AN INTRODUCTION TO PROPERTY VALUATION

by

A.F. MILLINGTON

BSc(Estate Management), FRICS, IRRV,FIMgt

One time: Professor of Land Economy, University of Western Sydney, New South Wales, Australia; Dean of the Faculty of Business & Land Economy, University of Western Sydney; Professor of Land Economics and Head of the Department of Land Economics, University of Paisley, Scotland; Lecturer in Land Management at the University of Reading; Lecturer in Valuation at the College of Estate Management. Fellow of the Australian Institute of Valuers and Land Economists, 1988–1996.

Estates Gazette

A division of Reed Business Information
ESTATES GAZETTE
151 WARDOUR STREET, LONDON W1V 4BN

FIRST PUBLISHED 1975
SECOND IMPRESSION 1976
REPRINTED WITH AMENDMENTS 1979
SECOND EDITION 1982
SECOND IMPRESSION 1984
THIRD IMPRESSION 1986
THIRD EDITION 1988
SECOND IMPRESSION 1989
THIRD IMPRESSION 1990
FOURTH EDITION 1994
FIFTH EDITION 2000
REPRINTED 2002 ✓

ISBN 0 7282 0350 2

Printed and bound by Bell and Bain Ltd., Glasgow

Contents

Preface to Fifth Edition

It is now 25 years since the first edition of this book was written, and the objectives of the fifth edition remain the same as those of the first edition, that is to provide "an introduction to and general background reading for the subject of property valuation". It is directed not just at would be surveyors and valuers, but at all those who may be interested in getting an understanding of property valuation.

The second edition saw the addition of a chapter on discounted cash flow, whilst in the third edition the book was adapted to make it more easily understandable and more useful to international readers. The author's experiences gained whilst working in Australia and for short periods in various parts of South East Asia and the Pacific Region indicated the value of comparative studies in a book of this nature, and this emphasis has been continued in subsequent editions. In the fourth edition a further additional chapter was added which sought to highlight a number of specific considerations to which valuers would need to pay special attention in the future.

With each new edition the major emphasis has been upon seeking to improve the product; there has been substantial updating, and an expansion of topics covered in earlier editions to take account of developments in the world and in the property professions in particular.

Over the 25 years since the first edition there have been considerable market variations, including booms and recessions, and the current edition seeks to take account of recent developments in economies and markets and to update the book in general and the examples in particular. There is also expansion of the topic of valuation standards, considerable revision of the chapter on taxation and property investment, and further development of the chapter covering special considerations for valuers. With the passage of time and the rapid development of modern technology, the valuation scene is very different from that of 25 years ago, and accordingly it is the author's view that these considerations become increasingly relevant and important to valuers with the passage of time. It is an extremely dynamic world, and no valuer today can make valuations with the expectation that the underlying market will be basically unchanged in 25, 10, or even 5 years time.

In writing a book such as this one is inevitably indebted to others for encouragement, knowledge gleaned from them, and direct assistance. In acknowledging assistance with this fifth edition, I would also repeat thanks expressed to some of those acknowledged in earlier editions, including:

My wife Imogen and sons Christopher and Andrew for their encouragement and assistance;

My mother for encouraging and assisting me to enter the profession of the land;

My teachers during my education at the College of Estate Management;

My past employers and work colleagues, and my friends in the property professions for knowledge and ideas obtained from them;

The late Mr Roger Emeny, one-time lecturer at the University of Reading, and Professor Will Fraser of the University of Paisley for their valued observations on the chapter on discounted cash flow techniques; Mr David Lloyd, lecturer in valuation at the University of Western Sydney, New South Wales, for his assistance in providing the comparison of valuation approaches contained in the Appendix;

My son Christopher, who is London based and who has supplied me with up-to-date United Kingdom information and without whose help I would have found completion of this edition far more difficult;

The Estates Gazette Limited and Mr Colin Greasby in particular for keeping me regularly supplied with copies of the Estates Gazette, which regular source of so much invaluable information is much appreciated;

Mr. Colin Greasby and his editorial and production staff for their valued assistance in the actual production process of this edition.

A.F. Millington
February 2000

Chapter 1

Introduction

The purpose of this book is to provide general background reading for students of surveying, estate management and land economics, who have not previously studied valuation. The lack of such a book was first noted by the author when he was a student, and the need became more apparent to him while he was a lecturer in valuation at the College of Estate Management and the University of Reading.

The book is written principally for students studying for examinations recognized by professional bodies connected with the land (such as the Royal Institution of Chartered Surveyors), including those studying for degrees in estate management and related subject areas. It is also hoped that it may prove of interest to students in other disciplines, including possibly economics and banking, and the author would be more than a little flattered if it proved to be of interest to any qualified practitioners.

No book can hope to be a completely authoritative work on any subject, and this does not even attempt to be that. It is hoped that it will be used primarily for background reading, and there will inevitably be gaps in it where it does not cover a specific examination syllabus. It is not intended to do that, but is meant to be used as a basic primer. More detailed and intensive studies of many of the topics should be obtained from other textbooks such as *Modern Methods of Valuation* published by The Estates Gazette, London.

The author hopes it will prove to be the sort of book that can be read in bed or in the bath, and, although he would like to think that much of it will be remembered it is not intended to be read and learnt "parrot-fashion".

If, having read it, a student has acquired a good background to valuation and a general understanding of the basic principles of the subject, the author will feel his efforts have been worthwhile. He does not profess or attempt to "answer all the questions". If, on reaching the end of the book the reader feels that there are many questions unanswered he offers no apologies, but on the contrary will be pleased, as it will suggest that interest has been aroused and the appetite whetted for further study of a fascinating subject.

Every effort has been made to ensure that errors have been eliminated and apologies are given for any that may have crept through.

For the benefit of those with a legal grounding, it must be mentioned that as the chief purpose of this book is to discuss the valuation of real property and leasehold property, the word "property" is used throughout these pages in a somewhat loose way to refer to these items, to the exclusion of items of personal property other than leaseholds.

Whilst examples are given using figures in £ sterling, for readers whose currency is different it should be noted that all references in the mathematics of valuation section to the Present Value of £1, the Amount of £1 etc., can be altered to the Present Value of $1, the Amount of $1 etc., or any other unit of currency which may be appropriate, without in any way altering the validity of the concepts or the functioning of the multiplier concerned.

Efforts have also been made to ensure that the figures for such things as rent, yields and building costs are regularly updated. However, these can vary considerably over quite short periods of time or between different areas, and it is emphasised that the main objectives of examples are to illustrate principles and methods which do not tend to change. The actual figures for such things as rents are in general incidental to the main objectives, and in cases where they have become out of date or in localities in which they would be inappropriate, readers can gain useful practice by reworking examples themselves using more appropriate figures.

The book is in no way intended to be taken as giving professional advice on specific matters, and neither the author nor anyone connected with its publication can accept responsibility for the results of any action taken, or any advice given, as a result of reading it. Indeed, if it teaches nothing more than that there are no hard and fast rules that can be applied in property investment and general investment, and that there are rarely, if ever, two identical situations in property and the property world, the book will have taught something useful.

Chapter 2

Value

The word "value" has already been used in this book, but what does it really mean? When a valuer uses the word it will normally mean "market value", which can be defined as the "money obtainable from a person or persons willing and able to purchase an article when it is offered for sale by a willing seller". There is no compulsion on either the vendor or the purchaser to enter into a transaction. The vendor will only sell if he or she obtains the sum required and the purchaser will only buy if he or she can do so at what is considered to be a satisfactory price. They are both willing parties to the transaction because they both consider the deal to be to their own personal advantage. The expression "market value" can be defined in many different ways, but the important thing is to grasp the concept rather than to remember any particular form of words.

Scarcity gives rise to value, and, generally speaking, when scarcity increases so will value increase. If articles in general demand become scarce, their value will normally increase; if there is no scarcity of an article there is likely to be little or no value attached to it. It is unlikely that one would be able to sell a bucket of water in India in the middle of the monsoon, but a person who had been lost without water in the middle of a desert might well be prepared to give all his or her worldly goods for a similar bucket of water. Similarly, air is free on the surface of the earth, where it is usually plentiful, but in a mine, where there is a lack of fresh air, it is invaluable, and much money may be spent to ensure that a regular supply of fresh air is available within the mine.

The same basic principle applies in the property market and, all other things being equal, as the supply of a particular type of property increases so will the market value decrease and vice versa. In the real world all other things rarely remain equal, and it is invariably found that both supply and demand are changing at the same time. The concept of value is nevertheless important, as the value of an article at any point in time shows the price at which supply and demand are equal – the price at which buyers and sellers who are prepared to do business are equal in number.

The value of an article therefore gives an indication of both the degree of scarcity of that article and of its utility when compared with other articles. Thus, in theory, one would expect people to pay more for articles which are very useful than for those which are less useful, although in practice this may not always be the case, if only because theory and practice often seem to be poles apart. Practice may often be based on imperfect knowledge, which results in such apparent anomalies arising as people paying higher prices for less useful articles, while theory may only consider the ideal circumstances which rarely occur in real life.

If a whole range of prices paid in the market is studied a pattern of consumer choice can be seen, and in the housing market the fact that purchasers pay high prices for houses in certain areas shows that consumers prefer to live in those areas, and are prepared to pay for the privilege. In examining the prices that have been paid for properties the individual valuations of all the purchasers are considered. Before purchasing an article each potential purchaser will normally put his or her own valuation on it. If the asking price is the same or less than his or her valuation, he or she will consider it worth their while to buy it. If the asking price is above their subjective valuation then they will not feel they would get value for money and will not purchase the article. A subjective value is the value to the subject or person concerned.

It may be that the person whose initial subjective valuation is below the asking price subsequently revises it in an upward direction, and does in fact purchase the article concerned. This is not unusual, neither is it necessarily unwise. As an article may have different values to different people, so may its value to one individual vary according to circumstances and needs. Just as a pint of beer may have no value to a teetotaller but some value to a regular drinker, so may a pint be incredibly valuable to the person who has just played nine vigorous sets of tennis, but of no value, or even of negative value, to the same person if he has already consumed 10 pints. Circumstances may well change, and the wise person will take changes into account in the decision making process.

The property market, as any other market, is composed of a whole range of subjective valuations, each of which may be regularly changing. Those of both buyers and sellers combine and interact to give the general level of values. The price of an article in the market will depend upon this interaction and will tend towards the point at which the number of potential purchasers and that of potential sellers is equal.

Chapter 3

General

The ownership of property is a form of holding money and thus when we consider the valuation of property we are concerned with money. It is therefore essential for the newcomer to valuation to get to know the way investors think. Both the student and the experienced practitioner should understand what is happening in the world of finance generally, and with this in mind the financial pages of a good newspaper should be read regularly. The more you know about the way monied people and institutions think and act, the better equipped you will be to become a skilled valuer. What are people currently spending their money on? What is the trend? What changes in investment policy are taking place? What is being done with money? Why are these things happening? These are all critical questions for the professional adviser to ask.

For property matters in particular it is necessary to become a regular reader of *Estates Gazette*, which is the main magazine of the United Kingdom property world, being published weekly. Similar journals are published in other countries with active property markets and should be read as appropriate. Initially, much of the contents may mean little to the new reader, but with regular reading he or she will soon find that they are getting the feel of the property world, and within a short time a good understanding of activities in the market will be obtained. He or she should not try to learn the contents but should read it as one would read any other magazine, paying more attention to the items which one finds to be particularly interesting, because in this way the branch of the property world which interests the reader most will be identified, and this will be a great help if at a later stage a specialisation has to be chosen.

Indeed, it is probably timely at this stage to emphasise that it is wrong to try to "learn" a subject such as valuation. Concentration should be directed towards trying to understand articles which are read and topics which are discussed. Learning without understanding is pointless. If there is understanding, learning will tend to follow automatically, and, even more important, opinions will be formed which will enable topics to be discussed. Moreover, a subject which is understood will almost inevitably be more interesting than one which is not understood.

The approach in this book will be to look very generally at property and valuations on the assumption that the reader has no knowledge of the subject and is in very much the same position as a person who is learning a new language. Some of the chapters will be almost entirely mathematical in content, but readers with non-mathematical minds should not put the book down at this stage, nor should they avoid the particular chapters concerned. The mathematical content is very basic indeed, and no-one need have any fears about not being able to understand it.

Valuation is not simply a mathematical process. It is much more than that, and probably the larger part of the valuation process depends upon the valuer forming opinions. He or she has to look at a wide range of facts and to try to predict the future. It is almost necessary to become a crystal-gazer. One has to weigh up all the facts in a particular situation, and, having done so, then form opinions upon which a valuation will be based.

There are probably many definitions of "valuation". It can be defined as: "The art, or science, of estimating the value for a specific purpose of a particular interest in property at a particular moment in time, taking into account all the features of the property and also considering all the underlying economic factors of the market, including the range of alternative investments". There will be different definitions, and the student will note that this definition itself is somewhat all-embracing and vague, even though it is long.

It is worth noting at this stage that there are circumstances in which a valuer will have to operate to very strict definitions of the terms "value" and "valuation". In some situations, such as the compulsory acquisition of interests in land, there may be definitions laid down by statute which dictate their precise meanings, while in others there may be definitions determined by professional bodies. There is, for instance, a definition of "market value" approved by the International Assets Valuation Standards Committee (IVSC) and The European Group of Valuers' Associations (TEGOVA) which is: "Market Value is the estimated amount for which an asset should exchange on the date of valuation between a willing buyer and a willing seller in an arm's-length transaction after proper marketing wherein the parties had each acted knowledgeably, prudently and without compulsion".

Definitions such as this should be checked regularly by a valuer as they may be subject to amendment from time to time, although this particular definition is fairly fundamental and unlikely to change substantially. This definition is intended to be used by valuers when they have to value the assets of companies, and it does not in any way alter the real world or the real market in which, unfortunately, parties to transactions do not always act "knowledgeably, prudently and without compulsion", and in which interests are not always marketed properly. In this respect the definition for assets valuation purposes requires a valuer to operate in a specific way envisaging a hypothetical market situation, and there will be other situations in which the valuer has to operate in a similar way observing guidelines laid down by law or a professional organisation.

However, much of a valuer's work will involve assessing "real world" market value, that is the figure for which the ownership of a property interest is likely to be exchanged in a free market situation, and this will rarely, if ever, be a simple task.

It is sometimes said that valuation is an art, and sometimes that it is a science. In fact it is a mixture of both, and in some instances the scientific content will be the greater, while in others the process will be almost entirely an art. The scientific part of valuations is the analysis of data and the mathematical calculations of value; the art is the skill of knowing which information to use to assist one's valuation, and the process of making judgements and forming opinions. Whatever it is called, valuation is not a simple cut-and-dried process, and although it may appear so when an experienced valuer is at work this is only because he or she has previously gone through all the processes of training and is now so familiar with the job that it appears simple and rapid.

In most instances the mathematical content of a valuation will be very simple. Valuation could also be defined as "The art of expressing opinions in a mathematical form in order to arrive at the value of a particular interest in a particular piece of property at a given moment of time". It is getting to the stage of being able to put opinions into a mathematical form which presents the problems – searching for all the facts concerning the property interest and the area in which it is situated, considering all these facts, and subsequently forming opinions.

In order to study the background to valuation, it will help to consider the first definition at some length.

The definition mentions the range of alternative investments, and this is all-important. The ownership of property is a form of holding money, and it is necessary to consider what people who have money do with it. All the items on which it could be spent should be considered. Money is not normally spent without the alternatives open to the spender being considered. There will always be the spendthrift who confounds this theory, but, generally speaking, money is only spent after careful thought, the more so in the case of large amounts. People wish to get value for money, and the valuer will endeavour to advise a potential purchaser of a figure which will ensure that this happens. A property which is purchased should be a good investment to the buyer or they would be better advised to spend their money elsewhere on something which gives better value for money.

People who have money can always simply spend it on what could be termed instant enjoyment. An expensive and rapid sports car could be bought, visits could be made to night clubs and popular places of entertainment, an expensive wardrobe could be bought and meals be enjoyed at the best restaurants. Expensive and luxurious holidays could be taken and the spender could generally "live it up". And what is wrong with that? Although personally we may disapprove of such action, who are we to say that another should not spend money in such a way? If the person concerned gets the greatest satisfaction from such expenditure then it is difficult to say that it is unwise, although it may well be that in later years, if he or she lives long enough, the spender will have formed different opinions and will have different needs, which may cause them to regret earlier excessive expenditure on instant pleasure. Nevertheless, when considering the options open to an investor, it should always be borne in mind that any expenditure on investment should give the spender at least equal value to expenditure on immediate needs and requirements.

It is possible to spend money on more durable consumer goods such as articles for the home, the home itself, tools of a trade, or possibly on the education of children. All expenditure of this type gives longer-lasting returns than does expenditure on "instant pleasure". The returns come principally in the form of the use given by the articles, but there may be an eventual money return of a somewhat remote form, in that the home and the articles in it have a very definite effect on the well-being of the occupiers and on their ability to work efficiently in their jobs, while tools of a trade and education assist in earning a living. So we see that when money is spent in this way returns come over a longer period of time than with expenditure on pleasure, and there may even be indirect financial returns.

There is a wide range of investments in which money can be placed such as stocks and shares, government-sponsored saving schemes, gilt-edge securities, debentures, local authority loans, building society shares and deposit accounts, assurance and endowment policies, unit trusts, investment trusts, property bonds, or, if a person is interested in landed property, either shares in property companies or investment directly in property itself. Of a more material nature than many of these items, but probably requiring smaller capital funds than landed property, there are investments such as silver, antique furniture, porcelain, works of art and the like.

With such a wide range of investment opportunities, the alternatives open to the investor should be studied before money is tied up in bricks and mortar. Unless such a study is made, how can an investor really be sure that the right decision is being made if a property interest is purchased? Some of these outlets for money will be considered later at greater length.

Chapter 4

Qualities of Investments

The potential investor's final choice should logically be that investment with features which make it nearest to his or her ideal investment. Different investments will vary considerably in their features and it is necessary to consider those which might appeal to different investors.

It is possible to list many qualities, but there are some which will be more important than others and which will be regular considerations for most investors. These will include security of capital, liquidity of capital, the security of income, the regularity of income, the ease of purchase and sale, the costs of purchase and sale, the divisibility of holdings, the security in real terms, capital appreciation prospects, and the costs of investment management.

Security of capital is a most important feature, because few investors will want to place their money in an investment if the prospect of losing that money is high. Only the gambler, or the person who has so much money that he or she is able to risk losing some, will be prepared to put their money into a risky venture and, even then, only if there is some possibility of a large gain if all goes well. The vast majority of investors will only wish to place their money in an investment if there is a strong probability that they will be able to recoup their capital at any point in time should the need arise. The greater the chances of their being able to get their original money back at any point in time (or the greater the security of their capital), the greater will be their willingness to invest. If the chance of getting the original money back is slight, then an investment will be considered insecure and relatively unattractive.

Security of income is another important consideration, and it must be remembered that, in investing money, an investor is giving up the immediate use of that money and is allowing its use to pass to some other party. If people give up the use of any other articles which they own, such as cars, they expect the lenders to pay hire fees, and it is just the same with investors. In return for giving the use of money to someone else, they require payment, and this payment will be the interest the money earns. It is the reward for forgoing the use of money, and before an investor is prepared to give it up he or she will wish to be reasonably certain that they will get adequate payment for such use, and that there is a high degree of certainty that the payment will in fact be made.

Coupled with the certainty that adequate interest will be paid, will also be the wish in most instances to receive *regular payments of interest*. If regular payments are made the borrower is less likely to get into arrears, while the receipt of interest at regular intervals enables lenders to phase such receipts to meet regular expenditure which they may be incurring, and this will assist their own budgetary control.

Investments which are *easy to purchase and sell* will be particularly attractive to investors who may be likely to require repayment of their capital at relatively short notice. If such a need is likely to arise it would be foolish to place money in investments on which notice has to be given before repayment of the capital will be made, unless such notice is very short indeed. Frequently, where notice is required it is in terms of months rather than days, and an investor will be at a considerable disadvantage if, having placed money in an investment which requires two to three months' notice before withdrawal of capital can be made, it is suddenly needed to meet unexpected debts within a matter of days. If such a situation is likely to arise it will be better if money is invested in something from which it can be withdrawn at a day or two's notice. Whether ease of withdrawal is important to investors will depend upon their own particular circumstances and needs, but other things being equal it will be an added advantage to invest in something from which money can be withdrawn readily and with little fuss.

Not only is the speed with which capital can be realised important, but the *cost incurred in investing and withdrawing money* is also important. With some investments there is little or no cost incurred in depositing money and subsequently withdrawing it, but with others there may be considerable expenses to be met, such as the payment of professional fees and stamp duty, and sometimes even the repayment of a certain amount of interest as a penalty if the withdrawal of capital is premature. Obviously the cheaper it is to invest and withdraw money, the more attractive will be an investment.

It may be that investors will wish to withdraw part of their money, but not the entire sum. If so it will be a great advantage if it is possible to sell only part of the holding, but this is not always the case.

For instance, it may be impossible to subdivide and sell a portion of a property investment, and the question of *whether an investment is divisible into smaller lots* which can be sold as and when convenient may be of considerable importance to an individual investor. This may be so even if the investor does not really expect to have to take such action, as the extra flexibility which the possibility of subdivision gives is an added attraction in an investment if all the other desired qualities are also present.

The first quality which was discussed was the security of the capital, and in times of inflation, when the purchasing power of money is rapidly decreasing, it is not only important that the capital which was originally invested should be capable of recoupment at any point in time, but also that the investment in which the money has been placed appreciates at a sufficiently rapid rate to keep pace with the changing value of money. Investors will wish to have their money in something in which capital appreciation enables them to withdraw it after a period of time and to purchase with that money at least the same amount of goods and services they could have purchased when the money was originally invested. If an investment is secure in this way it is referred to as being *secure in real terms*, as the real value of the money has been maintained. If the capital appreciates even more rapidly, the investment is of course still more attractive.

If the converse applies and it is not possible to purchase as many goods at the end of the period of investment, then the investor will not really be as well off as originally, and the investment will not have been secure in real terms. When an investment is secure in real terms it is often referred to as being a good hedge against inflation, or as being inflation-proof. In recent years this quality has perhaps been the most desirable quality of an investment, and the fact that property has generally satisfied this requirement has tended to make it a very attractive investment for those who require security in real terms. A cautionary word should be added at this stage to point out that not all types of property have fallen into this category.

Nevertheless, it is generally true to say that good property investments have possessed this quality to a very great degree despite the problems experienced by many property purchasers in many parts of the world during the economic recession of the late 1980s and early 1990s. Many of the problems experienced by people who purchased houses in the United Kingdom in the 1980s at prices which could not be recouped on subsequent sale following the crash in the housing market at the end of the decade, were largely attributable to the fact that many buyers paid particularly high prices on purchase and also borrowed heavily to do so. Problems which can result from "over borrowing" are discussed at length in chapter 22. The ability to be able to retain property ownership over a long period of time may be essential if capital growth is to be realized, and a sensible level of borrowing may be a precedent to long-term ownership. A forced sale of property is likely to mitigate against capital security if such results in sale during a property market recession or depression.

The overall risk attached to property will be a reflection of all the various qualities mentioned in the preceding paragraphs, and possibly also other qualities which the particular investor may consider important. Not all investors will place emphasis on all these features and, to some, certain qualities will be more important than others. One particular investment will not contain the same element of risk for all investors, and a person who wishes to have a high income yield may not be as worried about security of capital as another investor who must at all costs avoid losing any capital, but who has adequate income from other sources to satisfy daily needs. Individual investors will decide their own preferences, and the risk will depend on a combination of those preferences related to the features of the particular investment.

Generally speaking the riskier an investment is considered to be, the higher the yield investors will require before being tempted to place their money in it. In layman's terms, this means the riskier the investment the greater the return the investor will require. This is not unnatural, and is not dissimilar from the position a person might be in if asked to do a dangerous job. He or she would undoubtedly require higher pay than that expected for a safer job, and the higher yield that an investor requires from a risky investment is to some extent danger money for placing savings in an investment which may result in the loss of capital.

Apart from the danger money aspect of investment, there will also be a desire to recoup the capital as rapidly as possible and in receiving a higher return investors will be more rapidly decreasing their original outlay than if a small annual return is received. Although this may only be a rough and ready way of judging the quality of an investment, it is none the less a practical consideration, and if a 25% return can be obtained from a risky investment, by the end of the fourth year the original capital will have been recovered and so, to this extent, the risk will have been somewhat decreased. Obviously, to consider the situation in this way is to ignore the fact that the annual payments are really earnings for giving up the use of money and are not capital repayments – although periodic payments may also include an element of capital repayment, as is usual with mortgage repayments – and so strictly speaking this is not a mathematically correct way of assessing the quality of an investment, but it is nevertheless a useful rule of thumb method.

The costs of investment management and holding costs may also be important considerations for an investor; whilst an investment may produce an acceptable periodic income, such income may be less attractive if it is countered by high management or other holding costs. It may well be that an investment bond offers an apparently attractive annual percentage yield, but a high annual management charge may create a situation in which a lower yielding investment with no management or holding costs actually produces a higher net return. In this respect property investments often compare unfavourably with other investments, for it is generally necessary either to employ qualified consultants to manage an investment property, or to spend a considerable amount of personal time doing so. Either solution may have a high cost impact, whilst the investment owner may also be responsible for the payment of other holding costs such as rates, taxes, insurance, and maintenance costs, all of which may result in the net return from a property invesment being considerably lower than the gross income.

The reader may well be able to think of other qualities which may be of importance in different situations, but as this book is principally a consideration of property and property values, it is worth considering which of the above qualities are particularly evident in property investment, and also to consider their degrees of importance. There is little doubt, that over recent years the security of capital in real terms has been a particularly important feature. Indeed, in many instances capital has not only been secure in real terms, but marked capital appreciation has also been evident in many property investments. This has resulted in considerable capital gains being made on them, and this has naturally proved an attractive feature.

In good property investments the security and regularity of income has also been evident and where good quality property has been owned of a type for which demand has been high, there has been little likelihood of the income flow drying up or of rent not being paid on time. For many years, when such a misfortune occurred, property owners in most instances were able readily to relet properties with little or no interruption to income flow.

On the other side of the coin, it is not easy to sell property at short notice, and it may take many months, even years, to realize capital, as a property will not sell unless it fulfils the requirements of at least one potential purchaser. Because properties vary so much, it may be a considerable time before the purchaser who requires a particular property arrives on the scene, and this is a considerable disadvantage with property investment. It is also a fact that property is very often not divisible into smaller units, which is again awkward if only part of the original capital outlay needs to be recouped.

In recent years this indivisibility of property interests has proved a great problem with very valuable properties, so much so that the "unitisation" of property interests has occured: see chapter 9.

Investment in property is therefore likely to be of interest to the investor who will not need to recoup his or her money over the short term, and who places a high priority on security of capital, particularly in real terms, and on a regular receipt of income.

The relative stability of the property market and the fact that most freehold interests in property (and even some leasehold interests) have shown considerable capital growth over the years in real terms as well as in absolute terms, make property interests attractive security for those lending money. This is an important feature of property and enables owners of interests to borrow money using those interests as security (collateral), subsequently employing the borrowed money for business ventures. The ability to use property in this way is very important both to borrowers and lenders and is a very positive benefit of property ownership.

However, in the late 1980s and early 1990s events in the property markets of many countries world-wide cast doubts on the value of property as a good investment medium.

This raises the question "what makes an investment a good investment?", which was a particularly relevant consideration in the 1990s in view of the many undoubtedly bad property investment decisions made worldwide in the 1980s. Unfortunately, many property investments made in the 1980s proved to be very bad investments by the early 1990s when the almost worldwide economic recession resulted in a decrease in the demand for property by would-be users. Property in itself has no inherent value, rather is it the use to which property can be put which gives it its value. An officeblock or a factory have no value unless there are people who wish to use them, and the same applies to any property. Indeed, if there are no would-be users a piece of property can have a "negative value" as expenditure will be incurred on such things as unavoidable rates and taxes, basic and essential maintenance, and essential management tasks, while with no users no income will be produced to counter those outgoings.

Chapter 5

Investment Opportunities

There are many ways in which money can be utilised and the possibility of *expenditure on pleasure* has already been considered. Money which has been spent on pleasure is gone and cannot provide a future income flow in the form of earnings, but if one's resources are sufficiently large for a regular income flow to be of limited importance then this is not necessarily a great disadvantage. A person with large money resources may well decide that he or she can get the greatest enjoyment from life by simply dividing the money by the number of years they reasonably expect to live, and spending the resultant figure each year. They may have no dependants to worry about, and may well find this an exceedingly satisfying way of life. Although such a course of action may to some seem misguided, it must be remembered that there may be many who think otherwise, particularly if their philosophy of life is that today is more important than tomorrow.

Articles bought for use

These are of a more permanent nature and are likely to give returns over a longer period of time, but, even so, in most instances they will be wasting assets. They will wear out with time and will be less efficient as the years pass. They will also in all probability require regular expenditure on maintenance, and even on partial replacement.

Even so, many articles of this nature give great returns over substantial periods of time, and in that respect prove excellent investments. From household goods such as washing machines, to motor cars which give prolonged service both to individuals and businesses, to the plant and machinery essential for industrial production, all are articles bought for use which give great service over long periods, justifying both the initial capital expenditure and the essential periodic expenditure on them.

Stocks and shares

If *stocks and shares* are purchased, a small part of a company is in fact purchased, the size and part owned being represented by the relationship between the shareholding bought and the total number of issued shares in the company (ie the number of shares in the company that have been sold in the market). So if 1,000 shares in a company are bought and there are 500,000 issued shares, 1/500 of the company is owned by the purchaser. This does not in itself give the shareholder the right to tell anyone in the company how to work, or to make any decisions which will bind the company. The day-to-day running of the company will be in most cases carried out by paid employees and a paid management staff, and the policy decisions will be made by a board of directors, who will be elected by annual meetings of shareholders. However, as a part owner of the company, a shareholder will be entitled to a share in the company's profits, that is if it makes profits, and if the board of directors decides to distribute any of them. The board may decide to retain some of the profits within the company for such reasons as to provide for future taxation or for future expansion of the company's activities, but whatever profits it decides to distribute will be shared among the part owners according to the size of their shareholdings. The return to a share is known as the dividend, and this is one form of return that an investor will hope to get from stocks and shares, although the size of dividends and the regularity of payments will be uncertain and will depend upon the performance of the company concerned and the decisions of the board of directors.

If stocks or shares are bought in a progressive and prosperous company there is also likely to be capital appreciation. If a company commences with a given amount of capital and subsequently expands and builds up its capital resources, the value of the shares in the company will invariably increase. So not only will dividends be obtained from the company, but the capital will be secure and will increase in value. People want to be shareholders in successful companies, and as it becomes known that a particular company is very successful, so the desire of people to purchase shares in that company will push up their price in the stock market. Because of the increased demand, shares can be sold for considerably more than was originally paid for them. Obviously a share which pays good dividends and which also offers the prospects of a capital gain will be more attractive than one which merely pays a good dividend, unless there happens to be such a rare bird as an investor who does not want a capital gain.

Capital appreciation does not always result from investment in stocks and shares. It is just as easy to get capital depreciation in the Stock Market – probably easier. This is not unlike backing horses in some respects; for every capital gain made there is probably someone making a capital loss, or at least not doing particularly well out of a transaction. That a winner has been backed does not alter the fact that others may be backing losers.

There have been very well-known companies which have run into difficulties in many countries, which has highlighted the risks inherent in investment in Stock Markets. There is a risk in investment in stocks and shares, but it is probably true to say that unless there are risks there will be little prospect of high returns. If one plays safe all the time one will not make a fortune, just as the batsman who always plays cautious strokes is not likely to hit boundaries; if he does occasionally do so there will in any case probably be no spectators to watch him and he will not be a financial success, just as the ultra-cautious investor is unlikely to die a millionaire. Backing the two to one odds-on winner at the races will not enable one to make a fortune, and because the risk is far greater there will be more money to be made from backing the 100 to 1 against runner, always assuming it wins. The greater the risk the greater the returns one can hope for, but also the greater the possibility of complete failure.

Perhaps the biggest advantage about investment in stocks and shares is that generally they are very easily marketable. A telephone call giving instructions to a bank manager or stockbroker is all that is needed, and providing the quoted price at which a sale or purchase can be made is satisfactory, the transaction can be rapidly completed. There are now stockbrokers who do deals for clients at very competitive commission charges, some as low as about £25 to £30 per transaction, whilst shares can also be traded via the Internet at very competitive rates. Payment for shares purchased, together with commission charged for carrying out the transaction and any stamp duty which has been incurred, has to be made generally within about a week of a transaction being made. The net sum due to an investor following a sale will also be paid within a very short period of time, and it can therefore be seen that one of the main virtues of investment in stocks and shares is that an investor can quickly and easily realise money, should the need arise.

Following the purchase of shares the share certificate may not be received until several weeks later, but once a contract has been completed for a transaction one is the owner of the relevant shares. They can be sold again before the share certificate is received, and, indeed, shares may well be resold within minutes of being purchased. This will normally only happen if a quick profit can be made, but it might also happen if an unfortunate blunder was made in purchasing and the investor is attempting to cut his or her losses by immediately reselling. If both purchases and sales are made within a short period of time, the investor will simply receive a cheque from his or her broker for the difference between the net sums realised on sales and the total costs of purchases, or alternatively the difference may have to be paid if purchases cost more than sales realised.

However, one of the disadvantages of the Stock Exchange is that it is a volatile market, and, although it is easy to sell, a sale cannot necessarily be arranged at the right price. There is no guarantee that the original expenditure can be recouped. If a person is likely to want money back quickly and cannot afford to lose any of it, then they should not invest on the Stock Exchange. Stock Exchange investments should only be made if the investor can afford to lose money, as such a misfortune can easily occur. If it is known that a specific sum of money has to be available some time in the future it should not be invested in stocks and shares, because the capital could be worth considerably less at that future date than the sum originally invested. It is impossible to predict accurately what is going to happen on the Stock Market, so the investor must always be prepared for the unexpected, and must not complain if it happens. However, if money is available and its loss can be risked, and a careful study can be made of the market or a particular sector of it, then the Stock Market can provide rewarding and convenient investments.

Government sponsored savings schemes

National Savings Certificates are issued by the British Government, other governments having similar savings devices. When certificates are bought the precise purchase price is known and also the rate of interest which will be paid while they are held, while the normal redemption value will be stated at the time of purchase. They are therefore convenient for the more cautious type of person, or for the person to whom circumstances dictate that caution is necessary, for there are times when even the most confirmed natural gambler should exercise a degree of caution. As they can be purchased in small units they are also attractive to people of limited means.

Such investments normally can also be redeemed at an early date at a sum which can be accurately calculated, so they have the advantages of being a relatively liquid form of holding capital with little uncertainty attached. Against this they do not generally give very high returns and there is little capital appreciation to be obtained.

In order to overcome the disadvantage that investment in National Savings Certificates is likely to result in capital loss in real terms during a period of inflation, some certificates have been index-linked to the cost of living in recent years. Initially in the United Kingdom, index-linked certificates could only be bought by those who were 60 years old or older, but such rules can be and are varied with the passage of time and changes in circumstances.

As indicated above, governments in many countries sponsor such savings schemes. In doing so they probably have a range of objectives including that of raising money for government expenditure, the encouragement of saving and investment among the population in general, the provision of an easy and cheap savings medium for a wide range of people, and the provision of a safe investment medium for many people who know little about investment and whose main requirements in any investment would be security of capital, security of income, low costs of investment and disinvestment, and ease of withdrawal of funds. In these respects such schemes provide ideal investments for many small investors.

Gilt-edged securities

These are Government Stock, or Loan Stock which have been issued by a government as a method of raising money to fund its own activities or in an attempt to regulate the economy. Governments offer stock in exchange for a loan from the purchaser, and in return promise to pay a certain annual rate of return, and in many cases to repay the original capital at a specific future date. In other cases the stock may be undated and there is no promise to repay the capital at any specific time, if at all. In such cases the investment becomes what is virtually a permanent loan, if the Government cares to treat it as such. An investor can nevertheless realise capital by selling the stock on the Stock Exchange, but in this case they will not necessarily get the value of the original loan and will have to sell at the market rate for the particular stock at the time it is sold.

If the stock is dated, the Government promises to make repayment at some future date which may be quoted as a specific year, or a specific period of years, which would give the Government the option to choose the most favourable repayment date to them during that period. In the United Kingdom examples of the two types of repayment dates are 11.5% Treasury Stock 2004 and 8.75% Treasury Stock 2017. With the first, interest of 11.5% pa will be paid, and the nominal value of the stock will be repaid in 2004, whereas the second pays a rate of interest of 8.75% and has a more distant repayment date sometime during 2017.

A major difference between gilt-edged securities and such investments as National Savings Certificates is that gilt-edged securities can be bought and sold on the Stock Market at prices which fluctuate, depending upon the prevailing market conditions. Consequently their value at any future date cannot be certain, and there is therefore a relative insecurity of capital attached to such an investment. With a dated stock there is security of capital in the long term, although not necessarily in real terms, and there may be insecurity in the short term before the redemption date is reached.

Debenture stock

This is more usually referred to as "debentures" and is in fact loan stock issued by companies. A company may need to raise money and it may decide to do so by issuing debenture stock in return for loans from private investors who are willing to lend money to the company. Such a move provides an alternative source of fund-raising for the company as opposed to going to a bank or other institution, and, providing the company is healthy and flourishing, it will provide an investor in the debentures with a steady and reliable income flow at a known rate of interest. This will be at a rate quoted by the company when the stock is issued, and will have been pitched at a level which the company hopes will attract investors' money. For a debenture issue to be favourably received by investors it must therefore be issued at or about the current market rate of interest at the time of issue. Any hope of capital appreciation for the investor must depend upon market rates of interest moving in a downward direction, in which case the yield on the debentures will be above the market rate and a sale of the stock will be possible at a premium.

The holder of debenture stock is not a part owner of the company; he or she does not hold shares in the company, they merely lend money to it. If the company runs into difficulties they may possibly lose all the money they have lent, but on the other hand they will rank before the ordinary shareholders for the repayment of their loans; in such circumstances they will have a "prior charge".

Local authority loans

In the same way as companies may wish to raise money by borrowing, so also may local authorities. The terms of such loans are regularly advertised in the financial pages of the national press, when it is usual to see details of the rate of interest to be paid, the minimum size of loan accepted, and the minimum loan period. As regards their quality as investments, they are very similar to debentures in their attributes, although it would be hoped that the possibility of a local authority being unable to repay loan stock would not exist.

Insurance policies

An insurance policy is taken out to provide cover against the loss which could arise on the occurrence of an unlikely and unwanted event, which it is obviously hoped will not in fact occur. A premium is paid to the insurers, who agree to pay certain sums of money on the occurrence of the events insured against, the full details depending upon the full terms of the policy. In the simplest form of policy there is no return at all unless the event occurs, for example with a car policy there is no payment unless the car is damaged or stolen or someone is injured as a result of the use of the car.

An Assurance Policy is taken out to provide cover against an event which will assuredly occur, namely death. It is similar to an insurance policy in that there are no returns until the event occurs, when it will be someone other than the person suffering the damage who will receive the benefit of the insurance money.

An Endowment Policy is one in which, in return for paying a series of premiums over a period of time, or a single premium generally at the beginning of the period, the policyholder is guaranteed the payment from the insurers of a certain sum at the end of that period. The sum payable may be a specific amount, or it may be expressed as a certain amount plus profits. In the latter case the quoted amount will normally be lower than if there is no share of the profits, for in addition to the lump sum quoted in the policy, the policyholder will also be paid a proportion of the profits made by the insurance company during the term of the policy. In any event, the lump sum guaranteed by the insurance company will usually show a very low annual rate of interest when related to the premiums paid to the company, largely because payment is guaranteed. It may be that another reason is that many people who take out policies do so in ignorance of the low yield they are really accepting on their investment. Certainly, in the light of very high inflation in the period 1965 to 1990, many endowment policies taken out some years ago have proved to be incredibly bad investments. Nevertheless, there is a large degree of certainty about them, as it is unlikely that most of the leading insurance companies will go bankrupt, although this has occurred to some very well-known companies in the relatively recent past.

A popular form of insurance policy is that which links life cover with an endowment. A quoted sum, which will comprise the money they pay to the company – generally plus interest on that money – is guaranteed to be paid to policyholders when either the date quoted for the endowment has been reached or death occurs. This is perhaps the best type of policy to have, as with a little luck the insured person may enjoy the money himself. In return for a series of annual payments a young married man might be guaranteed the sum of £100,000 on death or in 35 years' time. In 35 years' time he will receive the money, but if in the meantime he dies, his widow will receive the money on his death and premium payments will cease. If the death occurred when only one premium payment had been made the receipt of the lump sum would yield an incredibly high rate of return on the investment, a fact which would be of great help to the widow, but of little help to the insured.

Although the indemnity cover of insurance is well worth having, it is debatable whether policies have much to recommend them as pure investments, judging by the way the purchasing power of the sum payable can be eroded by inflation.

Linked insurance

This is a type of investment which has come to the fore in recent years. Financial institutions realised the shortcomings of insurance policies which did not give adequate security in real terms, and various packages have been devised and offered on the market which give both insurance cover and participation in equity investment. Part of the premium provides for insurance, and the remainder is invested in equities, usually of a specified type or in a specified range of companies, the long-term hope being that the equity portion will give the investor growth which will at least keep the capital secure in real terms, and possibly even give growth in real terms. This then is an investment which gives both the security which insurance offers and the prospect of capital growth which the ownership of equities offers, although it must be remembered that the dual-purpose nature of the investment must inevitably result in each of these qualities being less pronounced than if an investment were made solely in either insurance or equities. It does tend to be a long-term investment with a commitment to the payment of regular sums, and it may be that with the passage of time and changes in circumstances, such an investment may cease to be attractive or may become inconvenient. Also, the quality of the investment will depend upon the quality of the trust management, which will vary from trust to trust, and may even vary within one particular trust over a period of time.

Unit Trusts and Investment Trusts

These are somewhat similar, in that money is paid to buy shares in a Trust which in turn reinvests the money in stocks and shares. Both offer a convenient way in which small investors can spread the risk when investing in the Stock Exchange; they obtain a block of shares from the Trust which will hold the shares of many leading companies, probably selected from a variety of industries, and consequently they become what could be described as owners at second hand of a few shares (possibly even fractions of shares) in a great many companies. Without the facility of the Trust an individual would require a far larger amount of capital to be able to acquire holdings in such a large number of companies.

By reinvesting the money in a variety of companies the Trust reduces the risk of investors losing all their money, but at the same time it is probably true that the possibility of making very large gains is also decreased, as large capital gains on one share may well be countered by smaller gains and even losses on other shares. So investing in Trusts increases the security of capital and probably the security of income also, and enables the person with limited resources to obtain an interest in a wide range of companies.

There are several differences between Unit Trusts and Investment Trusts, but to the small investor the most important difference is probably the fact that capital can only be realised by selling holdings in Unit Trusts to the Trust managers at the price quoted by them at the time of sale, while holdings in Investment Trusts can be sold on the Stock Exchange at the current market rate, which can be readily ascertained through the financial press, a stockbroker or a bank.

Property Bonds

These are similar to linked insurance, which was discussed earlier, with the investment in equities being restricted to property shares. Property Bonds therefore enable an investor to get insurance cover and an interest in property shares at one and the same time, the importance of the property element being that many investors have in the past regarded property, and consequently property companies, as being one of the safest and most rapidly growing types of investment available.

However, though it is easy to assume that an investment in property cannot go wrong, it is important to remember that not all property investments are good, a fact evidenced by events in property markets in many countries during the economic recession of the late 1980s and early 1990s. Again, the success of Property Bonds depends to a very large extent upon the quality of the management, and as the quality of bond managers and the quality of property investments approaches the margin between good and bad, then so will the quality of Property Bonds vary.

Property shares

Some investors may decide to invest only in the shares of property companies, many of which have quotes on the Stock Exchange. Many of these companies in the past became almost household names, some because of the rapidity of their growth and the quality of their management and products, others because of adverse press publicity, some of which is justified, much of which is not justified, and a large proportion of which is simply ill-informed and misguided.

Because of the speed and size of growth many of these companies became much favoured by investors, and the purchase of equities in property companies offers the small investor, who may not have sufficient resources actually to involve himself in property development and investment, the chance to share in someone else's involvement, and also, it is hoped, someone else's success. Whereas a considerable sum of money is needed before an interest in property can be purchased or a development scheme started, it is possible to purchase a holding of property shares for a relatively small figure. However, in the United Kingdom in 1999 and 2000 many leading property company shares had become relatively unpopular with investors. The recession of the late 1980s and early 1990s followed by a sustained period of low inflation in the economy, resulted in low capital growth of the assets of many property companies. One of the major attractions of property investment – the prospect of capital gains – was therefore absent, whilst the failure of many property companies to pay attractive dividends to their shareholders, not unexpectedly, also increased their unpopularity with investors.

Property

From investments of a general nature we have progressed to those with a property flavour, and as this is obviously a very important topic it will be covered in a later chapter at somewhat greater length than the other forms of investment.

At this stage it is sufficient to note that there are many types of property available for the would-be investor, and a variety of interests in such properties. There may be freeholds, short or long leaseholds, ground rents, offices, shops, factories, warehouses, farms and other types of property, and each of these interests or types of property will have features which make them more or less attractive to investors, depending upon circumstances and the requirements of the particular investor.

Several features of property investment which should be remembered when comparing it with alternative types of investment are:

1.   Property investment tends to be long-term in nature.
2.   The costs of purchasing and selling property investments are relatively high.
3.   The time involved in buying and selling is lengthy.
4.   Proof of ownership can sometimes be difficult.
5.   The amount of capital required to buy property is high, relatively speaking, and it is not often possible to buy part shares.
6.   There may be considerable management problems involved in owning property.
7.   Property seems to be prone to interference in the form of legislation by governments of all colours, and this can be extremely damaging to the value of an interest.
8.   There can be considerable prestige in being a property owner.
9.   Property values tend to appreciate with time, and property is therefore usually a good hedge against inflation.
10. The yield on property investments is generally much higher than the yield on a similar sum invested in other forms of investment, this being compensation for some of the disadvantages listed above.
11. An interest in property can often be used as security (collateral) against which loan finance can be raised.

The particular features of property investments will be considered again in chapter 9.

Property Investment and Underlying Factors of the Market

As there are many factors which can affect property values and which should be considered by property investors, it is essential for valuers to study the property market at considerable length, and also the underlying factors which affect it.

The international situation

Even the international situation can affect property values, and this was vividly illustrated by the events in the United Kingdom market during the years 1972 and 1973. In 1972 residential property values rose dramatically almost daily, to levels which would previously have been considered ridiculous, while there was a considerable reversal in 1973. Although prices did not fall to the levels of early 1972, there was nevertheless a considerable fall from the peaks of 1972, and many would-be sellers found difficulty in disposing of their properties. One of the major reasons for this was the shortage of mortgage funds in 1973 as opposed to the seemingly limitless supplies during 1972, and also the higher cost of borrowing as a result of worldwide increases in interest rates. So the international financial situation during this period had an important effect on the property market; as it did again in the period 1979–1981 when high international interest rates helped to make it expensive to borrow money.

The effect of international events was particularly evident in the early 1990s when international factors played a very important part in property markets worldwide. In the 1980s many national economies appeared to ride on the crest of a wave, with no apparent sign of any future downturn. Property development and property investment flourished in many developed countries, and in developing countries also, and there was much international property investment with the cash-rich Japanese being prominent.

With the advent of the 1990s (and a little earlier in some countries) economic growth faltered in many countries, rapidly resulting in recession in some. The deteriorating international economic situation was exacerbated by a number of international developments which caused uncertainty and thereby lessened the confidence of investors large and small on both national and international scenes. The Gulf War in 1991, the break up of the USSR, the bitter conflict which engulfed the old Yugoslavia in an internal war, and a number of major physical disasters throughout the world, all helped to cause further problems for and uncertainty in economies worldwide. Whereas since the Second World War the economies of Japan and Germany had seemed to be immune to the problems regularly experienced by other economies and had appeared to grow ever stronger, this was not the case in the early 1990s. Germany was troubled by the problems of unifying West Germany with the much poorer East Germany, a task which proved to encompass great problems including many of an economic nature, while the worldwide recession was so great that even the economic strength of Japan failed to render it immune to the problems experienced by other countries.

The net result of all these factors was destabilised economies worldwide, a downturn in consumption and production in most developed countries, and high levels of unemployment in many countries, with some countries such as Japan experiencing a growth in unemployment for the first time in many years. The downturns in production naturally affected the demand for property by would-be users, and in the United Kingdom, Australasia and North America in particular there was a demand situation which resulted in many unlet and unsaleable properties appearing in the market place. So high were the vacancy rates in many sectors of the commercial, industrial and retail property markets that no prospect existed of many properties becoming occupied and used without a major upturn occurring in national economies world-wide.

National property markets were in some cases not only adversely affected by the poor performance of their own national economies, but by the withdrawal of international property users and investors whose confidence, and sometimes their actual ability to invest, had been weakened not only by poor economic conditions in the foreign country but in their own country also. One thing that was very clear indeed in the early 1990s was that national economies in the modern world are so interdependent that most economies are adversely affected by downturns in economies elsewhere. As such factors affect production levels they also affect the demand for properties for use and therefore their market values also.

It is essential in the modern world for the property valuer to be aware of developments and trends in the international scene, as adverse developments may possibly even adversely affect local house prices in due course, or vice versa. There is little doubt that such prices are affected by the level of interest rates and the level of consumer confidence, both of which are very much affected by international factors.

Not surprisingly, similar effects were evident in property markets in other countries as a result of the same high interest rates. There will also be occasions when the market will react to moods of optimism or depression on the part of investors which may well be caused by such things as the international political situation. The property market does not normally react in quite so volatile a manner as does the stock market, as property investment is essentially a long-term investment and investors are therefore influenced more by long-term prospects than by short-term ups and downs. However, there is much to suggest that even the latter can have repercussions in the property market, and, of course, the longer or stronger the mood of optimism or depression, the greater will be its effect.

From about the mid 1990s to 2000 many developed countries have enjoyed periods of low inflation, sustained economic growth, decreasing unemployment, and low interest rates. Even troubles in some of the developing economies of South-East Asia failed to destroy the generally encouraging economic conditions with the result that property markets in many countries have operated to a background of low interest rates and strong demand fuelled by high economic activity. In 2000 there appears to be the danger in some economies that these conditions may be leading, yet again, to considerable price inflation in some property markets. Consequently, it may well be that some governments may decide to increase interest rates from levels at which it has been possible to secure housing finance at less than 6% per annum, in an effort to contain increasing levels of inflation, for it is recognized by most that relatively low inflation levels are preferable to some of the extremely high rates of inflation which existed in the period between 1965 and the 1990s.

The national situation and finance

National affairs are also very important in influencing the mood of investors, and the state of a national economy is very important. If a boom period appears to lie ahead, investors will be confident, while if future prospects appear gloomy, they will be far from confident, and such factors must make a terrific difference to the type of investment into which investors are prepared to put their money. Apart from affecting the thinking of investors, the economic situation can also affect the amount of money in circulation at any particular moment of time. The stringent credit restrictions in the United Kingdom in the late 1960s prevented many people from going ahead with schemes which appeared promising, purely and simply because they could not obtain the necessary finance. There was none to lend, and without money there could be no scheme. The quality of the investment as a risk was irrelevant, as even very reputable people could not obtain finance.

In 1971–1972 the situation was quite different and lower interest rates made borrowing cheaper. With plenty of loan money available in the economy, there were more would-be buyers with funds available, and this was a major factor in the surge of property prices in 1972.

The amount of money available and the cost of borrowing can also affect the preferences of investors, as well as determining whether they can afford to purchase or not. Having decided that he or she can obtain finance to make a purchase, if borrowing costs are low and if a large sum can be borrowed, an investor may well decide to invest in bigger and better investments than if the converse had been the case. Likewise, if everything in the garden appears rosy with plenty of money available, an investor may well be so optimistic as to go for riskier investments whose potential pay-off, if all goes well, is very high. If things generally appear tighter, the investor will be more likely to be attracted by the lower yielding and safer investment, and, if he or she is likely to need money back at short notice, liquidity of capital will probably become the most important factor.

In the period 1978–1981 there was a serious world economic depression. The British economy, like many others, suffered with high interest rates, high rates of inflation, depressed industrial production and rising unemployment. These factors influenced investment in property both in terms of the amount of investment which actually occurred during that period and the investment preferences of those investing.

The expected trend in the economy is also important in influencing investors. If current yields are high but in the long term they are expected to become lower, an investor may be keen to tie money up in a high-yielding, long-term investment. However, if current yields are low but in the long term are expected to rise, then it is better for the investor to put money into an investment from which it can readily be withdrawn for reinvestment when interest rates rise. In early 2000 it appears that the latter may be the state of affairs in a number of countries, and with the passage of time those who are able to arrange long-term finance at low rates of interest may find that they have made a very sound decision.

Government policies

Government policies are extremely important to investors. In many instances governments have given direct aid, such as grants to farmers for various purposes, and improvement grants, which have been given to encourage the improvement and modernisation of older residential properties. There is little doubt that such aid influences investment policy, as with the assistance of a grant, a scheme which previously was unremunerative and unattractive may suddenly become a paying proposition. Likewise the removal of a system of grants will probably have the converse effect, unless, having become involved with a particular line of investment, the investor has obtained sufficient knowledge and expertise to realise that even without a grant it is still worthwhile. Government policies of this type will inevitably influence the property market, even when they relate to other forms of investment, as there may well be a diversion to or from property investment as a result of factors affecting other investments.

In the United Kingdom budget of 1980 the Chancellor of the Exchequer, Sir Geoffrey Howe, announced the introduction of Enterprise Zones. Briefly, these were to be a number of selected areas, relatively small in size, where it was thought preferential treatment would be desirable to encourage industrial and commercial regeneration. Amongst the benefits for those developing in these zones would be the removal of planning restrictions and requirements, and preferential tax treatment. It is clear that such measures would be likely to make the Zones more attractive for investment and development than neighbouring and other areas which do not have such advantages. As a result their creation is likely to have a considerable effect on the decisions of those contemplating property investment and development in the localities of Enterprise Zones.

A similar type of government intervention occurred in Australia in the 1970s when the Government of Mr Gough Whitlam introduced development corporations. Two of these were in the Albury/Wodonga and Bathurst/Orange areas, both having the objective of seeking to bring about economic development in those areas. Infrastructure capital was provided for the development of the areas and various incentives were introduced to encourage industrial and other employers to move their activities to them.

Such schemes can generate a lot of worthwhile economic activity in the areas they benefit, but it is possible those activities may only persist as long as financial inducements are given to firms through such things as tax concessions, subsidies, capital grants, or other forms of assistance. While they are effective such schemes will distort property markets as they are likely to cause decreases in values in the localities from which firms move, likewise probably (but not necessarily) causing increases in values in the areas they are persuaded to move to. Whether such increases will actually occur will depend upon property tenure systems in the development areas and the long-term economic prosperity of those areas. It may be that once government financial assistance is reduced or withdrawn the local economy will suffer from the withdrawal of organisations which were in effect only induced to go there in the first instance by the strength of "government bribes".

The valuer should be aware of the effects on local property markets of such government policies and activities, and in particular of the likely effects of their possible introduction or removal.

Anticipating the effects of proposed or actual changes in government policy is an important task for a valuer. If changes are likely to affect the way in which property users think and act then they may well affect property values also. The introduction of a Uniform Business Rate in the United Kingdom varied the considerations to be taken into account by potential users of business accommodation. Previously businesses in some areas were faced with extremely high annual taxes on occupation because the properties they occupied were in areas of high property values in which the local tax in the pound was also high, the total taxes payable per annum being a product of the annual value of a property as assessed for rating purposes (such annual value being related to the annual rental value of a property) and the rate in the pound. In other areas business occupiers were faced with lower annual taxes in respect of the occupation of similar areas of accommodation because property values were lower, resulting in lower assessed annual values, whilst annual rates in the pound were also lower.

The decision as to where to locate a business could as a result be very much influenced by the fact that in one locality very high annual rates would be levied, while in another area the annual rates would be very much lower. Coupled with differences in rent levels, which in themselves might be partly a result of the different levels of taxation, businessmen could be influenced to locate in (or even to move their location to) one area as opposed to another. Such a pattern could result in the move of employment to low property value areas creating economic and employment problems in the high value areas. In order to reduce variations in the total costs of occupation the British Government changed the rating laws in 1990 and introduced a Uniform Business Rate, which resulted in the rate in the pound for business occupation being at a uniform level throughout the United Kingdom.

While such a change would not remove the differences in costs of occupation between different localities because variations in rent levels and assessed annual values for rating purposes would still remain, it would reduce the cost differentials, which reduction might influence business occupiers in their choice of location from which to operate their businesses. Valuers need to be aware of such developments and need to both anticipate and monitor the effects of such developments on the thinking of existing and would-be property users and therefore on property values, particularly on the relative values of properties in different localities. Such developments will not alter the principles to be employed by a valuer in assessing values, but they are underlying factors influencing value in the market place, which have to be taken into account in the valuation process.

Another obvious way in which government policy affects investment, and property investment in particular, is in respect of statutory policy, and this will be discussed in a later chapter.

The local economy

The state and trends in the local economy will be of prime importance. An investment in property in a depressed area is hardly likely to be as attractive as a similar investment in a thriving area, unless the depressed area currently offers investment possibilities at bargain-basement prices and it is considered that future prospects for the area are good, or unless, as in Enterprise Zones or Development Corporations, inducements are offered to make investment worthwhile. Investment in a brand new sports stadium and entertainments centre generally would be inadvisable in an area of high unemployment, particularly if the limited employment which did exist in the area provided a low average income. Such an investment would be best reserved for areas of high employment and high incomes.

Geographical factors

There are many geographical factors which influence the values of properties, and although many or all of them may appear obvious, it is nevertheless important to consider them, as it is often the obvious that is overlooked.

The *latitude* of the area may be important if the use of a property is likely to be affected in any way by the warmth of the climate or the length of seasons. A farm in the south-west of England will be suitable for a different type of farming from one in the north of Scotland, while an hotel in the former area will normally expect a longer summer season than one in the latter area. Likewise, the existence of a swimming pool in the grounds of a house is likely to cause a higher increment in value if the house is in a warm location rather than if it is in a colder area.

Topography – whether an area is flat or hilly, the existence or absence of rivers, and similar factors, may well have an important effect on value. The existence of geographical obstacles such as hills and rivers might well determine the catchment area of a shopping centre, while they could also create quite substantial price differences in the housing market. In an age of increasing affluence and leisure the existence of river frontage, particularly if it has good fishing rights, can make a house extremely valuable, whereas in earlier, less affluent times, and before the various local river authorities were so active in cleaning out rivers, it could well have been a positive disadvantage because of the danger of flooding and also the unattractiveness of polluted water.

The *aspect* of a property may be important, particularly in residential areas. Advertisements for houses in the United Kingdom often emphasise the fact that they are "facing south" or "built on the southern slopes of . . .", while if the objective were to try to grow vines in the United Kingdom a southern-facing slope would be a great help, if not essential. Another way in which aspect is important is in respect of the outlook of a house, any property which overlooks a pleasant area obviously being more attractive than one which overlooks unattractive surroundings.

Local climatic conditions may vary considerably and the value of properties will probably vary with them. Cotton and woollen industries have in the past often developed in particular areas because of certain features of the local climate, which favoured their establishment. In hill areas there are often considerable differences in annual rainfall within the space of a few miles which may greatly influence the pattern of residential settlement. At Minehead in Somerset, England, the average annual rainfall is 35 inches, while a few miles inland at Dulverton it is 60 inches, and it is 80 inches on the nearby hills of Exmoor.

Other factors

The existence of good *communications* and accessibility have always been important in influencing value and it is no less so today. Even though modern vehicles can travel fast, it is still an advantage to be close to good means of communication.

Some time ago I inspected a house set in a remote moorland area. The advertisement and the sales particulars both sounded very interesting and the property itself proved to have many attractive features, being set in a really beautiful and peaceful location close to a pretty trout stream and having several acres of land. However, to reach it involved a drive of about one mile across an unmade moorland road and through a ford which was quite deep even in the middle of a dry spell. The probability was that it would be cut off for several weeks each winter, and any prospective purchasers would need to equip themselves with a cross-country vehicle and a large deep-freeze in which to store provisions, apart from having the type of temperament which would allow them to stand real solitude.

The property did not sell that summer, and it is possibly still empty. The one major disadvantage was its inaccessibility and distance from good means of communication and this effectively "killed the market".

The closure of local railway lines and stations may well result in a change of values, particularly in an age when so many local bus services also seem to have closed down, or, where they still exist, are expensive to use. On the other hand, the development of new railway lines or roads may well facilitate the development of land previously considered unsuitable for development because of its inaccessibility.

Good communications are vital with commercial properties, particularly industrial and warehouse units, and it is common to see the proximity to a motorway (or freeway) access point stressed as the most important feature. A warehouseman storing perishable goods will not want them to go rotten in transit to the retailer, while in a competitive society any advantage of accessibility may enable a businessman either to undercut rivals or to make a higher profit by virtue of the lower costs incurred. Indeed, both may in fact be done.

An office user often needs to be close to or accessible to others with whom business is done, and it is also necessary for offices to be so located that staff can travel with ease to and from the premises, so making it easier to attract staff to work there. If staff can travel to work easily, and there is not the harrowing journey to work that is often a feature of modern life, it may even be possible for an employer to pay comparatively low wages and still have a happy work force, particularly if the office block is pleasantly located as well as accessible.

There is the possibility that the development of microprocessors and telecommunications systems may result in more office work being done by employees from their own homes, with a central office being used mainly for co-ordinating the work of the various "out-workers". It will be interesting to note to what extent such a trend develops, as it could well result in the accessibility of an office block being a less-important factor in determining value than previously. In effect, offices which are at present relatively inaccessible by road or rail could become relatively more accessible for communication by microprocessor.

Whereas in the past rivers, roads, canals and railways have at one time or another been the most important means of communications, it is probably true that the road network, and particularly the motorway network, is now the most important. The existence of good port facilities is important for international trade, as also are airport facilities to a growing degree, and the Channel Tunnel between England and France has made a substantial difference to road and rail transport communication between Britain and the rest of Europe.

Fashion and local demand are important, even though they are rather vague and difficult to assess. The stage has been reached in certain areas of the housing market in which buyers simply are not interested in a house which does not have a second bathroom, whilst even the type of shower fitment may be an important market consideration A few years ago the possibility of such a situation arising would have been considered ridiculous, but with increasing affluence it has become fashionable, and convenient, to have a second bathroom. The house that lags behind the change in fashion will inevitably command a lower price than those which have kept pace with it.

Similarly, in affluent areas in the United Kingdom a second garage and central heating have almost become essential, although it is noticeable that this tends to be the case more in the south-east of England than in many other areas of the United Kingdom.

It may well be that a property is difficult to market because it has a design or appearance that has become unfashionable. Some of the semi-detached houses erected in Britain between the two World Wars or the over-ornate properties erected in the latter part of the last century often fall into this category. Many properties erected more recently have elevations which are bedecked with a wide variety of finishes – shiplapping, varnished timber, tile-hanging, and many others – and while such finishes are in fashion they will prove good selling features. However, if the fashion changes, and it may well do so, particularly when high maintenance costs begin to rear their ugly heads, such finishes may eventually prove to be a great disadvantage.

Fashion can dictate the popularity of residential areas. In recent years it has become fashionable to live in parts of London, such as Chelsea, Islington, and Fulham, which were previously regarded as less desirable, and in many areas there are now wealthy people living in modernised cottages which were once the homes of labourers, and which may even have been legally unfit for human habitation before they were modernised. Indeed, many wealthy people now live in mews cottages which were once the homes of horses.

In many provincial cities large areas of housing have been demolished because the fashion of buying up old and poor quality housing and improving and modernising it never caught on. But, in many areas of London similar houses now provide excellent and desirable residences because some far-sighted people saw the potential of such properties in areas where there were acute housing shortages. Indeed, in recent years official planning policy has recognised the potential for improvement in many older properties to the extent that policies of wholesale clearance and rebuilding have been forsaken for policies of improvement involving the renovation of existing properties to improve both the quality of the individual properties and the locality concerned.

In other cities of the world similar trends have developed, and there are areas near the centres of Melbourne and Sydney in Australia where the gentrification of previously run-down city areas has occurred, and this trend continues.

The *individual features of a property* are of importance in determining its value. Does it have a good or a bad design, is it functional, is it adaptable? A factory into which it is difficult to install modern equipment because of the difficulty of access, the small size of the doors, the low load-bearing capacity of the floors or the restricted clear working height, is obviously relatively unattractive in these days. A factory which is nice to work in is more likely to maintain a high value, just as is the case with a house.

In some instances a period design adds a huge sum to the value of a house, particularly if it is in an area which is noted for such houses and which is fashionable. However, the same features in a high street shop in a market town could be a positive disadvantage. When modern marketing techniques dictate that alterations should be made, if the property is a listed building it may even be impossible to bring it up to modern retailing standards.

Several years ago in the United Kingdom there was great demand for new houses of modern design, and such properties sold readily while older properties stood on the market for many months and eventually fetched much lower prices. Later, purchasers realised that the majority of older properties were built to more spacious dimensions, and it became the fashion to purchase such houses and modernise them, while trying to retain their inherent character. The fact that the features of spaciousness and character became popular resulted in houses which possessed such features regularly commanding high prices on the market, thus illustrating the importance of fashion and trends in demand. Following the advent of instability in the Middle-East and the resultant energy crisis in the 1970s, the cost of heating increased as a result of which houses which were unduly spacious became relatively unattractive to purchasers, not only because of the increased cost of heating but also because of the high cost of rates on space which could not necessarily be fully utilised. In countries in which property rates are levied on site value rather than on the value of buildings the latter factor would not necessarily be a consideration which affected the cost of occupation.

The *state of repair* will influence a purchaser, for if a property is in bad condition he or she cannot contemplate a purchase unless the price will enable them to cover the cost of reinstatement and, preferably also make a profit. If this is not the case the potential purchaser would be well advised to purchase a similar property which is already in better condition. The property which is in good condition will invariably sell more readily and at a higher price than that which is in bad condition.

The term *"services"* covers such items as mains gas supply, mains electricity, mains water, mains sewers and to a lesser extent the existence of a telephone connection, and the presence or absence of these items must influence a potential purchaser. The idyllic country cottage will probably seem a little less attractive when it is discovered that at present there is only a chemical closet at the bottom of the garden, water is from a spring which has been known to run dry, lighting is by oil lamp, and mains electricity will cost a considerable sum to connect, while a telephone connection will also be expensive.

Although it is possible to live without all these modern conveniences, it is doubtful whether it is possible to carry on industrial activities without them, and it could well be that land which was physically suitable for industrial development is completely blighted by the lack of services. If services are available near to a property but connection to them is likely to be expensive, then it is almost inevitable that the market value of the property will be lower than if those services were already installed.

The *potential* of a property can greatly affect its attractiveness on the market. It may be in poor condition, but if it is capable of improvement it may still sell readily at a good price. This is what happened to many cottages in certain areas of London and also to inferior cottages in rural areas which have now become the sought-after residences of the affluent. If, however, there is no possibility of improvement or the cost would be excessive, then the property will be unattractive to purchasers.

This factor of potential may be relevant in the industrial as well as the residential market. An industrialist may require a factory of 5,000 m² in a particular location. He may well be unable to obtain a unit which satisfies both these requirements, and in such a case a 4,000 m² factory in the right location with adequate room for expansion to the required size may be more attractive than the right-sized factory in the wrong location.

The capability and space for extension of a property may also be a great selling factor in the residential property market.

It often happens that old factories too large for the requirements of modern producers are disposed of, and if it is possible for them to be split up into a number of smaller units of the size demanded by would-be tenants or purchasers, the total value of the property may thereby be greatly increased.

Whenever one has to value a property or advise a client one should consider whether the property is currently being used to its greatest potential and, if not, the possibility of putting it to a higher use should be given careful consideration.

The *time element* may be critical, particularly in periods of rapid inflation such as the late '60s' and early '70s'. A price which seems high at the commencement of a development scheme may be just about right by the time the scheme is completed, and a year or two later it may seem ridiculously low. An example of this is the case of contemporary-style bungalows which were built in a coastal resort, the price on completion being advertised as £13,500. This seemed very high even though they were built to a high specification, as other bungalows offering a similar amount of accommodation were on the market for about £8,500. The properties were eventually sold for figures in the region of £12,500 about 18 months after first being advertised. Within another two years similar properties were changing hands at between £20,000 and £25,000, while the passage of another 10 years saw these prices trebled. Time made all the difference, many of the underlying market factors having changed in the interim period.

Time is also important to the valuer, as it is often easy to fall into the trap of not moving with the times. Valuers with many years' experience have been known to make comments such as, "It went for £160,000, but it's never worth it". But the property must have been worth this sum to someone, otherwise that figure would not have been obtained. The valuer possibly made the mistake of not realising that value may change with time, and he may have had in mind the sum that the property obtained on an earlier sale some years previously. Alternatively, he may have made the mistake of associating his idea of value with the costs which he knew were incurred in originally creating the property, or of placing his own subjective valuation on the property rather than an objective assessment of what would be paid in the open market.

It should not be forgotten that properties may lose value with time, although in inflationary times this is the exception rather than the rule. Nevertheless, the further passage of time could result in the property of the last paragraph only fetching £150,000 on a subsequent sale. The doubting valuer might then consider his doubts well justified and the original purchase to have been at an ill-advised figure. This will not alter the fact that at the earlier date £160,000 was the market value.

Because of the time-factor it is always most important for the valuer to state clearly the date on which he or she makes a valuation. Property prices could vary incredibly in the space of a week because of exceptional circumstances. In Reading, England, a development of new houses sold with little difficulty, but within a week of the sale of some of them a scheme for a motorway (freeway) link road was announced. This was to go straight through some of them and extremely close to others. Most of these houses became virtually unsaleable overnight and the value of those that were to remain decreased immediately. Time is important and may witness great and rapid changes.

It is always as well to remember that because of the *lack of market information* which exists in many property markets, properties often fetch prices which, with better information, a purchaser would not have paid. However, many property transactions in many countries are of a confidential nature, which explains the reluctance of valuers to reveal details, although much could be done to make more information available without violating confidentiality. It sometimes happens that for various reasons a person must have a particular property, and to forestall other possible purchasers a price is paid which in reality is above the figure for which it could have been obtained. Should the purchaser subsequently have to sell the property because of a change of circumstances, there is no guarantee that the same figure will be obtained if buyers then in the market do not have the same pressing reasons for purchase. Such a possibility should be considered fully, particularly in valuing for mortgage purposes.

It is not suggested that the underlying conditions and factors considered in this chapter are completely comprehensive, and the reader will doubtless be able to think of other points. Nevertheless, those mentioned should be sufficient to illustrate the complexity of the background to property values, without, it is hoped, putting the would-be valuer off for ever.

The relevant factors will be different in different cases, but in all instances it will be important for the valuer to consider carefully all the features of a property, both its advantages and disadvantages, before any decision on its overall merit and value can be reached.

Chapter 7

Reasons for Valuations

In the earlier definition of "valuation" it was stated that it was the art or science of estimating the value for a specific purpose . . .". The expression "specific purpose" could refer to the fact that properties may be used for a whole range of different purposes such as residential, commercial, or industrial uses. However, the real significance is that there can be a wide variety of reasons for requiring a valuation, and it is possible to have a whole range of different values for one property at one particular moment in time, dependent upon the purpose of the valuation.

A valuer may be asked what a property will fetch if sold, and he or she then uses his or her skill and experience to advise the client as to the likely selling price. This is a *valuation for sale*. In making it the valuer will consider all the likely purchasers in the market at that time and all the alternative properties available to them, and having done that will estimate what price would result from the competition between the various potential purchasers.

A client may request a valuation because he or she is considering the purchase of a property. A *valuation for purchase* should not be quite as far-ranging in scope as one for sale, for although the range of people interested in the property may be just as great, the valuer's considerations must be directed towards the requirements and finances of one person – the client. In such circumstances the needs and personal characteristics of the client may result in a different decision from that which might result from considering the entire market. The client's subjective requirements may result in the value being lower to him or her than to others in the market whose subjective requirements are different. Conversely, they may result in them outbidding the market for a property which fits their requirements exactly, but which would be a compromise for other potential purchasers.

It may be that a valuation is required for *mortgage* purposes. At this stage the reader may not know what a mortgage is, but, briefly, it is a loan for which the security is an interest in property. Most people who purchase property have insufficient cash to pay all the purchase price. They pay as much as they can afford out of their own funds, borrowing the balance from some person or institution which has money to lend. The lender will not wish to lend money unless there is some sort of security which makes it relatively certain that their money can be retrieved should anything go wrong with the transaction. The loan is "secured" upon the property which is being purchased and the lender will be able to claim the money lent out of the value of the "mortgaged" property should the borrower default in any way.

The major objective in a mortgage valuation is to decide what sum a property is likely to fetch if a forced sale is necessary. When this is the case it is not always possible to put a property on the market at the most favourable time, or in the most favourable circumstances. The valuer therefore has to take a cautious view and has to decide what figure might be obtained in adverse conditions. Such a figure may differ considerably from that applicable to a valuation for sale, which in the majority of cases would only envisage a transaction being made in favourable circumstances. A valuation for mortgage purposes is a valuation for a lender who will not wish to take undue risks.

A valuer may be required to value a property for *rental* purposes because a client wishes to rent it. Account must be taken of what the client can afford to pay for the use of the property, and a figure of annual value will be required as opposed to a capital figure. The client must be advised what rent is appropriate and consideration must be given to what has been obtained for similar accommodation; in doing this the terms of letting of those comparable properties in comparison to those for the subject property must be carefully considered. It may be that a client is only able to obtain the property by paying what is considered to be more than he or she can afford or, alternatively, more than the property is worth to them. In such cases they will undoubtedly be advised only to do so if they can obtain a lease at the high rent for a relatively long period of time, as with time inflationary tendencies in the property market may in fact result in the original rent becoming low.

So the valuer will be influenced by the terms of a prospective lease as well as by physical factors such as the quality of the area and the property itself. Tenants will expect to pay lower rents if they are responsible for repairing the property than if the landlord is going to do those repairs for them. If there are onerous restrictions governing the use of the property potential tenants will similarly decrease their bids, and it is therefore important for the valuer to be fully aware of the purpose of the valuation and all the factors which are liable to be important.

A valuation for *insurance* demands a completely different approach. The market value of the property becomes to a large extent irrelevant, and the valuer has to calculate what it would cost to replace the bricks and mortar should the building or part of it be destroyed. The building is measured, its size calculated, and appropriate costings are applied to the various building works which would be entailed in replacing it, and in this way the cost of replacement, or the "value" of the building for insurance purposes is found. The actual value of the building if offered for sale on the market may be well in excess of this figure, if for no other reason than the fact that the market value includes the site as well as the building. It may alternatively happen that open market value is below the insurance "value", and this may occur where a building is rather ornate for its purpose, or is built of extravagantly expensive materials or to an expensive design.

Valuations are sometimes required for *balance sheet* purposes, when a company publishes its annual accounts, or for *redevelopment* purposes when someone thinks there may be redevelopment potential and wishes to have their views either confirmed or rebutted. Property may require valuing as an *asset* of a business if it is used for business purposes and the value of the business, including the property, has to be established. As noted earlier, the valuer must in such circumstances comply with the standards laid down either by statute or agreed by professional groups. In many countries this will entail compliance with the standards of the International Assets Valuation Standards Committee (IVSC). When a public authority wishes to acquire a property it is necessary to value it for compensation for *compulsory purchase*. A valuation for *rating* purposes may be needed where local taxation is concerned. When someone dies it may be necessary to value a property for *probate* purposes, that is to find the value upon which any tax payable at death is to be assessed (inheritance tax in the United Kingdom). In each of these cases there will be different considerations to take into account, many of the valuation criteria being dictated by law in some cases. As many of these valuations entail a considerable knowledge of advanced valuation methods and a not inconsiderable legal knowledge, they will not be discussed further in this book.

Although it is apparent that there may be a whole range of different values in any one property at any one time, depending upon the reason for the valuation, it should be remembered that there will only be one *market price*, which is the figure at which the property would change hands if offered for sale in the open market.

Chapter 8

Legal Interests in Property

It can be seen that the study of valuation is not quite as straightforward as might at first be imagined. Each different type of valuation may involve different techniques and different technical knowledge, and it is necessary to make a searching study of many subjects which affect property and property values.

One of these subjects is law. It was stated earlier that "valuation is concerned with a particular interest in property at a particular moment of time". Property itself cannot be owned; at law an interest in property is owned, which is somewhat different from owning the property itself. If the latter were possible there could normally only be one owner of a particular property at any one time (unless there happened to be joint owners). With interests in property it is possible for there to be many different interests in the same piece of property, each interest owned by a different person, and so there may be a hierarchy of interests at any one time, eg freehold, leasehold, subleasehold. The freeholder is the person who is usually regarded by the layman as the owner of a property, but in law each of the persons with a legal interest in the property is an owner.

A freeholder may not wish to use a property himself or herself but may have purchased it as an investment, and they may decide to split it into several parts physically and to let the various parts to different lessees. Each of them will own a legal interest in the property if all the correct legal formalities have been attended to. In time one of the lessees may decide that he or she no longer wishes to use the property and, providing the terms of the lease permit such an action, they may sublet part or all of the property which they lease. In such a case there will be three legal interests in that part of the property. As far as the entire property is concerned, the legal interests created will have resulted in it being divided on what might be referred to as both a horizontal basis and a vertical basis, ie separate interests have been created both at one level and at inferior levels. If the freeholder had purchased the property with the aid of a mortgage the mortgagee would also have a legal interest in it. It is not unusual to find on investigation that the legal interests in even relatively small properties are numerous.

When a lessee disposes of his or her entire legal interest to a third party, the transaction is known as an assignment of the interest, and the purchaser is known as the "assignee" of the lease.

From the valuer's point of view, if there are several interests in one property at one time, each interest will almost certainly have a different value, barring coincidence or absolutely identical circumstances, and the former will probably occur more often than the latter.

The valuer will therefore have to ensure, before making a valuation, exactly what interest is to be valued, and what are the terms on which the relevant interest is owned. Whenever possible, the valuer should personally examine the deed or lease, as details may be noticed which are vital as far as value is concerned, but which might not be thought important by a solicitor who is trained to notice legal detail and not to consider matters of value.

The valuer should also take care to specify quite clearly in the report what interest has been valued, and to include in the report a summary of all the important legal points so that there can be no misunderstanding as to the basis on which a valuation has been made.

The law of property is a subject in itself and it is not the purpose of this book to cover it. The student would be well advised to read the subject in a good textbook, as a thorough grasp of this branch of law will prove invaluable in a career as a valuer.

Chapter 9

Features of Property and the Property Market

Probably the main feature of *property interests* is that, to use an economist's expression, they are *heterogeneous*, or, in plain English, they are all different. Apart from joint interests in the same property, there will probably never be two interests which are exactly the same, and even though there may be row upon row of semi-detached houses which are apparently similar, they will all be on different sites. Although this may seem a minute distinction between them, it nevertheless means that each of the properties will be different.

Often such a small difference may be irrelevant in market terms, but in some instances it will make a considerable difference to the market value. The fact that certain houses are close to a school or a bus stop may be a distinct advantage, whereas others may be remote from these conveniences, although similar in all other respects. Other houses which are immediately adjacent to the school or bus stop may in turn suffer from the nuisance caused by such neighbours. So, along a street of basically identical houses, there may be differences in value resulting simply from their closeness to or distance from other property users.

Even when a new estate of basically identical houses is constructed, within a very short period of time differences between the properties will invariably develop. One person may move in and add another garage to a house, another may build an extension on to a kitchen, another may add a shower room, while yet another owner may install double glazing or roof insulation. The variations are almost limitless and the stage is very soon reached in which nearly all the properties have many detailed differences.

Similarly, with shops there are rarely two identical properties. Many built-up areas have shopping streets constructed around the turn of the century in which, from the street itself, the properties all appear to be similar. From the rear or from internal inspection many differences may be revealed, and it often turns out that nearly all the shops have been substantially altered during the passage of time. A shopkeeper who requires much storage space and much showroom space may build an extension over the rear yard of the property. Another shopkeeper may knock out partition walls in the main building to make it more suitable for his purposes. Other shopkeepers may also have built extensions at the rear, some smaller than that built by the first shopkeeper, others larger. An inspection of the various alterations and extensions may show that even where similar areas have been added, the type and quality of construction may vary. A number of shops which from the front appear similar may, therefore, on inspection all prove to be different to a significant degree.

Another important feature of *property* is that, relatively speaking it is *durable*. Many goods wear out and deteriorate rapidly, but property normally lasts for many years with a very slow rate of deterioration. Even when a property is so old that the deterioration is considerable, it is often found that the value of the land on which it stands has appreciated at a greater rate than the rate of depreciation in the value of the building. The result is that although the quality and physical condition of the building has deteriorated, the overall value of the land and buildings has been maintained, and quite regularly it will have increased. In this respect property is quite different from most other investments; it is very durable over time and generally *affords very good security for money over a long period*.

In comparison with other goods a *large amount of capital is usually required to purchase property*. Even the meanest interest will usually require a considerable amount of money for its purchase, and allied to this is the fact that *property is often in units which cannot easily be divided into smaller units*. The result is that the minimum sum required for property purchase is usually quite substantial, and it is necessary for a potential investor in land and buildings to be able to raise such a sum before he or she can in fact realise their ambitions. If a piece of cheese is purchased, should the original piece which the grocer weighs be too heavy for the customer, he merely cuts some off to leave a lower-priced piece. More often than not, such a solution is impossible with a property if it proves to be beyond the means of the would-be purchaser.

Sometimes it is possible to split buildings up into smaller units by using a little imagination and possibly by making minor structural adaptations. This may enable persons with smaller funds to purchase them, but often such a course of action is not possible and a would-be purchaser has to raise sufficient funds to buy the entire property otherwise a purchase may not be possible at all.

The market value of some properties has become so great that it is difficult for even very large organisations to purchase them. Not only is it difficult to raise the large sums of money required, (in March 1987 the long leasehold interest in the Regent Hotel, Sydney sold to EIE International Corp for $145 m at which figure there could only have been a limited number of potential purchasers, but this price appeared cheap in comparison to the $270 m paid for it in June 1989 by Nisshin Kisen Kaisha Ltd), but the purchase of a property worth many millions of dollars or pounds may result in too great a proportion of an investment portfolio being held in one property. With the passage of time even the figures quoted above seem small in comparison to the Bluewater Park retail development near Dartford in England, which was developed in the late 1990s by the Lend Lease Corporation and which opened in March 1999 with an estimated value for the completed development of £1.1 billion. Having too much capital invested in one property may be an unattractive proposition and the very high value of some properties may in any event make them almost impossible to sell rapidly if good management policy so demands. The solution whereby very expensive properties may be purchased by syndicates is not necessarily a good solution, as the portfolio management needs of syndicate members may vary considerably at any time; whereas, for example, one syndicate member may wish to sell a property, other members of the syndicate might wish to retain it. Also, they might not wish to have certain possible purchasers of shares as syndicate members.

There has, therefore, recently been consideration in a number of countries where such practices do not currently exist of the possible unitisation of property interests, and indeed various unitisation schemes have been put into practice in some countries. With unitisation it is possible for a legal interest in a property to be divided into a number of units of ownership, just as the value of a company is represented by a number of shares in it. Units in a property can then be bought and sold thus allowing each owner of units to pursue independent portfolio management policies. It also enables the value of very expensive properties to be broken down into units which are within the purchasing range of a great many more organisations, with the result that sales of interests in such properties are not dependent upon the ability of a small number of potential purchasers to raise incredibly large sums of money. Such a development should provide a more active market in interests in the more valuable properties and should prevent ownership being an unwieldy burden which few are able or willing to accept.

However, such schemes are not without their problems among them being the disadvantages which are associated with part ownership of any item. Property management decisions may not be within the powers of any individual unit holders who may have either little or only limited control over the quality of management for a property in which they own a unit. Also, individual unit holders may have no control over who else becomes a unit holder in the same property, and it may happen that a situation arises in which one unit holder is, so to speak, in partnership with another organisation with which they would not normally wish to be associated. Problems can arise when one unit holder goes into liquidation, as occurred with a major unitised property in Sydney, which could put a blight on or result in the devaluation of other owners' units for reasons completely unrelated to the real value of the property. For a number of reasons, therefore, it is not clear that the unitisation of large-value properties is the complete answer to making them more marketable as there remain unattractive aspects of only being the owner of a unit or units as opposed to completely owning an interest in property.

Proof of ownership of property can cause problems and it is often a quite difficult, lengthy and costly process to prove ownership of a particular interest satisfactorily. The owner of a pen who wishes to sell it proves his ownership principally by possession and the fact that he carries the pen around with him. If he can also produce a receipt showing when and where he purchased it, this will usually be more than adequate to satisfy any would-be purchaser. Even with larger articles such as a motorcar proof of ownership is still relatively simple. With a car the mere fact of possession is generally a good indication and if it can be backed up by possession of the registration book and a receipt for the purchase money this will invariably be sufficient.

Where property is concerned, possession is not always indicative of ownership, and it is obviously impossible to pick up a piece of property and carry it around. Many properties have several interests existing in them, each of which is capable of a transfer of ownership, and to establish ownership of any particular interest the possession of Title Deeds or some other proof of ownership is necessary. These will not always be simple and straightforward to understand, and there are sometimes omissions in them. It may be that on a transfer of ownership there will be lengthy work involved in the investigation of the precise details of ownership, some of which may never be cleared up to the complete satisfaction of all concerned.

As a result *the costs of transfer are invariably relatively high* when interests in property change hands. A skilled solicitor should be employed to undertake such work (it can be done by individuals themselves, but this is generally inadvisable), and quite understandably a skilled professional will require adequate remuneration for his or her services.

Because of the need to look into the legal history of a property and the complexities which may arise, *much time is usually involved in transferring interests*. A transaction with most types of goods can be carried to completion within, at the very most, a matter of days. Although in theory it is possible for transfers of property also to be completed within a few days, in practice this rarely occurs, as ideal circumstances seldom exist, and the time involved in property transfer tends to run into weeks, or even months.

Another feature of the property market is that normally it does not consist of one large market, but of a series of smaller markets, each of which is local in nature. Even within local markets the knowledge of property transactions tends to be far from perfect, and purchasers and vendors suffer from a shortage of information concerning past transactions. *Property markets* are what economists call *imperfect markets*, and the imperfection of knowledge becomes even more critical if an area in which a valuer has to work is not familiar to him or her. A person may have been familiar with many transactions in one locality over a long period of time, but if activities are moved to another area some way away they will be in the position of having to start from scratch in garnering background information about that new market. In such circumstances it will probably be much quicker and simpler for the investor to go direct to an expert in that local market, and to employ an experienced and competent surveyor as an advisor. This again will result in *expense on professional fees*, which would not be necessary if one was dealing with goods which have a relatively uniform value throughout the whole of the country. In the latter case the expert in one local market will also be sufficiently well versed to operate in another local market, but where property is concerned investors will normally have to allow for the costs of professional advice in calculating their overall expenses, unless they happen to be sufficiently expert to do their own work.

Property can provide a source of income and in this respect it is different from many other durable possessions. Because of this the ownership of property is often attractive to investors as well as to would-be users of the property. It provides an opportunity for a person to save money and having saved it to purchase something which then provides a regular income. This income is a direct result of, and the reward for, the original saving or the deferment of immediate consumption. Consequently property is often a particularly attractive investment for savers.

Another feature of the property market is *that the total supply of land is to all intents and purposes fixed,* although land reclamation may provide marginal increases in the total supply. Supply is not necessarily fixed in a particular use, but the total supply of land is fixed, and strict planning control in many countries has in effect tended to restrict the supply of additional land to a particular use. Although land may not be physically restricted to a use, planners have in fact often laid down rather rigid rules about what may be done and where it may be done. Because of this, and also because building is not normally a rapid process, an increase in the supply of a particular type of property can normally only take place over a relatively long period. Things happen more slowly in property markets than in most other markets. Consequently they will only attract people who can afford to let things happen slowly.

In the short run it may be possible to increase the supply of a particular type of property by conversion from one use to another. If there is a need for offices and there are not enough available, a spacious Georgian house could be converted into offices. But, again, it is normally only possible to take such action if planning permission for a change of use is obtained, and it may be that although market demand suggests that more offices are required, the local planning authority may not agree to their being provided.

It can therefore be seen that *the supply side of the property market cannot react very rapidly to changes in demand.*

Basic demand factors must also be considered. As *property fulfils a basic need of mankind* there will always be a reasonably strong demand for it. Everything that man does requires either land upon which to do it, or buildings and land. People live in houses; they work in buildings; those who work outside work on land, and even those who fly for a livelihood need land to take off from and to land on. So land fulfils a basic need of man, and those who own and manage this essential commodity are important people, more so in an advanced society. This is becoming more and more apparent with the passage of time, although it is only in the relatively recent past that land use has been considered on a long-term basis other than by the more enlightened estate owners. Land is a valuable asset and it is likely that great attention will be paid to land management policies in the future in most countries.

Chapter 10

Factors which Cause Changes in the Value of Property and Variations in Value between Properties

As value is a function of supply and demand, a consideration of the factors which are likely to affect either the supply of or the demand for property is merited.

Any *increases or decreases in population* will obviously affect property values. If there is an increase, all other things being equal, demand will increase. A larger population will require more housing in which to live, more buildings in which to work, and more buildings for leisure-time activities and all the other ancillary activities associated with modern life.

Population changes may be considered on a national basis or on a more local basis, and if there are movements of population from one area to another property values are likely to be affected in both areas. Demand will increase in one area and it will decrease in the other, unless any other factors counterbalance such changes.

Changes in the age distribution of the population may affect property values. For example, in one period a large proportion of the population may be aged under 40, and in another period the majority of the population may be over 40. Such a change in age distribution will affect the type of demand, as, whereas older people probably demand more bungalows and accommodation specifically designed for retirement living, younger people may be quite happy living in flats. It may be that the adaptation of the existing housing stock is sufficient to cater for such changes in demand, but it may equally well be that wholesale building is required. It is likely that variations in the age distribution of the population will also result in changes in demand for ancillary accommodation; young people may create a demand for gymnasiums, whereas the demand from older people may be for bowling greens and other leisure provision more appropriate to their ages.

Any change in the proportion of married people to single people will also be reflected in the demand for different types of properties, and hence in property values. Single people are more likely to be satisfied with one-room bedsitters than are married couples, who, even if they are extremely happily married, will sometimes wish to have the privacy of a room to themselves. Quite clearly they will wish to have a different type of accommodation from single people, and what is a suitable variety of property at one point in time may be an unsuitable variety when the ratio of single people to married people has changed.

Changes in the age distribution of the population and changes in the proportion of married to single people may also affect the size of *average disposable income*, which in turn is likely to affect the amount of money which individuals have available for house purchase. Population trends are therefore very important indicators of possible future changes in demand which might result from variations of this type. The size of disposable income will also be very much dependent on employment levels, the prevailing types of employment, and wage and salary levels in the prevailing industries and professions in any particular area.

Changes in fashion and taste also affect property values. The effect is much more obvious in a market such as the clothes market; at one point in time 15-inch trouser bottoms may be in demand, a year or two later a fashion for 19-inch may be in vogue, while a later change may be to bell-bottoms, before the whole cycle of change begins again. Fashion is important with property also, although in a less conspicuous way. Fashions change, as a study of older properties will reveal. Even though many pre-war houses provide good accommodation, the appearance of some is of doubtful quality. Often their unattractive design makes them difficult to sell, and many potential purchasers prefer to purchase a house of modern design and appearance, even though it may in other respects leave a lot to be desired. Many people want to be up-to-date with fashion and do not like to be left behind by others who have newer and more fashionable houses. Although the costly nature of property makes it difficult, if not impossible, for people to keep right up-to-date, changes in fashion over a period of years are evidenced in the price trends of different types of property.

Similarly, *the fashionableness of areas can change*. Areas of London such as Chelsea, Islington, and Fulham are outstanding examples of places which at one time were not particularly fashionable as places in which to live, but which later became so, with the consequence that the prices of houses in those areas rose. Many humble cottages in Kensington have become the highly-priced homes of the affluent because it is fashionable to live in Kensington. Large sums have been spent on modernising and rebuilding them, and yet in many northern cities similar dwellings have been cleared by bulldozers because they were in areas which were not fashionable, and there were no affluent purchasers prepared to spend money on improving them. There may well have been other factors which led to this happening, but the fact that the areas were unfashionable doubtless helped to speed up the process of demolition.

Not only may an unfashionable area become fashionable, but unfashionable designs of property may also become fashionable. It may also happen that a type of property which is fashionable in one area, may be very unfashionable in other areas.

Changes in the type of society are important, but in the United Kingdom such changes are only likely to occur on a limited scale nowadays, as the major changes of this type took place in earlier times. The change from an agrarian society to an industrial society is probably the most obvious example, and to a limited extent this change is still taking place today in that people are still moving from jobs on the land to work in industrial areas. In developing countries this trend may currently be very strong. However, a reverse movement is also taking place in the United Kingdom, in that whereas in the past people moved out of the country to live in towns and their former homes were left empty and decaying, nowadays when the farm labourer moves from the country into industry to obtain higher wages his empty cottage is bought by a town dweller for conversion into a luxury country residence. This is another example of a change in society which results in changes in property values.

In countries in which population is still expanding, such as Australia, or where there are substantial changes occurring such as the industrialisation of the economy, such as in Malaysia, property values will be very much affected by such major changes. Towns in many parts of both Australia and Malaysia are still expanding and population growth and movement help to sustain and even increase property values in areas so affected.

Changes in technology may also affect property values. In colder climates a house with central heating is invariably more valuable than a similar house which does not have central heating, but the difference in values has been decreased by the advent of improved installation techniques which enable complete central heating systems to be installed rapidly with a minimum of fuss and at relatively low cost.

Where factories are concerned, changes in technology may have far-reaching effects on values. It may be that new machines are designed and manufactured which carry out industrial processes more efficiently and more cheaply than existing machinery. If the new machinery is, however, taller than existing factory buildings, workshops with greater headroom will be required, and existing workshops may consequently become obsolete and less valuable. With warehouses in particular, modern mechanical and electronic storage methods have facilitated storage to about 35 feet or even higher, which has caused many older warehouse properties to be less attractive to possible users.

Changes in building methods may affect property values. Improved building techniques will not necessarily result in changes in value; they may simply result in builders making higher profits if there is no increase in supply as a result of the implementation of the improved techniques. However, if the new techniques do result in an increased supply of new buildings to the market, unless there is an increased demand, or an increase in the money supply, the values or prices of that type of property should fall. Similarly, *changes in building costs* may affect values if the result is that the supply of new properties to the market is also changed. The *money supply* has already been mentioned in passing. If there is a change in the amount of money available for the purchase of property, without there being changes in any other supply and demand factors, this should result in changed prices in the property market. If more money becomes available for house-purchase, rivalry among potential purchasers may result in prices being bid up, while if there is a shortage of money the lack of purchasers with cash available may result in prices falling back. It should be noted that if there is an overall increase of money in the economy, any resultant price-changes may not represent changes of value in real terms, as the purchasing power of money may have fallen considerably where an increased supply of money is available to purchase only a fixed stock of goods.

Not only does the ease or difficulty of obtaining finance affect property prices, the *cost of such finance* is also very important. Even though plentiful funds may be available for property purchase, if the cost of borrowing is high potential purchasers may be deterred from entering the market. Borrowing £50,000 for house purchase at 8% pa results in interest charges of £4,000 pa, which may be well within the range of many people. However, if the interest rate increases to 16% a great many people would be unable to afford the increased interest charges, which represent almost another £77 per week, every 1% increase in the cost of such a loan representing almost another £10 per week in funding costs. Changes in the cost of borrowing therefore affect property prices, as they affect the amount that potential purchasers can afford to borrow, although short-term fluctuations are not likely to substantially affect the market. However, if for instance increases in the cost of borrowing are substantial and prolonged they may result in people "trading-down" from expensive to cheaper houses.

Proximity to good means of communication is a great advantage with virtually any property, and if there is any *improvement or deterioration in the means of communication* it is likely to have a marked effect on the value of properties which either benefit or suffer as a result of the change. Country cottages reasonably accessible to a motorway access or a main-line railway station will usually fetch higher prices than those which are more remote, while with industrial and commercial properties the resultant price differences are likely to be even more marked. Few industrialists will wish to be too remote from their markets, particularly if transport costs form a large proportion of their overheads. Indeed, if the latter is the case, excessive transport costs could make their products uncompetitive in a competitive market. Employers of large work forces will benefit from good travel facilities in that the ease of travelling to work will make it simpler for them to recruit labour, and they may even be able to do so at lower wages than if they were situated in a location which involved employees in high travel costs. A study of property advertisements will readily reveal the emphasis placed on the existence of good means of communication, even with the humblest type of residence.

Planning control probably has a greater effect on property values than any other single factor, possibly even greater than all other factors combined. The power of the planner in today's world is very great indeed, and the decision of a planning committee can result in huge increases in value arising, or alternatively in such increases being denied. The farmer who obtains planning permission to develop his land as a residential or industrial estate may find that its value has increased 50-fold as a result of obtaining the favourable decision. Had planning permission been refused, the value would have remained as for agricultural purposes with perhaps a slight incremental value reflecting the hope that permission might be obtained at some future date.

Some of the big differentials which generally exist between the value of land with planning permission to a landowner and the value to the owner of similar land with no planning permission will be partly eliminated in circumstances where gains arising from the development value of land are taxed.

Nevertheless, particularly where the potential purchaser is concerned, there will be a considerable difference in value in the market, dependent upon whether planning permission for a valuable use has been granted or not.

The above factors are some of the more obvious which may affect property values, but there will also be others which the reader will think of or encounter in practice. One of the essential items in the valuer's armoury should be an alertness to all the possibilities, with the resultant ability to detect the vital factors in a given situation.

There are many factors which will cause variations in value between what might otherwise appear to be similar blocks of land or similar properties, and some of them have been referred to in previous chapters or earlier in this chapter (for example, proximity to or remoteness from good means of communications). Apart from the individual features of properties referred to in chapter 5, there is a range of specific considerations which are likely to be relevant to the value of any particular property.

The *situation of a property* within a particular locality may be very important particularly if there are variations in the quality of the environment within the locality.

Proximity to local amenities can be a positive benefit unless a property is so close that it suffers disadvantages, for example from excessive traffic loads generated by a nearby shopping centre or from the periodic noise emanating from and occasional traffic congestion caused by a local school.

The adequacy of *access on to a site* and the length of *road frontage* can be important factors in determining value. A property with inadequate road frontage can be disadvantaged while at the other extreme excessively long road frontages can result in excessively high fencing costs and possibly also security problems which may in fact adversely affect market value.

Site area is an important consideration as unless site size is sufficiently large for the current or intended use of a property, market value is likely to be depreciated. At the other extreme, value is likely to be enhanced if the size of a site is sufficiently large to offer potential for the extension of existing uses or for the further subdivision of the site and the subsequent sale of part of it.

The overall *shape of a site* and the relationship between its width and depth may affect both site value and the value of any property placed on it. Awkwardly shaped sites can result in the use of parts of them being severely restricted, while where sites are either too narrow or too wide in relation to their overall area there may be serious design constraints which reduce total value below that which would apply to a more appropriately shaped site of the same area.

The *slope of a site* or the *relative position of a site on sloping land* can be an important determinant of value. If there is an excessive slope to a site it could substantially increase initial development costs, while it could subsequently put off potential purchasers of a completed property. Similarly, it may well be that houses built towards the top of a site may command higher values than similar properties lower down a slope as purchasers often prefer elevated locations, particularly if they also benefit from a good view. However, if the more elevated location is accompanied by problems of access because of the slope or by greater exposure to inclement weather conditions, the reverse may be the case.

With residential properties in particular the existence or non-existence of a *good view* may, as hinted above, result in substantial variations in value between otherwise similar properties, and as indicated in chapter 5 a southerly aspect in the northern hemisphere or a northerly aspect in the southern hemisphere may be desirable features with resultant bonuses in value.

The possible *exposure* of any type of property *to such things as flooding, subsidence or problems from soil erosion* is likely to have a considerable effect on value. The difference in elevation of a few feet could result in one property being prone to occasional flooding and another being immune to it, which would normally affect their relative values, while any possibility of damage from land subsidence would have a detrimental effect on value.

A valuer should always take care to *note detrimental features of any kind* in respect of a property, some of which may not always be apparent from a physical inspection. Adjoining owners may for instance enjoy rights of light or rights of way over a property, which in some cases might only be identified by an inspection of title deeds or registration certificates, and failure to identify such encumbrances could result in the over-valuation of a particular property.

There will often be a range of features of the above type which cause variations in value between otherwise similar properties in a particular locality, and an important part of the valuer's job will be to identify such features and subsequently to take them into account in the valuation process.

Chapter 11

The Role of the Valuer

What is the role of the valuer and is it necessary to teach valuation? Statements to the effect that there is no need to consult a professional property valuer, and that old George who quietly sups his ale in the bar of the village pub knows as much about property values as anyone are often heard. Buy him a pint and he will tell you what prices different properties have fetched, what their advantages and disadvantages have been, and how properties currently on the market compare with these past sales. He may well be very good on local values, and with excellent local knowledge the advice which he gives may be first-class. In earlier chapters however, we considered many factors which should explain why old George will not always be a competent valuer. First, properties come in all shapes and sizes, and although run-of-the-mill semi-detached houses (duplexes) may be easy for the competent amateur to value, others will be much more complex and may create valuation problems with which only the trained specialist can cope.

Legal complications can only be allowed for if a valuer has a good legal education. This may be necessary for him even to appreciate that a legal problem exists in some circumstances. Even simple legal factors can create problems for the valuer. Is a freehold or a leasehold interest to be valued? If a leasehold interest, what is the length of the unexpired term and how will a term of 35 years differ in value from a term of 28 years in a similar property? How large is the rent or ground rent payable under a lease? The higher the rent which has to be paid, the less valuable will be the leasehold interest, all other things being equal. What are the other rights and restrictions under the lease, apart from the length of the term and the level of rent? Is the tenant responsible for repairs, are there restrictions limiting the user of the property? Who pays the insurance premium? These and many other points will be important legal factors affecting value, and a layman would find it difficult, and in many cases impossible, to cope with such valuation problems.

Third parties may have legal rights over a property. There may be rights of way or rights of light owned by adjoining property owners, and such rights may restrict an owner's user of his or her property. Such rights may have a particularly adverse affect on value if redevelopment is under consideration and the most valuable scheme cannot be put into effect because of their existence.

Finding out about such rights is part of the job of the trained professional valuer. There was an instance in the United Kingdom in the 1970s of a couple who purchased a new house on a new estate, only to have the local ramblers' association walk up the garden path and through their house. The ramblers were exercising rights along a footpath over which the house had been built. The trained valuer should be alert to such dangers, although it does not need a trained valuer to appreciate the drastic effect on value caused by such a state of affairs. This is an extreme example, but even the right of another person to share the use of a drive can have an adverse effect on a house, and the experienced valuer should be able to make a fairly accurate assessment of that effect in money terms.

Even where basically similar units exist side-by-side, the trained valuer with a good grounding in building methods and a knowledge of building costs can make accurate allowances for differences in the standards of finish and maintenance. How much difference in value does the existence of double glazing, central heating or air conditioning make, and what is the cost of installing such items? How much extra value results from the addition of a two-room extension, and what would be the cost of adding a similar extension to other properties? The layman with no knowledge of building costs or building technology would find it difficult to answer such questions.

Values will not necessarily vary proportionately with size, nor will they necessarily always vary by the same amount in respect of the presence or absence of such features as central-heating. This may not be such an indispensable feature in a warm locality as it would be in a cold locality, and it may be that only the trained valuer with his intimate knowledge of market conditions will be able to make an accurate allowance for the absence of a central heating system. Value is not necessarily related to cost, although a layman might sometimes be tempted to think it is.

Variations in values with changes in locality are sometimes quite astounding and the untrained person might often think some differentials to be so large as to be unrealistic and not factual. The trained valuer, however, is used to what may sometimes seem inexplicable market variations, and far from being puzzled by them acknowledges them to be market facts for which due allowance must be made.

Whereas a layman may rapidly get out of his depth as far as values are concerned once he moves away from his own territory, a skilled valuer should be aware of the general factors which affect values and should be able to accustom himself or herself fairly rapidly to the tone of values in an area to which he or she is a stranger. They should be far-seeing and versatile enough to be able to value over fairly large geographical areas and should even be able to work from basic principles to value the type of property which rarely comes on to the market. The layman may be a good valuer with standard, uncomplicated properties in his own immediate locality, but the trained valuer should be competent with a wide range of properties over a wider area.

Liability for repairs has already been mentioned and one of the skills of the valuer is to be able to assess the cost of outgoings arising from property ownership. How much does it cost to insure a property and how is an insurance valuation made? What will it cost to paint and maintain cast-iron rainwater goods and how will this compare with the costs for vinyl or aluminium rainwater goods? If a landlord supplies services for tenants, how much will these services cost? How much does it cost to provide central heating, hot water, gardeners, hall porters and lifts?

These last few items are most likely to arise in the case of commercial properties, and it is with such properties that anyone but the skilled and experienced valuer is likely to find the greatest difficulty. Few commercial properties are even remotely similar to each other. Shops, offices, factories and warehouses are frequently built or adapted to the requirements of the current occupier. Only the skilled valuer will be able to make allowances accurately to reflect differences in properties of this type. He or she will be trained to know the effect on values of various features and will be closely in touch with the requirements of the users of such properties, and with the state of supply and demand in the market.

Where more specialised buildings are concerned only the trained valuer can hope to make a reasonable assessment of value. Public houses, petrol-filling stations, cinemas, breweries, bottling plants, sports stadia and the like can only be valued by a person who knows just what affects the value of such properties. Such knowledge can only result from a good training and a thorough understanding of the market.

Apart from property law, which in many countries has developed over a number of centuries, nowadays there is in most countries a large amount of statute law which affects properties, and, as a result, property values also. Many valuations can, consequently, only be done properly if the valuer is fully apprised of all relevant law. In addition, there is often a set of valuation rules laid down by professional accounting and valuation organizations which must be observed in respect of valuations for specific purposes, such as for inclusion in company reports. It is therefore essential for valuers to be fully aware of all law and all rules and regulations which are likely to affect value or to dictate the approach to be adopted when undertaking a valuation, and a study of such areas is an essential part of both the initial and continuing education of valuers.

It seems to be a popular pastime nowadays to criticise the professional man or woman and to decry their roles, particularly with those least qualified to understand and therefore to criticise. As with all professions there is a whole range of skills among valuers and just as there are very good valuers there are also poor and even incompetent ones. In this respect the valuation profession is no different from any other, but the fully-skilled valuer has much to offer both individuals and society. No one else can provide his or her particular skill and expertise, nor the intimate knowledge of the wide range of factors which affect the value of properties.

Chapter 12

Rates of Interest and Yields

The "yield" of an investment is a particularly important concept as it indicates to an investor the level of earnings of that investment, or the speed at which it will earn money. All other things being equal, the higher the yield the more attractive an investment will be. A 20% yielding investment will pay £20 pa for every £100 invested, whereas with a 5% investment the return will only be £5 for every £100 invested. Thus, all other things being equal, the former would be the more attractive investment, but the concept of yields is a little more complicated, for a yield also gives an indication of the degree of risk attached to an investment.

Some investments will involve greater risk than others, and it is only natural that a higher return will have to be offered on riskier investments in order to tempt investors away from safer ones. Unless there is a possibility of a greater return a person is hardly likely to invest his or her savings in a risky venture, but if there is the prospect of a large pay-out this may persuade them that the risk is worth taking. This is not dissimilar from backing horses, in that the possible winnings from backing an odds-on winner are very much smaller, but more likely, than those which it is remotely possible to win by backing a 100–1 against horse which has little chance of winning.

In considering two investments an investor may also be influenced by the concept of what is known as "the pay-back period". With the 20% yield mentioned earlier £20 interest is received each year for an investment of £100. After five years the capital has been "paid-back", whereas with the 5% investment the capital would not have been paid back until 20 years had passed. This concept ignores the fact that annual earnings are an essential requirement from an investment. It is nevertheless a useful indicator as, with the riskier investment, the investor, having recouped the original capital after five years can regard all future returns as the reward for risk-taking. If there is a considerable risk attached to an investment an investor is likely to sleep better at night if the pay-back period is short. It is, of course, true that the shorter the pay-back period, the higher will be the yield. The incidence of tax on interest earned may lengthen the pay-back period.

In deciding what yield is acceptable from an investment an investor will work on the basic principle that the greater the overall risk the higher will be the yield required. Conversely, as risks decrease he or she will be happy with lower yields, although it should never be forgotten that investors will always hope to obtain the highest yield possible from any investment.

The yield which can be obtained will be affected by factors other than the risk directly attached to the actual investment under consideration. The attractions of alternative investments have already been discussed, and any prudent investor will always view the entire range of investment possibilities and make calculations to assist in determining what course of action is best.

Before looking at the type of calculation which should be made it is necessary to explain the expression "nominal rate of interest". A company may decide to raise money to finance its activities, or to obtain capital from the market, as it is sometimes described. To do so it may decide to issue stock in £100 units at a nominal rate of interest which will depend upon the conditions in the money market at the time the stock is issued. This nominal rate of interest will have to be sufficiently high to persuade investors to part with their savings, otherwise the firm will not be successful in its efforts to raise capital. Not only will it have to be high enough to persuade investors that saving is more attractive than immediate expenditure, but it will have to be high enough to compete successfully with alternative forms of investment. Having considered these factors the company may decide that 10% pa is the correct rate of interest at which to offer the stock, and each investor who purchases a unit of the stock will thereafter receive £10 pa for every £100 of stock purchased. This 10% will be the nominal rate of interest and the stock will be in units of £100 nominal value, but this does not mean that each stock unit will always be worth £100 or that the actual yield will always be 10%. Both the value of the stock and the yield may vary considerably with time and changing conditions in the financial market, and it may be that the only occasion on which the nominal rate of interest and the yield are the same is on the actual date of issue of the stock, (the date when it is first offered for sale).

The nominal rate at which stock is issued will usually be closely related to the minimum lending rate of the major banks at the date of issue. It is normal for governments to have a major influence in the important policy decisions of the "government bank" (that is the bank which is either under the direct control of the government or whose connections with the government are such that it may be obliged to react to the wishes of the government of the day). Without such an arrangement it would be exceedingly difficult for any government to attempt to control economic conditions and to affect trends in an economy. As a result, the rate at which the government influenced bank or banks is prepared to lend money to a limited range of low-risk borrowers is an important financial indicator which reflects minimum lending rates in an economy to which all other lending and borrowing will tend to relate. In the United Kingdom the Bank of England has traditionally been the "government bank" in this respect and its Minimum Lending Rate, later referred to as the "base rate", which was for many years determined by the Chancellor of the Exchequer, has been the normal controlling factor in determining the cost of finance. However, in the late 1990s the Bank was given a greater degree of independence in that the task of determining base lending rate levels and changes in interest rates was moved from the Chancellor of the Exchequer to a committee of the Bank of England.

A few simple examples will be considered to show how nominal rates and yields work in practice.

Example 1

An inspection of the London Stock Market prices shows that on February 16th 2000 the following British Funds were listed.

	Price £	+ or –	52 week		Yield	
			high	low	Int	Red
Treasury 13% 2000	102.75	–.04	110.86	102.75	12.65	6.06
Treasury 5.5% 08–12	97.69	–.04	110.05	95.68	5.63	5.76
Consols 2.5%	49.33	+.16	55.69	46.78	5.07	–

The left-hand column names the stock, the column headed "Price £" gives the value of a £100 unit of the stock on the previous day, while the "+ or -" column indicates what change in value that represented from the previous market price. So with Consols 2.5% the current market price is £49.33 for a £100 nominal value unit, which represents an increase of £0.16 (or 16 pence) over the previous day's price. The "52 week high low" column indicates the highest and lowest prices at which a £100 nominal value unit of the stock has traded over the previous 52 week period, and it can be seen that the highest trading price for Treasury 13% 2000 Stock has been £110.86 and the lowest price has been £102.75. Yield data is provided in two separate columns headed "Int" and "Red". The "Int" column indicates what percentage yield the current interest payments represent related to current market price, while the "Red" column shows the yield which will be obtained if the stock is bought at the current price and held until it is redeemed by the borrower (until the borrower pays back the original capital borrowed). With Treasury 13% 2000 Stock a return of £13 on a nominal value unit of £100 is in fact a 12.65 return on the current market value of £102.75.

When the nominal value of the stock is repaid to the lender the stock is said to be repaid at "par". Sometimes stock is repaid above par, which in effect means that the holder of the stock at redemption date receives the nominal value of the stock plus a bonus payment from the borrower. So, if £100 stock units were repayable at three over par the stockholder would receive £103 on the redemption date. The receipt of the extra money results in the yield during the period of ownership being increased slightly and the redemption column shows the yield calculated to include any redemption bonuses to be paid. The fact that the redemption yield on Treasury 13% 2000 Stock is only quoted as 6.06% suggests that there may be only one half year's interest remaining to be paid on that stock before redemption or that the redemption yield is based purely on the "bonus" which will be paid by the government when it redeems the stock.

Not all stock has a redemption date neither does all redeemable stock carry an entitlement to a bonus at redemption date. From the above listings it can be deduced that Treasury 13% 2000 stock will be redeemed (that is the stockholder will be repaid the nominal value of £100 by the government) sometime during 2000, the 2000 in the title of the stock indicating the year in which it will be redeemed by the borrowers, in this case the British Government. The redemption yields quoted may vary from the nominal yield of a stock both because of the right to receive a bonus payment at redemption date and also because the current price of the stock is either above or below the nominal value of the stock which is the amount that will eventually be redeemed.

With Consols 2.5% (an abbreviated way of referring to 2.5% Consolidated Stock) there is no promise by the British Government to redeem the stock. In effect the original purchaser of the stock entered into an agreement to make a permanent loan to the government of the £100 which was originally paid for a unit of the stock which the market currently values at only £49.33. Such a price is reached in the market after investors have evaluated the risk attached to this stock relative to the risks attached to other investments, and after they have also compared the return of 2.5% pa paid on this stock with the current returns obtainable from other investments. With 2.5% Consolidated Stock at a market price of £49.33 an investor with £10,000 to invest could have purchased stock to the nominal value found from the following simple calculation:

$$\frac{\text{Amount to invest}}{\text{Purchase price}} = \text{Units of stock which can be bought}$$

$$\frac{10,000}{49.33} = 202.716 \text{ units or £20,271 nominal value of stock can be bought}$$

The realistic purchase would in fact be £20,200 of stock – costing 202 × £49.33 or £9964.66 – which with a nominal rate of interest of 2.5% would produce an annual return found from the simple calculation:

Nominal value of stock held × nominal rate of interest = total return

In this instance the investor's total return would be

$$\frac{£20,200 \times 2.5}{100} = £505 \text{ pa}$$

The actual yield to the purchaser's investment would be found from the formula

$$\frac{\text{Actual return}}{\text{Purchase price}} \times 100 = \text{Yield, or in this case } \frac{£505 \times 100}{9,964.66} = 5.0679\%$$

The above calculation indicates how the newspaper quoted yield of 5.07% was found for Consols 2.5%, and it also indicates that potential purchasers were prepared to deal at a price which gave them a return of 5.07% on money invested. This yield is considerably higher than the nominal yield of 2.5% because units of stock can be bought at much lower than their nominal value. If the current market price of a stock is above the nominal value, the actual yield on a purchase at the current value would in fact be below the nominal rate of interest paid on it.

The possibility of the market value of a stock being above or below the nominal value has been considered, so it will already be apparent to the reader that stock prices fluctuate. A study of the daily Stock Exchange list over a short period will soon confirm this fact. Fluctuations in prices are nothing more than the reaction of the price mechanism to supply and demand factors. If a stock suddenly becomes popular and investors wish to buy it, its price will rise, whereas if its popularity wanes and investors offer it for sale on the market then its price will fall, unless there is an abundance of would-be purchasers. The operation of the market fixes the price of stocks and shares just as with most other goods.

The reasons for a variation in supply or demand of a stock or share may be many, but, whatever the reasons, a change in price will also result in a change in the yield. As the price of a stock rises so will the yield fall (unless the price rise is balanced by a simultaneous increase in dividends paid on the stock). The more popular a stock becomes the more its price is likely to rise and the yield to fall and vice versa. Investors, in their eagerness to become owners of the stock, will bid up its price, and therefore a low yield is a reflection of the eagerness of investors to own a particular type of investment, while a high yield reflects their comparative lack of interest in a stock.

The relative regard of investors for stocks and shares can be seen from an inspection of the Stock Exchanges prices, the following random selection of figures again being as at 16 February 2000.

	Price	+ or −	52 weeks High	52 weeks Low	Yield	PE
British Land	343	+11.5	586.5	329	3.0	34.0
Marks & Spencers	229.75	−8.25	461	229.75	6.3	17.7
Amstrad	425	−4	556.5	43.75	0.2	56.7
Tomkins	165	−10	299.5	156.5	7.3	7.0
Manchester United	305	+10	305	170.5	0.6	51.7
Rank Group	149.5	−3.5	303	137	12.4	−
Williams	248	−5	462	248	6.6	5.9
Alldays	49.5	−	95.5	48	24.2	2.7

These few figures show a wide range of opinions, from investors who are willing to purchase Manchester United shares at a price which provides only a 0.6% yield and Amstrad at a price which gives an even lower yield of 0.2%, to those who require a 24.2% yield before they are prepared to invest in Alldays. Overall the figures reveal that an interesting range of yields is available related to a range of different type of companies.

Two expressions which are frequently used in discussing shares are "price earnings ratio" and "asset value per share".

The price earnings ratio (or P/E ratio) gives the relationship between the market price of a share and the annual earnings of the company which are attributable to that one share. So, if the total number of shares in a company is 1,000,000, its earnings are £225,000 and the market price is 247.5p, then the price earnings ratio is 11:1. Each share earns £225,000/1,000,000 or 22.5p, and its market price is 11 times the earnings. The P/E column in the above table shows price-earnings ratios and it indicates that whereas a share in Alldays can be bought for 2.7 times its earnings, to purchase a share in Manchester United an investor would have to pay 51.7 times its earnings, and 56.7 times its earnings to purchase a share in Amstrad. It is also worth noting that whilst a purchaser of an Amstrad share will only receive a yield of 0.2% on the current market price, anyone who had been fortunate enough to purchase a share at the lowest market price in the previous 52 weeks would have seen the value of that share appreciate from 43.75 pence to £5.565, almost thirteen times the original purchase price in just one year. It should be noted that few purchasers of shares are that fortunate.

Sometimes the asset value per share is given, and this refers to the market value of the assets a company owns related to each individual share in the company. This should give an indication of the possibility of recouping investment funds if a company goes bankrupt. If the assets of the above company were considered to be worth £2 million on the open market, then the asset backing for each share would be £2. This suggests that anyone who buys a share at 247.5p could only lose 47.5p if the company became bankrupt and its assets had to be sold. Although this is a useful indicator of security, the investor should not place complete faith in it, as the value of the assets could well be subject to change or could be over-estimated.

To return to the concept of yields, these can be calculated at any point in time, and a calculation will show the return on an investment at that particular time. A similar calculation made at an earlier or later date may show a different yield resulting from changed circumstances, and although the actual annual return may remain constant, changes in capital values may result in there being a series of varying yields over a given period of time with any one investment.

Most property investors will be principally interested in what happens to their investment over a long period of time rather than what happens in the short term. Because of the time and cost involved, few people buy property with the intention of reselling at an early date, and generally the intention is to retain it as a long-term investment. For this reason property investors will not be particularly worried about minor fluctuations in yields which may occur from time to time, as they will be primarily interested in what yield is obtained over the entire period of ownership of an investment. Indeed, they may even purchase investments on terms which apparently reveal a very poor yield, but calculations in retrospect a few years later may reveal that over the long term the yield has steadily improved and that the overall long term yield is very favourable.

If yields are calculated monthly over a period of years it may be found that there are considerable fluctuations which throw up an uneven pattern of yields dependent upon underlying market conditions. Stock Exchange graphs showing performances of shares illustrate this point, both fluctuating capital values and fluctuating yields being revealed. In the stock market investors may well react to such fluctuations by immediately buying or selling if they think such a course of action advisable in the circumstances. Because the prime consideration of property investors is the yield obtained over the lifetime of an investment, they are only likely to take action if calculations reveal that the performance or likely performances of investments over a relatively long period are such that positive action is required. It would, however, be pointless for them to react immediately to changes in investment yields, because before they could go through the necessary legal processes to effect a sale or purchase of a property interest, conditions might have reverted to their original state, and the reason for action would have disappeared.

The property market is consequently much less volatile than the stock market, and the general tendency is for slow and gradual change over periods of months or years rather than for violent and rapid changes from day to day.

In other respects the property market operates similarly to the Stock Exchange in that prices result from the interaction of supply and demand. The general principles of yields are the same, with high yields being required for riskier property investments, and lower yields being acceptable from safer investments.

Over recent years there have been many variations in the yields applicable to various types of property, and there is little point in trying to suggest typical yields. Indeed, it could even be dangerous in view of the wide range of yields which might be found within even one particular category of property. However, a general indication will be given of the types of property which have found favour with investors in recent years, with suggestions as to the reasons for their popularity, and consideration will also be given to those types of property which have been unpopular.

In the early 1990s property markets in many parts of the developed world experienced some of the worst market conditions since the Second World War. Consequently, although the following general comments may be an accurate reflection of investors views for most of the period 1960–2000, they may not necessarily be applicable to conditions in markets at the start of the 1990s and to other periods of recession or depression. Nevertheless, the general observations will indicate the type of considerations which influence investors and the various features which make properties either more or less desirable as investments.

Generally, the most popular types of property have been those which appear to offer security in real terms. These are usually well-built properties situated in the best type of location for their purpose and constructed to a modern design which will not restrict the potential user of the building, and which is not likely to become obsolescent at an early date. The quality of situation is all-important, as a valuable site will remain even when the buildings on it become unusable, whereas if the site is inferior the quality of the investment is likely to deteriorate rapidly as the buildings deteriorate.

Modern shops in prime positions have been much sought after by investors and have sold at low yields. The presence of a reputable tenant, such as a well-known national multiple trader, has been regarded as an added attraction, strengthening the security of income from the investment. Older properties and those in inferior positions have been less attractive, although an advantageous location has often been sufficient to outweigh the disadvantages of an old building, particularly if modernisation and improvement can be effected without much trouble.

As the location of shops has become poorer, so has the willingness of investors to accept low yields decreased. The market for shops on the fringe of shopping centres, in neighbourhood centres and in poorer locations generally, has often been slack, with investors unwilling to put their funds into properties which are not in great demand by would-be occupiers, and in which there is little future potential. Investors have generally considered that demand will always be strong for prime position shops, but the changing shopping habits and modern marketing techniques will result in off-peak positions decreasing in importance.

The development of modern, self-contained shopping centres with plentiful car-parking space and facilities for refreshment and leisure, and the redevelopment and modernisation of older shopping centres to make more intensive use of existing shopping areas, have tended to centralise shopping activities more in recent years. The result has been that prime shopping locations have become even more valuable in relative terms, while locations remote from the new or redeveloped areas have become less attractive to both retailers and investors. Indeed, the development of one stop covered shopping centres, often referred to as shopping malls, has been one of the most significant features of the property development scene in the period 1960–2000. They have proved to be highly popular with shoppers and consequently with investors also, although the very high market values of many has meant that potential purchasers are restricted to large financial institutions only.

In some areas the development of large out-of-town shopping centres, or even of major new shopping complexes within existing retail areas, has had a major adverse impact on other longer established shopping areas. The valuer therefore has to be aware of proposed developments of this type when advising investors as retail locations which are currently much sought after may be much less attractive in the future. In some North American cities in particular, the increased use of the motor car by shoppers and the attractiveness of out-of-town retail centres have resulted in the deterioration of the traditional down-town retail areas to the extent that some town centres are suffering from great urban deterioration, with no demand for what were previously good retail investments.

There are strong indications in many countries of increased retailing by telephone or computer, and if there is a significant increase in that type of retailing, it could be that such a trend may, with the passage of time, cause as big a revolution in retailing practices as has the development of major self-contained shopping centres. In time this could cause a consequent decrease in the quality of these centres as investments.

Most of the comments concerning shops are almost equally applicable to offices. Modern office blocks in prime commercial situations have probably been even more attractive to investors over recent years than prime shops. Security in the long term seemed to be guaranteed if only because modern life, and government in particular, seems to revolve around an ever-increasing flow of paper and forms. There has therefore been a seemingly unending demand for well-located office buildings, especially those built to modern standards of design and possessing modern facilities. Well-located older blocks have also been popular, particularly if modernisation has been possible and feasible financially. The fear that the rapid development of modern telecommunication systems, which facilitate not only instant verbal contact but the incredibly rapid transfer of documents, would make central locations in major cities less essential and therefore less valuable in relative terms does not, as yet, appear to have become a reality. It will be interesting to monitor future trends as it may well be that the proximity of city centre office blocks to other city centre uses such as shops, theatres and transport systems may result in them remaining more in demand and more valuable than offices in other locations.

In many major cities there was substantial development of office accommodation during the 1980s because of its attractiveness to investors, to the extent, however, that many cities were in the 1990s very much oversupplied with such space. The substantial increase in supply was accompanied at the end of the 1980s and the start of the 1990s by severe recession in most countries, the result of the increased supply of space, and the recession enforced decrease in demand, being falling office rents and large amounts of unlet space in many cities. There was therefore a substantial change both in rent levels obtainable from office accommodation and in the yields accepted by investors, to the extent, for example, that major office buildings in Sydney in 1993 had capital values of about half the accepted capital values of 1988–1989.

This was not untypical and similar changes in the market occurred in the same period in major European and North American cities. Even with the passage of time some of the surplus accommodation has only just been let, frequently at lower rents than those originally envisaged by the developers. The impact in some cities of the oversupply of new offices in the early 1990s is still considerable, as the "ripple effect" has been that many older developments have suffered falls in rental and capital values as potential occupiers have been able to move to more modern developments at competitive rents.

In many markets there are likely to be considerable variations in yields between modern office developments and older developments, to the extent that many investment purchasers may only be prepared to purchase older offices and those in secondary positions if the purchase yield is attractive and considerably higher than the yield from prime offices.

Factories and warehouses have generally changed hands at higher yields than offices and shops, largely because of the greater rate of wear and tear on such buildings and the greater risk of early obsolescence with changes in technology. Quality of situation has again been critical, and it is commonplace to see advertisements stressing that a factory is close to good means of communication or easily accessible to markets. Older, multi-storey factories, and industrial buildings in poor locations have been relatively unattractive to investors, as such units have limited appeal to industrial users and security of income is therefore low.

Residential properties in the United Kingdom were for many years shunned by investors. For over half a century such properties were subjected to a constant stream of legislation, much of which decreased their attractiveness as investments by limiting the management power of owners and restricting the rental income that such properties produced. Investors consequently only showed much interest in those residences which were not affected by such legislation. Because of the existence of burdensome legislation and controls, and the fear that they would increase rather than decrease, investors would only purchase residential property if the yield was high enough to compensate for the very considerable risks entailed.

In countries where residential investment properties have not been subject to the type of legislative control which existed in the United Kingdom investors have in fact shown great interest in them. In Australia, for example, the purchase of a unit (a flat), or a duplex (a pair of semi-detached houses or bungalows) has for many years been a popular way in which individuals have chosen to invest superannuation funds to provide a steady income flow on retirement. Market forces have been allowed to operate in fixing rental levels, and owners could get possession if properties were required for their own use or tenants proved unsatisfactory. The result of the lack of onerous controls was a substantial stock of residential properties available for rent, but the introduction in 1986 of legislation which was more onerous for investors resulted in the early withdrawal of a considerable amount of accommodation from the rental market and a reduction in the appeal to investors of investment in residential accommodation.

This early reaction in the market to the recently introduced onerous controls resulted in the Government relaxing in 1987 some of the new legislative provisions which had deterred investors. The market again reacted quickly to the changes in legislation with investors returning to the residential market, more accommodation being offered for lease, and an early easing of the shortage of residential accommodation available for renting. These events clearly illustrated the reaction of investors to adverse statutory intervention and subsequently to the reintroduction of more favourable conditions. Valuers accordingly need to be aware of likely or proposed changes in legislation and the likely effect on market sentiments and property values.

The relaxation of controls on privately rented residential accommodation in the United Kingdom since the 1980s has resulted in a residential rental market developing similar to that in Australia. United Kingdom investors have shown a willingness to invest in residential property to let if reasonable financial returns can be obtained and if unduly onerous conditions are not forced on investors. In many areas there is now an abundance of residential property investment, to the extent that there are even indications of surpluses of accommodation to let in some localities.

For many years following the Second World War agricultural properties in the United Kingdom changed hands at incredibly low yields, in the region of 2% to 4%. There were several reasons for this, including the fact that agricultural products are a basic need of man and there will always be a demand for farms to produce them. Successive governments gave considerable aid to agriculture and this continuing policy doubtless strengthened the confidence of agricultural investors. There were for many years estate duty concessions on agricultural properties which increased their attractiveness as investments, although these concessions were curtailed on the advent of Capital Transfer Tax. Similarly, prospects of large gains resulted from the possible development of agricultural land. Such prospects were considerably reduced by the introduction of Development Gains Tax, and its successor Development Land Tax which in turn has been repealed. These factors, together with a period of low returns to agricultural production, resulted in yields on agricultural investments rising somewhat as market demand fell.

In the mid-1980s low returns to agricultural production continued, while the decrease of Government financial support and the imposition of controls by the European Community have resulted in agricultural producers experiencing hard times compared to the previous 30–40 years, and a reduction in the attractiveness of agricultural properties as investments.

In other countries similar trends have been experienced; the relative unattractiveness of Australian agricultural properties from the mid 1980s resulted mainly from world gluts of wool and wheat with resultant low market prices. The problems caused for Australian agriculturalists illustrates why for many years British Governments provided financial protection for farmers to protect them from world conditions beyond their control. The introduction and the removal of such assistance by governments can be of considerable importance in the property investment market as is also illustrated in the case of residential property.

Whatever the type of property, wise investors will always work from basic principles in order to determine what is an acceptable yield and will retain open minds and have flexible investment policies to enable them to adapt to changing conditions in the market.

The future will inevitably see some property investments becoming more popular as others decrease in popularity. The student should endeavour to keep up to date with the underlying conditions of the property market in order not only to be able to explain changes as they occur but hopefully to be able to predict them before they become reality.

Chapter 13

Methods of valuation

There are five conventional methods of valuation: the Comparative Method (or comparison), the Contractor's Method (or summation), the Residual Method (or the Hypothetical Development Method), the Profits Method (or the Accounts Method or Treasury Method), and the Investment Method (or Capitalisation).

The names in brackets are those used in different countries in lieu of the traditional English names for the methods, but the theory and practice remain the same wherever the methods are used.

The Comparative Method (Comparison)

This is probably the most widely-used method and even if one of the other four methods is used by a valuer he will still almost inevitably have recourse to comparison as well. The method entails making a valuation by directly comparing the property under consideration with similar properties which have been sold in the past, and using the evidence of those transactions to assess the value of the property under consideration.

Although this sounds simple and straightforward, there may be many pitfalls to trap the unwary. In using the method it is desirable that the comparison should be made with similar properties situated in the same area, and with transactions which have taken place in the recent past. The less the comparable property complies with these requirements, the less valid will be the comparison. Often a valuer is able to get evidence of sales which do accord with these requirements, particularly when a valuation is of a property such as a semi-detached suburban house (a duplex). However, the more uncommon a property is, and the more specialised the type of property, the less likely is it that the valuer will be able to find good "comparables", and it is not unusual for there to be a complete lack of evidence of sales of comparable properties.

Even when properties appear to be similar, close inspection often reveals that they are in fact different. A row of apparently identical houses may on internal inspection prove to have many differences, and the skill of the valuer will be required to make an allowance in money terms for such differences. Similarly, a skilled valuer will be able to quantify the difference in value caused by a different geographical location.

It is essential in using this method to have as much evidence as possible readily available, and good office records are invaluable. Any valuer should ensure that there is an efficient filing system which is regularly kept up to date, and which contains as much information as can be obtained of each market transaction which is recorded. Modern computer filing systems greatly facilitate good record keeping and the almost instantaneous recall of information. However, for helpful records to be kept, there must be evidence of suitable comparables available, and unfortunately, in the real world, this is often lacking.

The word recent is a relative term. At some points in time a sale which took place a year earlier may be recent enough to be a valid comparison if the market has remained relatively stable in the intervening period. If the market has been volatile or has changed in any way during this time, then market evidence might need to be much more recent for it to be a good comparable. Even if there have been no transactions in the recent past, a valuer may be able to get some guidance from considering market evidence over a period of earlier years or months, as it is possible that a clearly defined trend of values might be detected. The valuer may decide that this trend would have continued through to the date on which the valuation is being done.

In using the Comparative Method a most thorough inspection of all the underlying factors in the market must be made in order to decide whether there have been changes in conditions since other transactions took place. For instance, changes in the general level of interest rates over a period of time may mean that what may at first appear to be useful comparable evidence has to be rejected because of the substantially different market conditions in which they occurred.

The method involves few dangers if the market is stable and active. When it is not stable, valuers may encounter problems in its use, and this may also be the case if there are few comparables, or if there are no true comparables, that is if the range of properties sold does not contain anything truly identical to the property under consideration.

Some differences may be relatively unimportant, and differences in architectural design, for example, may often make little difference to value. However, if a particular design renders a building inefficient in use, the difference may be very important, and in using the method the valuer should always consider such possibilities. The age of a building can be important in that it may be such as to render it either more or less fashionable and in many instances the structural condition and state of repair will be directly related to the age of the building. Valuers should also pay considerable attention to the accommodation offered by different properties, and try to make allowances for variations in the amount of floor space and differences in the layout and the number of rooms provided. Allowance should also be made for differences in the quality of fixtures and fittings, and for differences in the size of plots on which properties stand. Location is always very important, and valuers should be wary of assuming that values should be similar simply because the size and accommodation of properties is identical. A slightly different location can make a vast difference to market value.

In using this method the valuer should always bear in mind the fact that property is heterogenous, and should always ask himself or herself whether any special factors affected the market value of the comparables being used, or whether any special factors are likely to affect the value of the property which is under consideration. The following is an example of the comparative method in use.

Example 1

A valuer has to value a four-bedroom house, which provides net usable space of 180 m^2. Research reveals that in the same street there have been sales of similar sized four-bedroom properties as follows:

> 12 months ago a house of 180 m^2 sold for £180,000; 8 months ago a house of 170 m^2 sold for £173,400; 6 months ago a house of 185 m^2 sold for £190,550; 3 months ago a house of 175 m^2 sold for £183,000; 1 month ago a house of 185 m^2 sold for £195,200.

After analysing these sales the valuer decides that whereas the first sale revealed a capital figure of £1,000 per m^2 of usable space, the last sale was at a figure of £1,055 per m^2 of usable space. The valuer also notices there has been a consistent rate of increase of £5.00/m^2 per month in the capital values of this style of house and therefore values the subject property at £190,800 (180 m^2 @ £1,060/m^2).

This is a very simple example of the comparative method in use and in reality its use will normally be more complicated. It would be necessary to check whether the various houses provided similar numbers of rooms in total, whether their locations were of equivalent quality, whether their fixtures and fittings were similar and of equal quality, and also to check on other features of each of the properties. Adjustments in the analysis of each piece of market evidence would be necessary to allow for variations in such factors, and it is likely to be rare in the real world that market analysis and use of the analysed information would be as straightforward as in the example above, in which there is general consistency in the level of values and in the rate of increase in value over time.

There is a great danger in the use of comparable evidence that the valuer may place too much faith in it, forgetting that while such evidence reflects what happened in the past his present task is to determine what the current value of property is. Current market value may be affected by different factors to those which affected past transactions and it could therefore be that past evidence might be of only very limited assistance or, in extreme circumstances, it might be positively misleading.

Perhaps the biggest danger in the use of the Comparative Method is the underlying and simple assumption that because in the past one person was prepared to pay a certain figure for a particular property, another person will also be prepared to pay a similar figure for a similar property. It may be that the purchaser of the comparable property had special reasons and specific personal circumstances which both prompted and enabled the purchase to be made, such reasons and circumstances being completely irrelevant to others in the market place.

An example of such a situation might be a person paying a high figure to buy a house near their own home for occupation by an aged or infirm parent who can contribute a substantial amount of capital towards the purchase. Other potential purchasers of nearby houses may be quite unable and also unwilling to pay a similar price. In the retailing world, a retailer selling high-value goods with big profit margins may be able to pay a high rent or a high purchase price for retail accommodation. However, if the only potential occupiers of nearby vacant and similar retail accommodation are traders who sell low value goods at low profit margins, it may be completely unrealistic to anticipate a high rent or high capital value for the vacant accommodation.

The simple assumption that, all other factors being the same, the market evidence of the value of one piece of property at a point in time is automatically a good indicator of the market value of another similar property, itself has to be treated with great caution. The valuer has to be extremely knowledgeable about the market for which valuations are to be made, and must make every effort to determine who is looking for property of the relevant type, what their precise needs are, and what their purchasing power is. Such factors could have a great influence on how comparable evidence should be used and whether it is indeed a reliable indicator of the prices likely to be paid in the current market.

However, in spite of the need for great care in the use of the Comparative Method and the frequent shortage of suitable comparable evidence, it is a method which the valuer will use regularly and which will give reliable results if used properly and in the correct circumstances.

The Contractor's Method (Summation)

This is used to value the type of properties which seldom change hands and for which there are therefore few or no comparables. It must at this point be stressed that cost and value are rarely the same, but this method of valuation is based loosely on the assumption that they are related. It should therefore be appreciated that it is a method used only infrequently, and which is something of a last resort. The basic theory of the contractor's method is that the cost of the site plus the cost of the buildings will give the value of the land and buildings as one unit.

With the majority of properties there is ample evidence to show that this proposition is not correct, but the contractor's method is used to value properties for which there is little general market demand and which are consequently rarely sold. The types of property for which it could be appropriate are hospitals, town halls, schools, libraries, police stations, and other such buildings. It will be noted that this list comprises principally public buildings, although the use of the method is not necessarily restricted to public buildings alone. Cost is normally only one factor of many which may affect supply and demand and which therefore affect value, but it is probably true that with this type of building it is a predominant factor. It would always be possible for the would-be users of such buildings to acquire alternative sites and to construct new buildings rather than purchase an existing property at a greater overall cost. Competition between rival potential users would be unlikely and it is therefore reasonable to assume that cost and value are not unrelated with such specialist buildings.

However, if an alternative building were constructed it would be a new property, whereas with an existing property it is obvious that there would be some wear and tear resulting from its previous use and there might also be a degree of obsolescence which had arisen since it was new. In using the Contractor's Method the valuer must therefore make a deduction to allow for both depreciation of the buildings and obsolescence of design. The basic valuation approach then becomes as below:

	Cost of site
Plus	Cost of building
=	Total cost of similar property
Less	Depreciation allowance and
	Obsolescence allowance
	Value of existing property

This method is most frequently used for rating purposes where rates are levied on the value of buildings and sites together and it is also sometimes used in valuations for compensation when property of a specialist nature has been compulsorily acquired.

The following illustrates the typical use of the Contractor's Method of valuation.

Example 2

A public library which occupies a site on the fringe of a town centre has to be valued. The building is about 100 years old, built in a rather ornate style, and shows some evidence of deterioration.

The valuer using evidence and his own judgement decides that an equivalent site would today cost £800,000, that a 500 m² building in the same style would cost £1,250/m² to build today, and that considerable allowances should be made for deterioration and obsolescence. He accordingly values as follows:

	Cost of site	£800,000
Plus	Building cost 500 m^2 @ £1,250/m^2	625,000
		1,425,000
Less	25% obsolescence allowance	
	(based on building cost)	156,250
		1,268,750
Less	15% depreciation allowance	
	(based on building cost)	93,750
	Value of existing property	£1,175,000

Expressed as a hypothetical example in a book, the method appears simple to use. In reality all the inputs to such a valuation are likely to be difficult to determine, even the building costs, which will be far from clear-cut in the case of the unusual types of building for which the method is used. In particular, the precise allowances to be made for depreciation and obsolescence are very difficult to determine and very much dependent upon the judgement of the individual valuer.

The Residual Method (Hypothetical Development Method)

This is used when a property has development or redevelopment potential. It is needed when there is an element of latent value which can be released by the expenditure of money on a property. Residual valuations are quite regularly made by people who purchase residential properties which they consider could be made more valuable if money were spent on improvement and modernisation. This would-be purchaser may look at a house and decide that it is worth £100,000 and that it needs expenditure of a further £40,000, after which it will have a market value of £180,000. A quick inspection of the figures shows that latent value of £40,000 has been released by the expenditure on improvements. This is a very simple example of the method in use, but even when much more complicated calculations are involved, the basic approach to the method is still exactly the same.

	Value of the completed development
Less	Total expenditure on improvements or development
	(including developer's profit)
	Value of site or property in its present condition
	(Residual Value)

The value of the completed development is sometimes referred to as the Gross Development Value, or as the Gross Realisation.

The use of the Residual Method involves considerable skill and it is first necessary to decide what is the best form of development suitable for a site or property, and then to predict the value of such a development after it has been completed. This is not an easy task, as apart from the skill involved in choosing the best use for a site, the valuer also has to estimate the value of a building which does not as yet exist. He has to cast his mind forward to some future date and imagine the building in existence on the site before he can begin to estimate the Gross Development Value.

Even when this has been done with great skill and accuracy, all the costs of improvement and development must then be estimated. These may include such items as the cost of site clearance, architect's fees, site engineer's fees, quantity surveyor's fees, and all other professional fees incurred in creating the development. The costs of building must be estimated, and these could well increase during the time-lag between the acquisition of the property and the completion of the development. Once a property has been developed it has to be let or sold, and the valuer must make an allowance for all costs which would be incurred in letting or disposing of the property envisaged as one day standing on the site. These costs will include estate agent's fees, advertising fees and the solicitor's fees and legal expenses. To purchase the site and subsequently develop or redevelop it will usually require a considerable amount of finance, and the costs of obtaining this must be deducted as a development cost. No one will wish to take all the risks involved in a project without any reward, and a developer's profit must be allowed.

There may be other items of expense to consider, and the more variables there are the more difficult it will be to maintain an acceptable degree of accuracy in the calculations. However, a skilled valuer with a specialist knowledge of the type of development for which he or she is valuing, who is in touch with the market and who is familiar with the costs of development, can use this method with a considerable degree of accuracy.

There will doubtless be instances in which, in retrospect, a purchaser will be seen to have paid too much for a property because the figures on which the residual valuation was based have changed in the period since the property was purchased. The method is nevertheless acceptable for finding market value, as it must always be remembered that this value is the figure which would be paid in the market at a definite point in time, taking into account future expectations. It is not unusual in any aspect of life for retrospective consideration to show that an earlier decision was based on false assumptions, and the fact that a residual value may subsequently be shown to have been optimistically calculated does not alter the fact that in the light of circumstances at the time of purchase the estimate of market values was correct.

The method is often criticised as being clumsy and containing too many variables, but there is little doubt that it is the only real method of valuation applicable when there is latent value in a property.

The following simple example illustrates the way the method is used in practice.

Example 3

A site with approval for the development of four houses of 160 m^2 each is to be auctioned and a developer wishes to know how much he can afford to bid for it. He estimates it will take one year to complete and sell the development, that he will have to pay 10% pa interest on borrowed money, and that the houses will cost £525/m^2 to build including all site works and the provision of services. He also decides that he will require a profit of £20,000 per house, each house having an anticipated market value of £160,000.

His residual valuation is as follows:

Four detached houses at £160,000 each

Gross Development Value		£640,000 GDV
Less Costs of development:		
(1) Building costs: 4 houses × 160 m^2 @ £525/m^2 =	£336,000	
(2) Professional fees (architect's, planner's, quantity surveyor's) @ 10% of building costs =	33,600	
(3) Cost of borrowing half of (1) + (2) @ 10% = $\dfrac{10}{100} \times \dfrac{369,600}{2}$ =	18,480	
(4) Legal expenses on sale of houses say	12,000 ·	
(5) Agent's fees on sale of houses say	10,000 ·	
(6) Advertising costs on sale say	3,000 ·	
(7) Developer's profit = 4 × £20,000	80,000	
Total development cost		£493,080
Residual Sum		£146,920

On the above calculations the developer considers the sum of £146,920 is available to cover the purchase price of the land, all the expenses he will incur on the land purchase, and the interest charges he will incur on holding the land from the date of purchase until the development scheme is completed and sold. He will calculate these costs and his bid price for the land will be reduced below £146,920 by the amount of these costs.

All the costs he uses in this valuation will be carefully assessed by the valuer in an effort to calculate a realistic figure for the land value. However, because they are all predicted figures the accuracy of the valuation will depend upon the skill and judgement of the valuer.

The selling price of the houses will be based on his market knowledge, while items (1) and (2) will be based on his knowledge of costs and perhaps even on the preliminary estimates of builders and professional advisers.

Until the development is built and sold the developer will incur the cost of interest on money borrowed to complete the development, or alternatively he will sacrifice interest earnings if he uses his own money. As a general rule, a realistic estimate of the amount of money a developer will need on average throughout the development period is half the cost of building and half the cost of professional fees, as calculated at (3). Clearly, such a calculation is only an approximation of interest charges, but it is nevertheless a useful approach for initial development appraisals.

In reality, most development schemes usually involve lower levels of expenditure in the early stages of the scheme than in the later stages, and this type of initial assessment may therefore tend to over-estimate the costs of finance. However, it is nevertheless a useful approach for initial appraisal purposes.

Items (4), (5) and (6) are costs which are likely to arise with any development and the valuer will assess them on the basis of recent experience of such costs.

The developer's profit is the return for his expertise and the reward for the risks he takes, and there is no fixed rule for the calculation of this figure. It is sometimes calculated as a percentage of Gross Development Value, sometimes as a percentage of Building Costs, or it may be calculated by other methods thought appropriate by individual developers, such as the profit per house used in the above example.

The residual sum found is based on the anticipated sale of the completed development at a future date, in the above example 12 month's time. If the developer bids for the land he must therefore allow for the professional fees and legal costs he will incur on purchasing the land, and also on the interest charges he will incur during the development period on borrowing the money for the land purchase and the associated costs.

From the above example such adjustments can easily be made to the figure of £146,920 which was calculated as being available to cover the price of the land, the costs incurred in purchasing the land (eg solicitor's charges, valuer's fees, and any taxes payable on purchase), and the interest charges incurred on the money tied up in these items during the construction period. If it is found that the costs of purchase would be 5% of the price of the land the following calculations would be appropriate.

Sum available for purchase of land, costs of purchase, and interest charges during the construction and marketing period of one year –	£146,920
× Present Value of £1 in one year at 10% per annum	× .909090
Sum net of interest for one year at 10% per annum	£133,563
Divide by 1.05 to allow for 5% costs of purchase	1.05
Price payable for the land	£127,203

These calculations reveal that £127,203 can be paid for the land on which purchase costs of 5% or £6,360 would be incurred, giving a total expenditure of £133,563. Interest on this sum at 10% pa for one year would be £13,356, creating a total expenditure in respect of the land of £146,919, which was the figure originally calculated as being available for the land purchase and associated expenditure, subject to a rounding error of £1 in the calculations.

The use of the Present Value of £1 for the appropriate time period at the relevant cost of borrowing merely reduces that figure to allow for interest accumulations, and the concept of the Present Value of £1 and other valuation formulae and tables is more fully considered in chapter 14.

The example is relatively simple, but even complex residual valuations are only developments on this basic approach.

The Profits Method (Accounts Method, Treasury Method)

This is based on the assumption that the value of some properties will be related to the profits which can be made from their use. It is a fact that the vast majority of properties will have no value unless they can be put to beneficial use, but not all properties will be used in a way which generates a calculable profit in money terms. There will also be cases in which a profit from use can be calculated in which the profit does not provide the best basis for assessing market value of the property being used. The Profits Method is therefore not normally used where it is possible to value by means of comparison, and is generally only used where there is some degree of monopoly attached to a property.

Such a monopoly may be either legal or factual, a legal monopoly existing where some legal restraint exists to prevent competition to the property user from the users of other property. Such a situation may occur when a licence is required for the pursuit of a particular trade. A factual monopoly may arise when there is some factor, other than a legal restraint, which restricts competition. An instance of a factual monopoly is the restaurant at the top of Mount Snowdon in Wales where there is no other property to offer competition and where none is likely to be built. Whenever there is an element of monopoly it is obviously not possible to use the comparative method of valuation, as there could be no true comparison to a property which enjoys a monopoly, and it is also a reasonable assumption that any rent which would be paid for the use of such a property would relate to the earning power in that use. It should be noted that with this method the valuer attempts to estimate the rental value of a property and not the capital value. Profits are made on an annual basis, and any figure obtained from them will also be on an annual basis. The basic equation on which the profits method is based is as follows:

	Gross earnings
Less	Purchases
	Gross profit
Less	Working expenses (except rent)
	Net profit

Part of the net profit which the business earns must be allocated to pay the tenant for his work in the business, a further allocation must be made to cover his risk and enterprise, and a final allocation to allow interest on the capital he has put into the business. Care should be taken that a deduction for the tenant's wages has not previously been made in calculating the net profit, as double counting must be avoided. After allowance has been made for these various items, a sum of money will remain which would be available to pay for the use of the premises. This figure will vary, depending upon the size of the deductions for other items, and skill is required, and a knowledge of trading returns, for a valuer to be able to make a reasonable assessment of the division of net profit between these various items.

Much care is also needed in the calculation of the net profit and in the use of the various figures involved. If a business is already in existence there will be accounts which can be inspected. The obvious approach is to base a valuation on those accounts. However, care is necessary, as a business may have been badly run and the use of the accounts figures in such circumstances would give too low a net profit. Alternatively, a business may have been run by an exceptional, astute and hard-working business man, and no other trader might be able to achieve the same level of profits. In such circumstances no other trader might be prepared to pay the same level of rent, whilst in any case the super-efficient trader would require a large share of the profits for himself or herself as a reward for their skill and management. If such a situation does exist, the valuer should not take too high a proportion of the net profits as rent, as they are only endeavouring to assess what normal market value is, not what a property may be worth to a super-efficient business person. It also sometimes happens that accounts do not give a true reflection of trade, either because they have been badly kept through ignorance or inefficiency, or because the business person has chosen not to put all items through the books.

In practice the method is normally used in the valuation of hotels, public houses, and sometimes for cinemas and theatres. It is regularly used in rating valuation, the annual value which is obtained being directly related to the value required for rating purposes in the United Kingdom and many other countries. In other instances a valuer may have to convert the annual figure into a capital figure. This will be done by means of the Investment Method, which is the fifth conventional method of valuation.

The following is an example of how the method might be used in practice.

Example 4

A valuer has to assess the rental value of a restaurant, which has a licence for the sale of alcoholic drinks and which took £358,000 in the last trading year. He inspects the books and decides the business has been run in a way other potential tenants would run it, and he produces the following valuation.

Gross takings		£358,000
Less Purchases	£95,200	
Running expenses	119,400	
		£214,600
		£143,400
Less Owner's remuneration	£50,000	
Interest on capital (furniture, kitchen fittings, etc) @ 12% on £74,000	8,880	
		£58,880
Divisible balance		£84,520

Using his judgement the valuer decides the divisible balance would be split 55% profit to business, 45% available for rent; therefore estimating the rental value to be approximately £38,000 per annum.

The Investment Method (Capitalisation)

This is based on the principle that annual values and capital values are related to each other and that, given the income a property produces, or its annual value, the capital value can be found. The method is widely used by valuers when properties which produce an income flow are sold to purchasers who are buying them for investment purposes.

Many properties are let to tenants and the income they produce is known. In other cases, although a property may not be let, it is possible to predict what its rental value is by comparing it with similar properties which have been let. If this is the case, or if a rent actually passes on a property, the only problem the valuer faces is determining the relationship of annual value to capital value. The way in which the conversion is made is by the use of a multiplier which is commonly known as "Years' Purchase", or, in abbreviated form, the "YP". As will be seen later, in valuation terms, this multiplier is more appropriately described as the Present Value of £1 per annum.

The Years' Purchase or multiplier is derived from the rate of interest which an investor decides he will require from a property, that is, the yield which he wishes to obtain. This yield reflects the quality of the investment in comparison with other property investments, and other investments generally. Consideration has been given to factors which may influence the investor in his choice of yield, and the valuer will obviously need to be conversant with these when using the Investment Method. Here again an element of comparison arises, as the valuer will be comparing both other investments and other properties. The choice of the yield, and its conversion into a suitable multiplier will be considered at length in chapter 15.

In addition to these conventional methods, valuations can be made using *Discounted Cash Flow* techniques. These are to all intents and purposes extensions of the Investment Method, and entail estimating all future items of income which an investment will produce, and converting these future sums into present day equivalents in money terms. These future cash flows will be the sums remaining after taking account of anticipated future outgoings, so that only the net income for each future period will be discounted back to its present value. Calculations are frequently based on periods of one year, but can in fact be made for any time period considered appropriate in specific cases. When the present value of the net income for each period has been calculated, the total of all the discounted figures will represent the present capital value of the investment.

There are several different types of approach to discounted cash flow calculations, but generally they all amount to a more comprehensive form of investment valuation, with more variables being included in the calculations and with each period's income assessed separately. As very versatile and cheap calculators and computers are now available, the objection that such techniques involve too much complicated mathematics is no longer valid. The quality of the results which emerge from the calculations will, however, depend to a very large extent on the quality of the inputs, that is, the figures which are estimated for future returns and outgoings, and the rates of interest used for discounting.

The techniques will be considered further in chapter 19.

The Mortgage/Equity Approach

As the majority of property purchasers have to raise finance by borrowing, the Investment Method of valuation in its simplest form may not provide an entirely correct way of assessing the value of an interest to a particular would-be purchaser. What can be termed a *mortgage/equity approach* should really be adopted, giving a basic calculation as in the following example.

Example 5

An investor wishes to determine what is the maximum he can bid for a property which produces an annual income of £40,000. He can put £200,000 of his own funds towards this purchase and requires an 8% return on that money. He can borrow the remainder of the necessary purchase funds at 10% per annum.

Total property income		£40,000
Purchaser's capital	£200,000	
Return required 8%	.08	
Purchaser's required income		£16,000
Balance available to pay mortgage interest		£24,000

Mortgage obtainable @ 10% pa
Interest payable = Mortgage capital × Interest rate
£24,000 = $\frac{MC \times 10}{100}$ 2,400,000 = 10 MC   MC = 240,000

Equity	£200,000
Plus Mortgage	£240,000
Maximum price payable for property	£440,000

The annual income being known, the potential purchaser first calculates the return required to his own money (his equity), and deducts this from the income to leave a balance which is available to pay interest on any money borrowed on mortgage.

As the interest rate charged on mortgage money will be known, he or she is then able to calculate how much mortgage capital can be borrowed for that amount of annual interest, and this is the second calculation.

The maximum he or she can then afford to pay for the property is the sum of their own money and the money they can afford to borrow. If they pay more than this sum they will have to borrow more money, the mortgage interest charge will increase, and the return to their own capital will then be too low. If they can purchase the property for less than this sum they will get a better return on their own capital.

The importance of this type of approach to individuals can be illustrated by considering the same property and another potential purchaser who can borrow money for 9% pa, and who only requires a 7.5% return on equity (that is the purchaser's own money) which again is £200,000.

Example 6

Total property income		£40,000 pa *10%*
Purchaser's capital	£200,000	
Return required of 7.5% ×	.075	
Purchaser's income required		£15,000 pa *Retn*
Balance available to pay mortgage interest		£25,000 *(40K-15K)*

Mortgage obtainable @ 9% pa

Interest payable = Mortgage capital × Interest rate

$$£25,000 = \frac{MC \times 9}{100} \quad £2,500,000 = 9MC \quad MC = £277,777$$

2,500k ÷ 4 =

Equity	£200,000
Plus Mortgage	£277,777 *– 240,000 =*
Maximum price available for property	£477,777

These calculations show that another potential purchaser who is prepared to accept .5% pa less return on equity and who can borrow for 1% pa less can outbid the original potential purchaser by £37,777 in this example. They also show that the original purchaser, by being prepared to accept a lower return on equity and by obtaining more favourable lending terms, could in a competitive market situation increase their bid by that same amount of £37,777 if necessary and if considered advisable.

This approach will be considered again in a later chapter.

Chapter 14

Valuation Tables and Valuation Formulae

Valuing property is partly an art and partly a mathematical process. The mathematical content of most valuation work is relatively limited and quite straightforward, and there is no reason for any newcomers to the subject to fear that they may be unable to cope with the mathematics. There is nothing magical or mysterious about the concepts employed, and this is also true of valuation tables.

In the past valuation tables were extensively and regularly used by valuers. In the United Kingdom the most commonly used tables are *Parry's Valuation and Investment Tables*, published by The Estates Gazette*, which are produced simply with the intention of making a valuer's life easier. They are one of the tools of the trade and, as with a handyman's tools, they help to take the drudgery out of jobs and enable them to be done more rapidly. They do not provide valuation answers in themselves; they merely provide rapid answers to many of the mathematical calculations which a valuer may have to make. Ownership of a set of tables does not turn one into a competent valuer.

The advent of cheap table top computers and very versatile financial calculators has resulted in them replacing valuation tables for use on a regular basis by most practising valuers, and there are now many computer programs which can produce all the figures given in valuation tables, and a lot more figures also, very rapidly indeed. There are consequently those who say that valuation tables no longer have any use for the valuer and need not be purchased or used. Even good pocket-sized financial calculators can produce most of the figures found in the tables and this is therefore an appealing argument.

However, it is the author's experience that many who produce figures from calculators and computer programs do not really understand what the machine has done and what financial

---

* All references in this book are to the Eleventh Edition of the Tables.

calculations have been made, or the true implications of the figures produced. It is extremely dangerous to use figures without understanding what they represent, and it is therefore a most important part of the process of understanding to study the valuation formulae and valuation tables in order to be able to apply the various figures at the appropriate times and in the appropriate ways. With such understanding a valuer should know what each figure produced by electronic means represents, and should also be able to recognize when figures produced in such a way are erroneous or unacceptable.

Additionally, the importance of variations in the rate of interest can instantly be recognized from studying the pages of tables which deal with the Amount of £1 (or the Amount of $1) in particular. While small differences in rates of interest over a short period of time may appear insignificant, the tables clearly illustrate the importance of such differences over longer periods of time, and also the significance of larger differences over any period of time. Where particularly large sums of money are involved it can be seen that even small variations in interest rates over relatively short periods of time can be important.

It is therefore considered that use of the tables and a good understanding of them is an extremely useful part of the education of students even in the computer age, and there are in fact still times when in practice it will be more convenient and even quicker to use tables as opposed to a computer program.

Practically every calculation made by a valuer involves compound interest theory, and the valuation tables are a series of compound interest tables with what can be described as variations around the basic compound interest theme. This basic theme is that when money is invested the original capital sum earns interest which, if it is added to this capital, will in subsequent years itself earn interest. This process will be repeated each year, and the capital is therefore regularly increased and, as a result, the interest earned increases with each period, or it compounds.

This basic theory is used in a number of different tables, each of which will be considered and the practical implications explained. It must be stressed that in the following pages the author uses only sufficient mathematics to give a basic understanding of the tables. Because of this emphasis on the minimum of mathematics those who wish to become more involved with this aspect should refer to either *Parry's Valuation Tables* or *Modern Methods of Valuation*.

There are other sets of valuation tables in use besides *Parry's*. These include *Rose's Property Valuation Tables*, *Donaldson's Investment Tables* and *Bowcock's Property Valuation Tables*. Which set of tables is used will depend to some extent on the preferences of the valuer and to some extent on the purposes of the valuation. Each set is constructed using slightly different variations on the basic compound interest theme, and answers obtained after using the different tables will therefore vary slightly. The valuer should take care that he or she understands the underlying principles in the construction of each table that he or she uses to ensure that a valuation is based on the correct assumptions and the correct mathematical process.

There are also numerous cheap and versatile electronic calculators on the market, and those who are proficient at mathematics can, as indicated above, often dispense with the use of valuation tables. Some of the more advanced calculators are programmed to give equivalent calculations to those provided by valuation tables, and it is therefore a matter of choice for the individual valuer as to whether he or she uses a particular set of tables or relies upon an electronic calculator.

The reader should note carefully that all the examples which follow have been done using the Eleventh Edition of *Parry's Valuation Tables*.

Readers from countries in which the £ is not the unit of currency can substitute their own country's currency in the remainder of this chapter, eg the Amount of $1, the Present Value of $1, etc.

The amount of £1 table

This table shows the amount to which £1 will accumulate if it earns a given rate of compound interest for a given number of years.

The student should note carefully that there is a single investment of £1 only which is invested to earn compound interest. The amount of £1 will always be greater than unity, as each figure will represent the original capital plus compound interest earned. It follows that the total interest earned is the figure found in the appropriate column of the tables, less the £1 originally invested.

The calculations on which the tables are based are as follows:

	Capital at start of year	Interest earned during year	Capital at end of year after interest has been added
Year 1	£1	i	$1 + i$
Year 2	$1 + i$	$(1 + i)i$	$(1 + i) + (1 + i)i$ or $(1 + i)^2$
Year 3	$(1 + i)^2$	$(1 + 1)^2 \times i$	$(1 + i)^2 + [(1 + i)^2 \times i]$ or $(1 + i)^3$
Year 4	$(1 + i)^3$	$(1 + i)^3 \times i$	$(1 + i)^4$
"N" years			$(1 + i)^N$

N = the number of years
i = interest earned by £1 in 1 year

The concept of "i" sometimes causes confusion with students, but it is really quite simple. If the rate of interest is 5% pa, £100 will earn £5 in one year. £1 would therefore earn £5/100, or 5p. Expressed in decimal form this is £.05.

A simple calculation can be done to show how the above calculations work when a specific rate of interest is chosen.

	Capital at Start of Year £	Interest at 10% £	Capital at end of Year £
Year 1	1	.10	=1.1
Year 2	1.1	$(1.1).1 = .11$	$1.1 + .11 = 1.21$
Year 3	1.21	$(1.21).1 = .121$	$1.21 + .121 = 1.331$
Year 4	1.331	$(1.331).1 = .1331$	$1.331 + .1331 = 1.4641$

So the amount of £1 in 4 years at 10% is 1.4641, but instead of doing the above calculations this figure could be rapidly looked up on p104 of *Parry's Tables*. The figure tells us that if £1 is invested to earn interest at 10% and it remains undisturbed for 4 years, at the end of that period there will be a total fund of £1.4641, representing the original £1 and £0.4641 accumulated interest.

It is interesting to consider a few figures in this table to understand fully the effect of the compounding of interest, and to see how rapidly the rate of accumulation increases as the time or the rate of interest increases.

		Total Interest earned by £1
Amount of £1 @ 2% in	3 years = 1.0612	£0.0612
	10 years = 1.2190	£0.2190
	20 years = 1.4859	£0.4859
	50 years = 2.6916	£1.6916
Amount of £1 @ 10% in	3 years = 1.3310	£0.3310
	10 years = 2.5937	£1.5937
	20 years = 6.7275	£5.7275
	50 years = 117.3909	£116.3909
Amount of £1 @ 20% in	3 years = 1.7280	£0.7280
	10 years = 6.1917	£5.1917
	20 years = 38.3376	£37.3376
	50 years = 9,100.4381	£9,099.4381

The figures need little explaining, but it is interesting to note that at 10% almost as much interest is earned in 10 years as is earned in 50 years at 2%. The 20% figures illustrate the almost staggering interest accumulations at high rates over a long period of time, and make one ponder what a marvellous old age could be enjoyed by a prudent youngster lucky enough to find a 20% investment.

The use of the table can best be explained by an example.

Example 1

If £25,000 is deposited and is left undisturbed for 17 years, earning interest at 6%, what sum will be available at the end of the time period.

Sum deposited	£25,000
Amount of £1 in 17 years at 6% (p102 of *Parry's Tables*)	2.6928
Sum available after 17 years	£67,320

The calculation is based on the simple proposition that if £1 becomes a certain amount after a given period of time at a given rate of interest, larger deposits will accumulate to larger amounts in direct proportion to their relationship to a £1 deposit.

To summarise and to stress the important facts:

(1)   The Amount of £1 Table provides the basis for the other valuation tables.

(2)   The Amount of £1 Table is based on compound interest calculations.

(3)   Amount of £1 figures are always greater than unity.

(4)   The excess over unity represents total interest earned.

(5)   The formula for calculating the amount of £1 is $(1 + i)^n$ where "i" is the interest earned by £1 in 1 year, and "n" is the number of years.

(6)   The amount of £1 is often referred to by use of the symbol A, eg $A = (1 + i)^n$

The Present Value of £1 Table

The Present Value of £1 Table gives "the sum which needs to be invested at the present time at a given rate of interest in order to accumulate to £1 by the end of a given period of time".

Alternatively it can be defined as "The present value of the right to receive the sum of £1 at a given time in the future, discounted at a given rate of interest".

The two definitions together explain the concept of the present value of £1 or "V", as it is frequently written. A person who has the right to receive £1 at some future date would be as well off were he or she to receive a lesser sum now which could be invested to earn sufficient interest to ensure that, by the time the future date was reached, that sum plus the interest earned would be equal to £1. The opportunity cost concept is involved, as anyone who had the right to receive £1 at a future date would only sell that right for a figure which left him or her no worse off financially today.

The figures found in this table again involve capital figures of £1 and are based on compound interest calculations. They are the reciprocals of those found in the Amount of £1 Table, as if £1 invested today will amount to $(1 + i)^n$ at a future date, the right to receive £1 at a future date must be worth $1/(1 + i)^n$ or $1/A$.

An inspection of a few typical figures from the tables may again prove interesting.

Present Value of £1 in		
	3 years @ 2% =	£.9423223
	10 years @ 2% =	£.8203483
	20 years @ 2% =	£.6729713
	50 years @ 2% =	£.3715279

Present Value of £1 in	3 years @ 10% =	£.7513148
	10 years @ 10% =	£.3855433
	20 years @ 10% =	£.1486436
	50 years @ 10% =	£.0085186
	3 years @ 20% =	£.5787037
	10 years @ 20% =	£.1615056
	20 years @ 20% =	£.0260841
	50 years @ 20% =	£.0001099

These figures illustrate how the present value of the right to receive £1 at a future date decreases the more distant that date is, and they show how this decrease is even more marked at high rates of interest. They also illustrate how much more highly the future is valued when the opportunity cost rate of interest is low. Where this is 2%, £1 in 50 years' time is today worth over 37p, but where it is 20%, £1 in 50 years' time is today worth about 1/100 of one penny.

Again, the use of the table in practice is best illustrated by examples.

Example 2

An investor has the right to receive £100,000 in 10 years' time. What sum would he or she be prepared to accept today in lieu of that future sum?

It is found on enquiry that money can today be invested to earn interest at the rate of 8% over the next 10 years.

Sum receivable in 10 years' time	£100,000
× Present Value of £1 in 10 years at 8%	.4631935
Present value of £100,000 in 10 years at 8%	£46,319.35

This calculation suggests that the investor will be as well off accepting £46,319.35 today as if he waited 10 years for the £100,000. The validity of the assumption can be checked by calculating what figure will be available in 10 years' time if he does in fact invest the money today.

Example 3

Sum available to invest	£46,319.35
× Amount of £1 in 10 years at 8%	2.1589
Total available after 10 years	£99,998.84

There is a small error revealed in the checking which could be avoided if more decimal places were used. The example illustrates both a common use of the Present Value of £1 Table, and also its relationship with the Amount of £1 Table as shown in the checking.

Example 4

A sum of £320,000 must be available in 12 years' time to meet expenditure which will be essential at that time. What sum should be put aside today to meet this future expense if 7% interest could be earned in the intervening period?

Sum required in 12 years' time	£320,000
× PV £1 in 12 years at 7%	.4440120
Sum to be invested today	£142,083.84

The reader can obtain practice in the use of the tables by carrying out a check similar to that in the previous example.

To summarise, some of the important facts relating to this table are:

(1)   The table is based on compound interest calculations.
(2)   The present value of £1 is the reciprocal of the amount of £l.
(3)   The present value of £1 is always less than unity.
(4)   The formula for calculating the present value of £1 is
$$V = \frac{1}{(1 + i)^n} \quad \text{or} \quad V = \frac{1}{A}$$
(5)   The difference between any given present value of £1 and unity represents the interest that could be earned by that present value of £1 if it were invested for that period and at that rate of interest.

The Amount of £1 Per Annum Table

This table gives the amount to which a series of payments of £1 invested annually at the end of each year will accumulate over a given period of years if compound interest is earned at a given rate.

The Amount of £1 Table involves a single payment of £l, but quite frequently money is saved or spent at regular intervals of time. The Amount of £1 Per Annum Table envisages a series of payments of £1 at the end of each year, each £1 going into an ever-increasing fund on which compound interest is earned.

It should be carefully noted that the payments are made at the end of each year, and this is almost the same as if the first payment were made a day later, that is at the beginning of the second year. If the concept of the Amount of £1 Per Annum is considered in this way any Amount of £1 Per Annum figure could be found by taking the Amount of £1 figures at the appropriate rate of interest for a time period of one year less than the required time period, adding them together, and then adding a further sum of £1 to the total representing the £1 deposited on the last day of the time period.

For a time period of 5 years at 10% the Amount of £1 pa could be found by adding the figures in the table on p110 of this book, and adding a further £1 for the last year, giving a total of 6.1051, as below:

End of Year	Capital saved	Period for which interest is earned	Accumulates to	Total at end of 5th year
1	£1	4 years	A £1 in 4 yrs @ 10% =	1.4641
2	£1	3 years	A £1 in 3 yrs @ 10% =	1.3310
3	£1	2 years	A £1 in 2 yrs @ 10% =	1.2100
4	£1	1 year	A £1 in 1 yr @ 10% =	1.1000
5	£1	No interest as paid into fund on last day of the 5 years		£1.0000
Amount of £1 pa in 5 years @ 10%				6.1051

It will be noted from the above calculations that the result comprises a total of capital invested of £5 (£1 for each year of the period considered) plus total accumulated interest of £1.1051.

Whenever the amount of £1 pa is required, the above calculation could be made, but it is far easier to find the answer in *Parry's Tables*, and this particular figure can be found on p122.

It was noted in the Amount of £1 Table that when long periods and high rates of interest were involved, the resultant figures were surprisingly high. With the Amount of £1 Per Annum Table in similar circumstances the results are even more surprising, and it is useful to consider a few figures from the table.

		Total capital	Total interest earned
Amount of £1 pa @ 2% for 3 yrs	3.0604	3	.0604
2% for 10 yrs	10.9497	10	.9497
2% for 20 yrs	24.2974	20	4.2974
2% for 50 yrs	84.5794	50	34.5794
10% for 3 yrs	3.3100	3	.3100
10% for 10 yrs	15.9374	10	5.9374
10% for 20 yrs	57.2750	20	37.2750
10% for 50 yrs	1163.9085	50	1113.9085
20% for 3 yrs	3.64	3	.64
20% for 10 yrs	25.9585	10	15.9585
20% for 20 yrs	186.688	20	166.688
20% for 50 yrs	45497.1905	50	45447.1905

The 20% figures cannot be found in *Parry's Tables*, but are shown above for consistency with previous examples. Some of the figures are almost unbelievable but they are mathematical facts and they clearly illustrate the rewards for regular saving, particularly if high rates of interest can be earned.

The formula for calculating the amount of £1 pa is $\dfrac{A - 1}{i}$

where A = the amount of £1 and i is the interest earned by £1 in 1 year.

It can be written more fully as $\dfrac{(1 + i)^n - 1}{i}$

For those who have difficulty remembering formulae, it may be helpful to remember that in this table the figures are large, and the denominator i, being small, will help to give appropriately large results.

As before, the use of the table is best explained by examples.

Example 5

A man of strong will and frugal habits manages to save £10,000 pa out of his salary. He invests this at the end of each year to earn interest at 7.5%. What sum will he have available at the end of 18 years?

Annual savings	£10,000
Amount of £1 per annum at 7.5% for 18 years	35.6774
Total available after 18 years	£356,774.00

The total capital invested is 18 × £10,000 or £180,000, and the balance of £176,774 is accumulated interest.

Example 6

Woodland has been planted which will take 40 years to mature, and it is estimated that £24,000 pa will have to be spent on maintenance. What will the true cost of maintenance have been when the timber reaches maturity, if the woodland owner has to borrow capital at 11%?

Annual maintenance cost	£24,000
Amount of 1 per annum over 40 years at 11%	581.8261
	£13,963,826.40

The actual capital spent will be 40 years × £24,000 or £960,000 and the balance of £13,003,826.4 represents the interest which would be payable on the money borrowed.

Alternatively, if the owner had not needed to borrow money the £13,003,826.4 would represent earning power he had sacrificed by investing his money in trees.

In summary, some important features of the Amount of £1 Per Annum Table are:

(1) The table concerns a series of annual payments of £1.
(2) Each payment is made at the end of the year, the first payment being at the end of the first year.
(3) The table is based on compound interest calculations.
(4) The last payment earns no interest.
(5) The formula is $\dfrac{A - 1}{i}$ or $\dfrac{(1 + i)^n - 1}{i}$

The Annual Sinking Fund Table

This table gives the amount which must be invested annually at the end of each year to provide £1 at the end of a given period, taking into account the accumulation of compound interest at a given rate during the period concerned.

As the sum which finally accumulates is only £1, each of the annual sums will be considerably less than £1, and the longer the time over which the money accumulates and the higher the rate at which interest is earned, the smaller will be the size of the annual instalments. A sinking fund is nothing more than a savings fund into which a series of annual payments are made with the object of ensuring that a specific sum of money is saved by a given future date. At that future date the fund will comprise all the annual payments that have been made into it plus the compound interest earned by each annual payment since the date it was first put into the fund.

An inspection of some of the figures will illustrate some of these points, and will give the student some idea of the type of figures involved.

Annual Sinking Fund to redeem £1 in	3 years @ 2%	.3267547
	10 years @ 2%	.0913265
	20 years @ 2%	.0411567
	50 years @ 2%	.0118232
Annual Sinking Fund to redeem £1 in	3 years @ 10%	.3021148
	10 years @ 10%	.0627454
	20 years @ 10%	.0174596
	50 years @ 10%	.0008592
Annual Sinking Fund to redeem £1 in	3 years @ 20%	.2747252
	10 years @ 20%	.0385230
	20 years @ 20%	.0053565
	50 years @ 20%	.0000219

These figures need little explanation and it is apparent that only very small amounts need to be saved annually if the fund is to accumulate over a long period, particularly if high rates of interest are involved. If it is remembered, for instance, that £.02 is 2p, it will be realised just how small a figure £.0000219 is. However, it is extremely doubtful if anyone would ever be fortunate enough to get such a favourable rate of interest as 20% for a sinking fund.

A building society mortgage is in practice somewhat similar to a sinking fund in that a mortgagor borrows a sum of money and then makes a series of equal payments over a period of time at regular intervals to ensure that at the end of the time sufficient money has accumulated to repay the debt. The essential requirements and the operation of a building society mortgage are similar to those of a sinking fund, although the student should be warned that the concepts behind the mortgage calculations are different, even if the mathematics involved gives the same result in money terms.

The Annual Sinking Fund is proportional to the Amount of £1 Per Annum. With the latter table a series of annual payments of £1 are made to accumulate to a larger future sum. With the Annual Sinking Fund a series of proportionately smaller payments are made to accumulate eventually to £1. So the Annual Sinking Fund formula is the reciprocal of the Amount of £1 Per Annum formula, that is:

$$\text{Annual Sinking Fund (ASF)} = \frac{1}{\text{Amt £1 pa}} \quad \text{or} \quad \frac{i}{A-1}$$

As with previous tables its use in practice will be illustrated by means of examples.

Example 7

£50,000 will be required to purchase a new car in 10 years' time. Money can be invested today to earn interest at 7.5% in the intervening period. What amount should be invested annually at the end of each year to ensure that sufficient money is available in 10 years' time?

Annual Sinking Fund to accumulate to £1 in 10 years at 7.5% (p91 of *Parry's Tables*)	.0706859
Total capital required	£50,000
Total annual payment required	£3,534.295

The total annual payment is merely a multiple of the figure found in the tables in the ratio that the sum eventually required relates to £1.

Example 8

The validity of the above calculation can be checked by use of the Amount of £1 Per Annum Table.

Amount to which £1 pa will accumulate in 10 years at 7.5%	£14.1471
Annual savings	£3,534.295
Amount to which £3,534.295 saved annually for 10 years will accumulate, interest being earned at 7.5%	£50,000.025

This proves that the Annual Sinking Fund will fulfil its objective, and it also illustrates its relationship to the amount of £1 pa. Example 7 involves 10 annual payments each of £3,534.295. At the end of the 10 years the £50,000 available will comprise 10 x £3534.295 or £35,324.95 of capital saved, plus accumulated compound interest amounting to £14,657.05.

In the property world this table is extremely useful if it is known that capital sums will be required at future dates, and there is a regular income, such as rents coming in, from which ASF payments can be made. The replacement of fixed equipment in buildings, the renewal of roofs, repointing and similar jobs could all be allowed for by setting up Annual Sinking Funds.

In summary, the more important features of the Annual Sinking Fund are:

(1)   The table concerns a series of annual savings each less than £l.
(2)   The sum eventually saved is £l.
(3)   The £1 saved comprises all the annual capital savings plus accumulated compound interest.
(4)   The last year's savings earn no interest, being placed in the fund on the last day of the last year.
(5)   The ASF formula is $ASF = \dfrac{1}{Amt\ £1\ pa}$ or $\dfrac{i}{A-1}$

The Present Value of £1 Per Annum

This table shows the present value of the right to receive an annual income of £1 at the end of each year for a given number of years, each year's income being discounted at a given rate of compound interest.

It is possible to find the value of such an income flow from the Present Value of £1 Table, and if an income of £1 were receivable for each of the next 6 years it could be valued as follows, the value of each £1 receivable being the PV of £1 for the number of years preceding the receipt of that £1.

PV of £1 in		
	1 year @ 8% =	.9259259
	2 years @ 8% =	.8573388
	3 years @ 8% =	.7938322
	4 years @ 8% =	.7350299
	5 years @ 8% =	.6805832
	6 years @ 8% =	.6301696

Present Value of the right to receive an income of
£1 per annum for 6 years discounted at 8% (PV of £1
Per Annum for 6 years at 8%)   £4.6228796
                                                      _____

The present value of £1 Per Annum for any period at any rate of interest could be found simply by adding the figures from the Present Value of £1 Table for each of the years involved. With long periods this would soon become very tedious, and the Present Value of £1 Per Annum Table obviates the need for such a time-consuming exercise. The £4.6228796 found above can be looked up direct on p34 of *Parry's Tables* where it is rounded off to 4.6229, but although it is therefore not necessary to refer to the Present Value of £1 Table it is nevertheless important to remember its relevance.

As with previous tables, an inspection of some of the figures from *Parry's Tables* may be helpful.

Present value of £1 Per Annum in	10 years @ 2%	8.9826
	20 years @ 2%	16.3514
	50 years @ 2%	31.4236
	Perpetuity @ 2%	50.0000
	10 years @ 10%	6.1446
	20 years @ 10%	8.5136
	50 years @ 10%	9.9148
	Perpetuity @ 10%	10.0000
	10 years @ 20%	4.1925
	20 years @ 20%	4.8696
	50 years @ 20%	4.9995
	Perpetuity @ 20%	5.0000

It is apparent from the above that when low rates of interest are involved the present value of £1 pa figure for any period is relatively high, and that when high rates of interest are involved the figure is relatively low.

If the reader refers back to these figures after completing this chapter he or she will note that when an income flow in perpetuity is valued at 2%, 32.7% of the total capital value is attributable to the first 20 years' income flow, when 10% is used 85% of the total capital value is attributable to the first 20 years' income flow, while when 20% is used almost 84% of the total value is attributable to the first 10 years' income flow. In the latter instance nearly all the value is attributed to the near future, and it is well to remember at this stage that 4.1925 represents the sum of the PV's of £1 at 20% for each of the first 10 years.

In the earlier calculation the right to receive £1 pa for 6 years was considered. The right to receive any larger sum for the same period discounted at the same rate would be proportionately larger and the use of the table can be illustrated by example.

Example 9

What is the present value of the right to receive an annual income of £3,755 at the end of each of the next 6 years, discounted at 8%?

Annual income	£3,755
Present Value of £1 Per Annum for 6 years at 8%	4.6228796
Present Value of the right to receive £3,755 for 6 years at 8%	£17,358.9129

Example 10

What is the value of an annual income of £6,875 for the next 27 years discounted at 7%?

Annual income	£6,875
PV of £1 pa for 27 years at 7% (p32 of *Parry's Tables*)	11.9867
Capital value of income of £6,875 for 27 years at 7%	£82,408.562

Example 11

What is the value of an annual income of £20,000 receivable in perpetuity at 9%?

Annual income	£ 20,000
PV of £1 pa in perpetuity at 9% (p35 of *Parry's Tables*)	11.1111
Capital value	£222,222

The last three examples are in fact simple valuations using the investment method of valuation, annual income flows having been converted to capital figures by means of multipliers. The present value of £1 Per Annum is the best title for this multiplier as it explains what it represents, but it is frequently referred to as the years' purchase, or YP for short.

The reader will note that in example 11 the income is receivable in perpetuity, that is there is no future date at which the income flow will cease, but that it will continue permanently. As in all other instances the YP is the addition of each PV of £1 for each of the years the income is receivable. Mathematically the figure can be found from the formula:

PV of £1 Per Annum in perpetuity or YP in perpetuity
$= \dfrac{100}{R}$ or $\dfrac{1}{i}$ where R = the % rate of interest per annum
and i = the interest earned by £1 in 1 year.

The validity of this formula can be checked by examples.

Example 12

What is the value of an annual income of £7,500 receivable in perpetuity discounted at 4%?

Annual income	£7,500
YP in perp @ 4% = $\dfrac{100}{R}$ = $\dfrac{100}{4}$ =	25
(p 31 of *Parry's Tables*)	
Capital Value	£187,500

If the above income flow is purchased for £187,500 we can check whether the YP has converted income into a capital value in the right ratio to ensure the correct return on the money invested.

A 4% return is required on £187,500.

$\dfrac{187,500 \times 4}{100}$ = £7,500, and the YP has fulfilled its function correctly.

Example 13

What is the value of an income of £40,000 pa receivable in perpetuity, being the rent from a good-class shop? The appropriate rate of interest is 7%.

Rental Income	£40,000
YP in perpetuity @ 7% = $\dfrac{1}{.07}$	14.285714
(p33 *Parry's Tables* – 14.2857)	
	£571,428.57

The functioning of the YP can again be checked, a 7% return on capital being required.

$$\frac{571,428.57 \times 7}{100} = 39,999.999$$

As the property produces £40,000 each year it can be seen that a 7% yield is in fact obtained if £571,428.57 is paid for it, and that the YP in perpetuity has correctly fulfilled the task of converting annual income into a capital value.

The use of the PV of £1 Per Annum Table or the YP will be considered again in the next chapter, which deals with the Investment Method of valuation, and a summary of the more important features of this table is:

(1)   The table concerns a series of annual receipts of £1 each.
(2)   Each £1 is received at the end of the year.
(3)   The receipt of each future £1 is discounted at an appropriate rate of compound interest.
(4)   The formula for the Present Value of £1 Per Annum in Perpetuity is $100/R$ or $1/i$ (the formula for the YP for a limited period will be considered in a later chapter).
(5)   The function of a YP is to show the relationship between income and capital, and to convert income flows into capital values.

The tables – general

It will have been noticed that all the above discussion relates to interest calculations which are made annually, and to receipts and payments at specific times, for example at the end of each year.

The formulae can be adapted to allow for variations in the pattern and it would be relatively simple for a YP to be worked out on the basis of each £1 being received at the beginning of each year or in four equal quarterly in advance payments, which are probably more realistic payment patterns in the modern world, particularly the payment of rent quarterly in advance. However, the current set of *Parry's Tables* is based on the descriptions as above.

If interest calculations are necessary at intervals of less than one year, it is possible to work out the correct figure by adaptation of the Amount of £1 formula, which is $A = (1 + i)^n$. If "t" is the number of interest calculations made in any year, the formula $A = (1 + i/t)^{tn}$ will give the correct Amount of £1. So if interest calculations are made half-yearly the formula becomes $A = (1 + i/2)^{2n}$ and if monthly calculations are made it becomes $A = (1 + i/12)^{12n}$. The meaning of "i" and "n" in the formula remains as before. As A is the basis of all the other formulae, by use of this adapted formula any of the other tables can be calculated for interest periods of less than one year. Although such adjustments will not always give absolutely correct answers, the resultant figures will be sufficiently accurate for most valuation purposes.

Chapter 15

The Investment Method of Valuation (Capitalisation)

This method has already been discussed briefly, and the purpose of this chapter is to look a little deeper into its use in practice, and to consider the role of the Years' Purchase.

As stated earlier, the method involves the conversion of an income flow from property into an appropriate capital sum. The capital value of a property is therefore directly related to its income producing power, which may give rise to actual, notional, or potential income. The income flow will be actual when a property is let on lease and the tenant pays a rent for use and occupation. There is a notional income flow when an owner occupies a property himself or herself, as although he or she will obviously not pay a rent, the notional rent is the figure which he or she would otherwise have to pay to acquire the use of a similar property. The value of the property to him or her as an occupier should also be at least as great as the market rental value, otherwise he or she would be better off letting it on the market. So, even if property is not let, the full rental value can be estimated, and this is also possible when a property is vacant and available to let. The Investment Method of valuation can therefore be used even when no rent actually passes on a property, the income producing potential being estimated by the valuer.

The *full rental value* of a property is the maximum rent for which it could be let in the open market on a given set of letting terms. The concept envisages that it is possible to let the property at that rent, and it follows that if the full rental value is known a valuer can use it in his or her valuation, confident in the knowledge that it would be paid by at least one potential tenant. The assessment of full rental value requires thorough knowledge of the relevant market, and skill in the application of that knowledge on the part of the valuer. The more onerous the terms on which a property is offered the lower will be the full rental value, and vice versa.

If a property is offered on the condition that a tenant would be responsible for doing all repairs, the full rental value would be lower because of the extra responsibility for repairs and the resultant extra costs to the tenant than if the landlord undertook that responsibility. If the terms on which a property is to be let are varied, there will normally be a complementary variation in the full market rent.

The assessment of a full rental value will invariably involve comparison with similar properties which have recently been let. In many circumstances direct comparison may not be possible because of the differences in size of different properties, and the valuer will frequently have to rely upon the use of *Units of Comparison*. If a factory of 1,000 m^2 has to be let and there is evidence of lettings on three factories of 700 m^2, 900 m^2 and 1,200 m^2, the valuer would reduce each piece of market evidence to a price per square metre, a square metre being the unit of comparison. He or she would study the property to be valued and the three comparables, and would choose a suitable price per square metre for the valuation. In choosing a figure adjustments would be made as considered necessary to reflect differences between the various properties, such as differences in quality of accommodation and location, and possibly even to reflect the existence in the market of a quantity allowance: see later.

The essential feature of units of comparison is that properties of varying sizes are reduced to a common unit of area to enable comparison to be made between them. Typical units of comparison are listed below, both metric and imperial units being quoted

	Units of Comparison Imperial	Metric
Building land	Per acre	Per ha
	Per sq yd	Per m^2
Residential building land	As above and also per unit of accommodation or per plot or block	
Factories and warehouses	Per sq ft	Per m^2
Offices	Per sq ft	Per m^2
Shops	Per sq ft	Per m^2
	Also in some instances per ft or per m of shopping frontage to the street.	
Agricultural land	Per acre	Per ha
Fishing rights	Per rod, or per ft or yd of river bank	Per rod or per m of river bank
Cinemas and theatres	Per seat or per full house	

There is no rule that says the above have to be used, and customs vary in different areas. As long as the valuer is consistent in his or her choice of unit a valid comparison can be made.

The actual *assessment of rent* is an important process, as any discrepancy at this stage will be multiplied many times once a multiplier is used to convert the rent to capital value. This is not to suggest that complete accuracy is possible in assessing rental values, as the best that can be hoped for is a reasonably accurate and intelligent estimate, the valuer's own judgment being as critical a factor as the mathematical analysis. The process of comparison and assessment of rental value will be illustrated by a number of examples.

Example 1

A single-storey shop with a frontage of 6 m and a depth of 13 m has to be let. What rent is it likely to command if a near-by single-storey shop with a frontage of 5 m and a depth of 13 m was recently let for £10,000 pa?

Analysis of recent letting
5 × 13 @ £10,000pa   = 65 m^2 @ £10,000
                                      = £153.8462/m^2

Rent assessment
6 m × 13 m @ £153.8462/m^2 = Estimated rental value
                                      = £12,000 pa

Alternative analysis
5 m frontage @ £10,000 pa   = £2,000/m frontage pa
Rental assessment
6 m frontage @ £2,000/m frontage   = £12,000 pa.

Note. Analysis by means of value per metre frontage can only be done if properties are of similar depth. When different depths are involved a process known as "zoning" is frequently utilised: see later.

Example 2

A site of 2.5 ha with planning permission for 65 semi-detached houses has recently been sold for £1,170,000. Another near-by site of 1.8 ha with planning permission for 54 semi-detached houses has to be valued for sale.

Analysis of sale

2.5 ha @ £1,170,000	= £468,000 per ha
or	
65 units @ £1,170,000	= £18,000 per house plot
Density = $\dfrac{65}{2.5}$	= 26 plots per ha

Valuation

The valuer will first note that the density is somewhat greater on the land to be valued.

54 units on 1.8 ha = $\dfrac{54}{1.8}$ plots per ha   = 30 plots per ha

In view of the greater density the houses, when completed, will probably sell for less. The valuer may therefore decide that each plot is worth £17,000.

Value of 1.8 hectare site = 54 units @ £17,000 each = £918,000.

Note. Although this valuation utilises a lower price for each plot, the total site value is greater than if a valuation had been made using the unit of comparison "price per hectare". The latter unit would not have allowed for either the greater density of development or the lower value per plot. An adjustment could have been made to it in an attempt to allow for these two factors, but such adjustments are much easier when applied to the smaller unit of comparison of the single plot.

Example 3

A factory of 2,000 m² has to be valued for rent purposes. It is similar to three other factories of 2,200 m², 1,800 m², and 1,500 m² each of which was recently let for £79,200, £72,900 and £64,125 pa respectively.

Analysis of comparables

1.	2,200 m² @ £79,200 pa	= £36/m²
2.	1,800m² @ £72,900 pa	= £40.50/m²
3.	1,500m² @ £64,125 pa	= £42.75/m²

It is apparent that the rental value per square metre decreases as the size of the factory increases. Such a trend sometimes occurs in the market, but this is not always the case. The valuer should take care to ascertain that such variations do not arise because of other factors, before he makes a "quantity allowance".

"Quantity allowances" may occur if for no other reason than that the management costs per square metre may be lower on a large area than on a smaller one, and the freeholder may therefore be prepared to accept a lower average rent. They do not always occur, and their existence will generally depend upon a number of supply and demand factors.

Valuation

In this case the valuer may decide that £38.25/m^2 is an appropriate value.

2,000 m^2 @ £38.25/m^2 pa. Rental value = £76,500 pa

"Zoning" was referred to earlier, and is a technique which is sometimes used to enable a rather more sophisticated comparison to be made between units of varying sizes. It is used principally with shops, as the area immediately adjacent to the street is generally considered to be more valuable than the space at the rear of a shop. The space at the front of a shop can be used for display purposes to attract passing pedestrians, and enticing the customers inside a shop is perhaps the most important step in promoting a sale. The area behind this can be utilised for sales purposes, while that at the rear of the shop is possibly only usable for storage. It is therefore common for shop rents to be "zoned", the front zone having the highest value, the next zone a lower value and subsequent zones even lower values. The depth of a zone and the decrease in value between zones are not fixed by any rigid rules, but are arbitrarily fixed as a result of market negotiations between landlords and tenants. However, it is often found that zones have been calculated at 20 ft depths, and are now calculated at 7 m depths. It is rare to use more than three zones in valuing or analysing transactions. It is generally found that values are "halved back" with zoning, for example Zone A (the front zone) at £X/m^2, Zone B at £.5X/m^2 and Zone C at £.25X/m^2.

Example 4

72, High Street, a shop with 7 m frontage and 18 m depth, was recently let for £16,500 pa. The rental value of 68, High Street, which has a frontage of 5.3 m and a depth of 16 m, is required.

72 HIGH STREET	68 HIGH STREET
£16,500 pa	
7 m	5.3 m

ZONE A	7m	ZONE A	7m
ZONE B	7 m	ZONE B	7m
ZONE C	4 m	ZONE C	2m

Analysis of 72, High Street

Zone A = 7m × 7m @ £X/m² =	49X
Zone B = 7m × 7m @ £.5X/m² =	24.5X
Zone C = 7m × 4m @ £.25X/m² =	7X
Total rental value £16,500 pa =	80.5X
£204.97= X	
The value/m² of Zone A =	£204.97
The value/m² of Zone B =	£102.48
The value/m² of Zone C =	£ 51.24

These figures would not have been used in practice to find the rental value originally, and figures of £204, £102 and £51 were probably used. However, for comparison with number 68 it is best to be consistent and to use the figures as analysed.

Valuation of 68, High Street

On the assumption that in all respects other than size the properties are truly comparable, the rental valuation is:

Zone A = 5.3m × 7m @ £204.97 =	7,604.39
Zone B = 5.3m × 7m @ £102.48 =	3,802.01
Zone C = 5.3m × 2m @ £51.24 =	543.14
Full Rental Value =	£11,949.54

In practice this would probably be rounded off to £12,000 pa.

For a further study of zoning the student should refer to books on valuation for rating purposes.

Once a rent has been estimated or is known, it must be reduced to a figure of net income before it can be converted into a capital value. An investor can only spend the money which remains after he or she has paid all the outgoings for which they are responsible, and any liabilities must be deducted from the gross rental to give the net rental value.

Outgoings which a landlord may have to pay include repairs, insurance premiums, general rates, water rates, and management costs. It is an important part of the valuation process to determine exactly what outgoings a landlord is responsible for and then to estimate the annual expenditure involved.

Repairing liabilities may vary considerably from property to property and from lease to lease. It is common for a lessor to attempt to make a lessee take on all the repairing liabilities. If this happens the landlord will generally have to spend nothing on repairs, but the valuer should always be prepared to make a contingency allowance for repairs if it appears that a tenant may be financially unable to meet his or her repairing commitments, or if he or she appears to be the sort who may disappear at dead of night, leaving expensive repairs to be done. If the lessor is liable for repairs, the liability may vary from being responsible for structural repairs only, or for external repairs and structural repairs, to being responsible for all repairs. The valuer should carefully determine the extent of the liability by examining the lease documents, and should then inspect the property and carefully estimate the cost of complying with any repairing liability. The anticipated repairs should be costed, and it is not acceptable to allow a percentage of the rent to cover repairs. The rent passing on a property does not necessarily give any indication of likely repairing costs which will be related to the situation and exposure of a building, the local climate, the design of the building, the materials of which it is constructed, the quality of past maintenance and the degree of wear and tear involved in its current usage.

The cost of *insurance* against fire and similar risks will be related to the cost of constructing a similar alternative building. As insurance is an indemnity against loss, the valuer must estimate the cost of replacing the present building in the event of its being destroyed. As with repairs, this is not necessarily related to the rent produced by a property, and insurance costs should not be estimated by taking a percentage of the rent. The area of the building must be calculated, the cost of constructing a building of that size must then be worked out, and the premium charged by the insurers will be directly related to that cost, and will be the figure to be deducted as an outgoing. In estimating the sum to insure for, the valuer should also allow for professional fees likely to arise on reinstatement, for demolition costs of the damaged building, for the costs of shoring adjoining buildings and land, for taxes levied on building costs and professional fees, and for loss of rent during the reinstatement period when let buildings are involved.

Rates, both *general and water and any other similar charges* are normally the liability of an occupier, and will not often have to be paid by a landlord. When they are the landlord's liability, the amount payable will be the Rateable Value of the property multiplied by the Rate in the pound for the area in which the property is located.

Management expenses will occur with virtually all property interests, even when the landlord has no repairing liabilities, and no other responsibilities. In the latter circumstances the landlord will still have to check that the tenant is complying with all his or her obligations, and he or she may also have expenses connected with collecting and banking the rent and keeping accounts. When the landlord does the management himself or herself, the cost will be his or her out of pocket expenses and the cost of his or her time. When managing agents are employed the cost will be their charges, which will be related to the work and expense involved, and which may be levied on the basis of a percentage of the rent passing on the property.

Once the net income has been calculated the capital value will be proportional to it in a ratio which will depend upon the quality of the property as an investment. As discussed earlier, this will be revealed by the yield, or the rate of interest at which it is valued, which will have been determined following comparison with alternative investments.

In recent years there has been considerable questioning by some valuers as to whether the valuation of properties by use of such yields is acceptable, it being argued that they are very rough-and-ready and insufficiently precise measures of the income producing quality of investment properties. The Investment Method utilises what is known as "the all risks yield", that is the yield has to be appropriately chosen to allow for all the bad features of the property as an investment and all the risks attached to those features, as well as reflecting all the good features of the property as an investment. It is indeed difficult to reflect all the pros and cons in the yield selected, and it is suggested by some that the Investment Method is no longer appropriate and that all investment valuation should be undertaken using a discounted cash flow (DCF) approach as discussed in chapter 19.

While a DCF approach to an investment valuation allows more precise attention to be paid to a range of considerations and variables, it is nevertheless a fact that many market valuations are still carried out using an investment valuation (or capitalisation) approach, and for that reason it remains important for students to study the approach thoroughly. It is also a fact that in some valuation situations the traditional investment valuation approach remains as good a method as any, which probably explains its survival in use. It should also be noted that yields quoted for other investments, such as stocks and shares, are in fact all risks yields, and the property all risks yield is therefore useful for comparison with other investments.

The reader will also see later that in reality the DCF approach is only a development of the Investment Method which permits the use of more information when it is available than is normally used in an investment valuation, and when it is desirable it also facilitates the incorporation of more variables than are normally used in the Investment Method.

To illustrate how the market operates, a number of hypothetical but not untypical, transactions will be considered.

1 Property	2 Net income	3 Capital value	4 Multiplier
Factory	£15,000	£135,000	9
Factory	£16,500	£150,000	9.09
Factory	£67,500	£615,000	9.11
Shop	£7,500	£93,000	12.4
Shop	£10,500	£130,000	12.38
Shop	£13,500	£170,000	12.59
Offices	£45,000	£650,000	14.44
Offices	£120,000	£1,700,000	14.17

Column 2 shows the rents being produced by the properties listed in column 1, while column 3 shows the prices for which these investments have been sold. Column 4 has been calculated from columns 2 and 3 and shows the relationship that each sale price bears to the income the property produces. It can be seen that although all the multipliers are different, there is a reasonable consistency between similar classes of property, with factories being capitalised with a multiplier of approximately 9.1, shops with about 12.5, and offices with just over 14. The consistency of multipliers within a particular class of property is not always as regular as that illustrated, but it is a market fact that within one class of similar quality properties, the multipliers evidenced by sales are reasonably consistent. The variation between nine and 14.44 in the table above reflects the different regard of investors for the different types of property, and indicates that offices are most favourably regarded as investments, while factories are least favourably regarded. In valuation terms the figures reveal that property investment valuations were done on a basis which approximates to the following pattern:

Factories	*Shops*	*Offices*
Net income	Net income	Net income
× 9	× 12.4	× 14.3
Capital value	Capital value	Capital value

Our consideration of these transactions is in retrospect, but the investors who purchased the properties would have decided the figures they were willing to pay by doing valuations similar to the above. They would have chosen a rate of interest which they considered would give them a yield which adequately reflected the risks involved, and from this rate the Years' Purchase would then have been calculated. This basic approach is used in most investment valuations, although the calculations may become more complex when there are variations in the income flow over time, or when terminable incomes are involved.

If the figures in the table are considered further they can be related to the formula for the Years' Purchase in Perpetuity which is YP = 100/R. With the first factory this becomes 9 = 100/R and R must therefore be 11.1. This indicates that this factory was considered to be a good investment if the income it produced was an 11.1% return on capital.

The price of the first shop can be analysed in a similar way, the Years' Purchase being 12.4, YP = 100/R becomes 12.4= 100/R, and R must therefore be 8.06. This indicates that the purchaser of the freehold considered it to be a good investment if an 8.06% return on capital was received.

The reader may care to calculate the yields on each of the other properties.

It can be seen from the table that the lower the yield an investor is prepared to accept, the higher will be the multiplier with which the income flow will be capitalised. It follows that if two investors are interested in one property, the investor who is prepared to accept the lowest yield will place the highest capital value on the property, and vice versa.

Another aspect of this fact is that the poorer the investment is, the higher will be the yield the investor will require, the smaller will be the Years' Purchase and the lower the capital value which results. The converse also applies.

When called upon to value other offices, shops and factories, the valuer who possesses the market evidence in the table will be influenced by it. If a property similar to one of these classes of property has to be valued, it will be reasonable to expect it to be valued on a similar basis and at a similar rate, unless there is good cause to suspect that investors have changed their investment criteria.

As the investment method involves capitalising an income flow, a valuer might be tempted to think that if that income is known, a visit to a property to inspect it is unnecessary. This is not the case, and an inspection should always be made, as it is otherwise impossible to tell whether the income flow can be expected to increase or decrease in the future. The quality of the property must be ascertained for this purpose and also to determine the correct yield and to estimate any outgoings which may have to be met. No matter how small the income flow or how humble the property, an inspection should always be made.

At this stage it may prove helpful to read again the section in chapter 14 dealing with the Present Value of £1 pa, or Years' Purchase. A reconsideration of the mathematical concept and the examples which show how the Years' Purchase fulfils its role may be useful.

The *analysis* of market evidence involves nothing more than valuation in reverse. The capital value or rental value is known, and analysis is an attempt to find out how such figures were originally calculated. The unknown will usually be either the rental value or the yield at which the property was valued. It is rather like attempting to reconstruct a crime, and it is almost inevitable that, when analysing, a valuer will have to make certain assumptions and deductions. In doing so the valuer must exercise as much skill as possible, but there will nevertheless be an element of doubt in most analyses. It is consequently dangerous to suggest that the results of analysis necessarily prove anything; at best they suggest what probably happened in a given situation.

The simplest way to analyse transactions is to adopt exactly the same approach as in valuing, and to use the symbol "x" for the unknown quantity. Some simple analyses are illustrated in the following examples.

Example 5

The freehold of a shop which is let at the full rental value of £18,000 pa was recently sold for £150,000. A valuer with a similar shop to value wishes to find the yield on this comparable.

Full Rental Value	£18,000 pa
Years' Purchase in Perpetuity	X
Capital value	£150,000

$18,000 \, X = 150,000$ $X = 8.3333$

The Years' Purchase in Perpetuity $= 8.3333 = \dfrac{100}{R}$

The rate of interest with which the property was valued is therefore 12%.

Example 6

A freehold factory which was recently let at full rental value has just been sold for £2,025,000. A similar factory has to be valued, and the valuer wishes to use this market evidence. The factory has 5,900 m^2 of usable floorspace.

Full Rental Value	X
Adopt a Years' Purchase calculated at 11% in perpetuity	9.09
Capital Value	£2,025,000

$9.09 \, X = 2,025,000$ $X = £222,772$

The full rental value as calculated is £222,772 pa, which represents a rent of approximately £37.75/m².

It will be noted that to do this analysis the valuer had to assume a yield of 11% as there were two unknowns, the full rental value and the yield. In adopting a yield, knowledge of the investment market for industrial properties will have been utilised, and the rate chosen is therefore a considered estimate rather than a guess. Obviously, the more unknowns there are, the more cautious the approach which has to be adopted, but even where there is more than one unknown, analysis can nevertheless be a very useful valuation tool.

Many analyses will be much more involved than these examples, which are simply intended as a basic introduction to the process. If the valuer is careful to determine as many known facts as possible prior to embarking on analysis, and is subsequently very methodical in approach, there should be few problems encountered in doing more complicated analyses.

It should also be remembered that the results of analysis are likely to be more useful and more reliable the greater is the number of similar properties which can be analysed. Unfortunately, in the real world actual market evidence is often most evident for its absence, but nevertheless the valuer must attempt to use what evidence there may be as intelligently as is possible.

Chapter 16

The Valuation of Varying Incomes

It is possible for the income produced by a property to vary at different future dates. This may occur when a lease has been in existence for a number of years, during which time rental values may have increased, the present rent consequently being out of date. When the current lease expires it is therefore often reasonable to assume that at least the present full rental value of the property would be obtainable on re-letting, and that this would produce a higher income flow than the lower rent paid under the existing lease.

It is suggested that the present full rental value would be obtainable at the future date, as capital values are frequently calculated by use of values which are known at the date of valuation. Any possibility of further increases in rental value between the present and the future is reflected conventionally by the choice of a lower yield for the valuation of a property, signifying that the risk entailed in investing is reduced by the prospect of an enhanced future rental value. Most values change with time if there is inflation in an economy (particularly in property values), and it is often suggested that valuers should attempt to calculate what future rental values will be by considering trends in value, and that valuations should then be made utilising anticipated future rents. Valuations done on this basis became much more of a practice during the 1980s as changes in rental levels became quite substantial over relatively short periods of time. Some valuers therefore argue that it is unwise and impracticable to value without taking into account expected future increases in rental income, as to ignore such increases will result in the under-valuation of interests in property.

If anticipated future rent increases are included in a valuation the valuer should consider what recent trends have been, whether they are likely to continue in the future, and whether values are likely to rise or fall, and if so at what rate and for how long. It is impossible to make accurate predictions of such matters, but the valuer has to remember that a current market valuation represents an estimate of the present value of all the future returns that will be produced by a property. Past transactions are historic evidence; the future could be very different and the valuer must therefore guard against placing unquestioning dependence upon comparables which are firmly based in the past.

Where an approach is utilised by the experienced valuer which entails building into a valuation a monetary allowance for future growth in rental value, the valuer should be careful not to adopt a low yield which also reflects the prospect of future rental growth, as if that were done growth prospects would be valued twice.

Nowadays, in certain valuation situations the valuer will be unable to base his or her valuation on the conventional approach which only allows for future rental growth through the choice of yield. Certain clients now demand valuations which include the estimation of future rental values and yields. In particular developers, who wish to know what possible developments will be worth when they are in due course completed, and financial institutions which lend money to finance development schemes, find valuations based on current market evidence of limited use to them. Their decisions are completely dependent upon estimates of values in the future and many of them demand valuations which actually incorporate predicted future rental values and market yields. However, currently the adoption of such a valuation approach is still not universal and many valuations are still done placing considerable reliance upon comparable evidence and current market rents and yields.

Events in the early 1990s in many property markets throughout the world illustrate the problems which can be encountered with valuations which incorporate anticipated future rental increases. The Sydney office market proved a good illustration of a typical chain of events, with rents for prime Central Business District office space being about \$400 per m² in the mid 1980s, rising significantly in the second half of the decade until some valuations were being done incorporating rental levels in the range of \$800 to \$1,000 per m². Although such levels were actually achieved in a number of lettings, the recession of the early 1990s completely changed property markets to the extent that rental levels dropped right back to those of the mid-1980s. In fact, the extreme difficulty in even finding tenants for office space resulted in valuable incentives (such as lengthy rent-free periods) being attached to lettings which were negotiated, the cost of such incentives to a landlord being so great that it is arguable that rental levels actually fell below those of the mid-1980s.

The result of these happenings was that in the period 1985–1990 the capital valuations of many major buildings increased by 100%

or even more, only for their valuations to be subsequently halved over an even shorter time period. This illustrates the quandary faced by valuers who could reasonably argue that the valuations done at any particular time reflected market opinion of value as at each point in time. On the other hand, some purchasers of property at high values soon found their purchases to be worth very much less than they had paid, which not unnaturally left them displeased and many of them sought to blame valuers for their predicaments. Such a state of affairs emphasises the need for valuers to determine at the outset of a valuation the precise purposes of the valuation and their client's objectives and requirements, and then to make quite explicit in their valuations and reports the underlying assumptions made by them and the various inputs to the valuations which have resulted in the valuation figures calculated by them.

The discounted cash flow (DCF) format of valuation is favoured by many valuers over the traditional Investment (or Capitalisation) format because it more easily lends itself to the incorporation of a wider range of variables and enables the valuer to make explicit a number of assumptions incorporated in the valuation, as previously noted in chapter 15. However, care has to be taken that the ease with which variables and variations can be incorporated in the DCF format does not encourage their inclusion where such inclusion does not improve the reliability of the result, where they cannot be predicted with reasonable precision, or where the state of the market over a period of time is likely to be so uncertain as to make long-term predictions (especially of growth) unwise.

Varying rental incomes may also occur because of arrangements between the parties to a lease which have been made for their own convenience. A tenant may take a lease of a property in which he or she wishes to set up a business. The freeholder may agree to occupation of the property for a number of years at a concessionary rent so that the lessee may have a better chance of establishing a successful business. This may be in the interest of both parties, as it gives the businessperson a greater chance of success, while the landlord will benefit from having a prosperous tenant. After the concessionary rent period, the businessperson would normally be expected to pay the full market rent for the property.

Because of the rapid changes in value which occur during periods of high inflation, landlords are reluctant to enter into long leases at a fixed rent, as this might result in their receiving an income which might become rapidly devalued by inflation. A series of short leases with new rents which kept pace with inflation would be preferable, but might not suit a potential tenant who would prefer the security of knowing that he or she had the right to occupy the premises concerned for a reasonable number of years. To satisfy the landlord's desire for a rent which does not rapidly become out-of-date, and the tenant's desire for security of possession of a property, leases are often arranged for a reasonably long term, with rent increases agreed for specific future dates. Also common are leases without pre-determined rent increases, but with "rent-review" dates fixed in advance, with agreement on a specified way of assessing the rents to be fixed under these future reviews.

In order to consider the problems involved in valuing varying incomes, the principles will be discussed by reference to a freehold property which is let for a period of 15 years. The rent for the next 5 years will be £10,000 pa net, for the second 5 years it will be £15,000 pa net, and for the final 5 years £20,000 pa net.

The conventional valuation approach to the problem posed by a varying income-flow is to value each block of income separately and to add together the separate capital figures so found, the total being the present capital value of the right to receive all the items of future income. In effect, the valuer is faced with a series of separate valuations, as illustrated below.

First 5 years	Rent reserved	£10,000 pa	
	× Multiplier	————	
	Capital value of first 5 years income flow		£ ————
Following 5 years	Rent reserved	£15,000 pa	
	× Multiplier	————	
	Capital value of second 5 years income flow		£ ————
Final 5 years	Rent reserved	£20,000 pa	
	× Multiplier	————	
	Capital value of final 5 years income-flow		£ ————
Thereafter	Rent obtainable on re-letting	£	
	× Multiplier	————	
			£ ————
	Present capital value of freehold interest = the sum of the present capital values of all the future income-flows		£ ————

This format is used to value most varying incomes, and it is apparent that the answer will be influenced by the multipliers chosen. On p120 it was shown that the Years' Purchase for a given number of years is in fact the sum of the Present Value of £1 figures for each of those years, and that the YP figures for specific periods at given rates of interest have for ease and convenience been provided in *Parry's Tables*. Using these tables, the capital value of the income flow for the first five years can be found as follows:

First 5 years	Rent reserved	£10,000 pa
	YP for 5 years at 10%	3.7908
	Capital value of first 5 years' income	£37,908

The only problem with this stage of the valuation is the choice of an appropriate rate of interest, which the valuer would select using market experience so that the risks of the investment were adequately reflected.

The income flow for the second period can be tackled similarly and converted to a present capital sum.

Second 5 years	Rent reserved	£15,000 pa
	YP for 5 years at 10%	3.7908
	Capital value of the right to receive £15,000 pa for the next 5 years	£56,862

The approach used is exactly the same as for the first five years, but a problem has arisen. A YP for five years is the equivalent of adding the present values of £1 for each of the next five years to obtain a multiplier. When applied to the annual income of £15,000 as above this presupposes that £15,000 will be received this year, the second £15,000 in the second year, and so on for five years. But the first five years' income flow is £10,000 pa and have already been valued, and the flow of £15,000 pa will not commence until five years has elapsed. The position is therefore exactly the same as if a capital figure of £56,862 were not receivable until five years had elapsed. To find the value of the right to receive a future capital sum, the Present Value of £1 table can be used, as on p113. The second stage of the valuation of the varying income then becomes

Second 5 years	Rent reserved	£15,000 pa
	YP for 5 years at 10%	3.7908
	Present capital value of the right to receive an income flow of £15,000 per annum for 5 years, commencing this year	£56,862
	× PV of £1 in 5 years at 10%	0.6209213
	Present capital value of the right to receive an income flow of £15,000 per annum for 5 years which commences after the elapse of 5 years	£35,306.826

The valuation of the third income-flow involves exactly the same principles as those used for the second stage.

Third 5 years	Rent reserved	£20,000 pa
	YP for 5 years at 10%	3.7908
	Capital value of income-flow of £20,000 per annum for 5 years	£75,816
	× PV of £1 in 10 years at 10%	0.3855433
	Present capital value of the right to receive an income-flow of £20,000 per annum for 5 years which commences after the elapse of 10 years	£29,230.35

In this third stage of the valuation a PV for 10 years has been used as that is the period which will elapse before the first year of the income-flow of £20,000 pa is reached.

A problem arises on considering what would happen after the 15th year. The freehold interest should continue to provide an income flow after the end of that year because the interest continues in perpetuity, and the property can therefore be relet to produce another income flow. The problem is, what rent can be expected from the property at that time? The value of the property is not confined to the first 15 years, and an effort must be made to assess the value attributable to the period thereafter, that is from the fifteenth year in perpetuity.

Earlier in the chapter it was pointed out that it is conventional to value by reference to known current values, and it is therefore normally assumed that after the current lease has terminated it would be possible to relet the property for the current full rental value. If on investigation it were found that the present full rental value of the property in the example was £25,000 pa the final stage of the valuation would become

Reversion after 15 years to Full Rental Value	£25,000
YP in perpetuity at 10%	10
Capital value of an income-flow of £25,000 per annum	
received in perpetuity	£250,000
× PV of £1 in 15 years at 10%	0.2393920
Present capital value of the right to receive £25,000	
per annum in perpetuity commencing after the elapse	
of 15 years	£59,848

The above calculation involves the use of two multipliers – 10 and 0.2393920. The same result would emerge if these were multiplied together and the product applied to the annual income flow.

Reversion to Full Rental Value		£25,000 pa
YP in perpetuity @ 10% =	10	
× PV of £1 in 15 years @ 10% =	.2393920	2.393920
Capital value		£59,848

The figure 2.393920 is referred to as the Years' Purchase of a Reversion to a Perpetuity deferred 15 years at 10%. A YP of a Reversion to a Perpetuity can be calculated for given periods of deferment at given rates of interest. A table containing such calculations commences on p45 of *Parry's Tables* and the figure 2.393920 can be looked up direct on p56, so avoiding the need to refer to two separate tables.

Two multipliers were also used in each of the second and third stages of this valuation, and it is customary to apply the two multipliers to each other and then to utilise the product in the valuation, by which process the second stage would become:

Rent reserved		£15,000 pa
YP for 5 years @ 10%	3.7908	
× PV of £1 in 5 years @ 10%	0.6209213	2.3537884
		£35,306.826

The valuation of each stage of varying income can be seen to involve the following basic process.

YP for period of income
× PV for period of deferment }

Income flow

Multiplier
Capital value

An easy way of understanding how long a period of deferment must be is to think of it as the period of waiting before the year in which an income flow will commence.

Before reviewing the entire valuation there is a further modification which can be made to the different stages in order to comply with convention. In working through the different stages it was decided that the current full rental value of the property is £25,000 pa net. The majority of valuers maintain that if at any time a property is occupied at less than its full rental value, that element of income is more secure than the receipt of the full rent would be. There is less risk of a lessee defaulting in the payment of rent if he or she is obtaining a property cheaply, as apart from the fact that a low rent is in any case easier to pay than the full rent, the tenant is hardly likely to run the risk of losing the profit which he or she is enjoying. Should he or she move elsewhere they would have to pay a higher rent for a similar property, so they are unlikely to do that. Should they not wish to use the property themselves, if the lease allows them to sublet, they could do so at the full market rent and make a profit. Consequently, whenever a rent paid is below the full market rent, it is generally considered to offer greater security to an investor. It is therefore valued at lower rates of interest, as it is reasoned that an investor would be prepared to accept a lower yield from that particular income flow. The author considers that this theory is not always completely valid, but it has been the convention for many years and the valuation will now be reviewed, using progressively lower rates of interest as the rent reserved is lower.

Valuation

First 5 years	Rent reserved		£10,000 pa	
	YP for 5 years @ 8.5%		3.9406	
				£39,406
Second 5 years	Rent reserved		£15,000 pa	
	YP for 5 years @ 9%	3.8897		
	× PV £1 in 5 years			
	@ 9%	0.6499314		
			2.5280382	

				£37,920.572
Third 5 years	Rent reserved		£20,000 pa	
	YP for 5years @ 9.5%	3.8397		
	× PV £1 in 10 years			
	@ 9.5%	0.4035142		
			1.5493735	
				£30,987.469
Reversion to Perpetuity				
	Full RentalValue		£25,000 pa	
	YP in perpetuity			
	@ 10% deferred 15 yrs		2.39392	
				£59,848
	Capital value of freehold interest			£168,162.041

A few further comments on this final valuation are necessary. It will be noted that where lower rates of interest have been used in the valuation a higher capital value results than when 10% was used to value the same stage. The greater security results in a higher capital value.

Some of the calculations have been made to as many as seven places of decimals. A valuer in practice may well not do this, for a variety of reasons. Prior to the advent of cheap electronic calculators, such calculations were time-consuming and therefore costly. As valuation cannot possibly be a precise science, it can also be argued that they are unnecessary. For the same reason it is also customary to round-off valuation answers to the type of figure which would be paid in the market, and the answer above would probably be "rounded off" to £168,000. Now that efficient calculators are available the author's preference is to do all calculations to about four places of decimals, confining "rounding off" to the answer. Where a series of figures are all "rounded off" there is always the possibility of cumulative errors being unacceptably large. This will not often be the case and the degree of accuracy to which mathematics is done is very much a matter of personal choice to each valuer, although all approximations must be kept within acceptable limits.

The Years' Purchases for limited periods of time can be derived from the formulae

$$YP = \frac{1 - V}{i} \qquad \text{and} \qquad YP = \frac{1}{i + ASF}$$

The mathematical proofs of these formulae can be studied in *Modern Methods of Valuation* but a comparison with each of them and a YP from *Parry's Tables* will be made.

(i)   YP for 10 years @ 8% = 6.7101
         (*Parry's Tables*, p34)

(ii) $YP = \dfrac{1 - V}{i}$

         V = PV of £1 in 10 years @ 8% = 0.4631935

         $i = \dfrac{8}{100} = 0.08$

         $YP = \dfrac{1 - 0.4631935}{0.08} = \dfrac{0.5368065}{0.08} = 6.7101$

(iii) $YP = \dfrac{1}{i + ASF}$

         ASF 10 Years @ 8% = 0.0690295

         $YP = \dfrac{1}{0.08 + 0.0690295} = \dfrac{1}{0.1490295} = 6.7101$

The above calculations have been done partly to show the reader that the formulae are in fact alternative and more rapid ways of working out YPs for limited periods, as opposed to adding together the PVs of £1 for the relevant time, and also to show that each of these three methods of finding the YP is equally accurate.

The Layer Method of Valuation

Earlier in this chapter the conventional approach to the valuation of varying incomes was considered and it was noted that it was customary to value income flows which were at less than the current full rental value as being more secure than the full rental value, as a result of which a lower yield rate was used to reflect that greater security. Each income flow was valued in its entirety for the period of years during which it would flow, higher income flows being valued at higher yields.

The Layer Method adopts a slightly different approach and its use assumes the indefinite continuation of the rent currently passing which is valued at an appropriate yield, only the increments of rent in subsequent time periods being valued separately and at higher yields. The underlying logic behind this approach is that it is only the increase in rent (or the added layer of rent) to which a higher risk attaches, whereas in the earlier method all of the higher income in the subsequent time period was treated as being at greater risk.

Use of the Layer Method is not without difficulty as the choice of an appropriate yield to reflect the level of risk attached to the incremental income is far from easy.

Applying this approach to the earlier example in this chapter would give a result similar to the following, the yield used for the valuation of the base income being the same as in that example.

Valuation

Base rent reserved		£10,000 pa	
YP in perpetuity @ 8.5%		11.7647	
			£117,647
Second 5 years:			
Incremental income		£5,000 pa	
YP in perpetuity @ 10% =	10		
× PV of £1 in 5 years @ 10 % =	.6209213	6.209213	
			£31,046.065
Third 5 years:			
Incremental income		£5,000 pa	
YP in perpetuity @ 12% =	8.3333		
× PV of £1 in 10 years @ 12% =	.3219732	2.683099	
			£13,415.4963
Reversion to perpetuity:			
Incremental income		£5,000 pa	
YP in perpetuity @ 14%		1.00069	
deferred 15 years			
			£5,003.45
Capital value of freehold interest			£167,112.00

This answer is slightly lower than that calculated in the earlier example in this chapter, but it is only £1,050 less. It is probably surprising that the two answers are so close and it is perhaps chance that the valuation has been done on the assumption that the incremental income layers should be valued at 10%, 12%, and 14%. Had higher rates been used the answer would have been even lower, but had lower rates been used to value the increments the valuation figure might have been higher than in the more traditional approach. The choice of the appropriate rates is left to the judgement of the valuer, although the reader may wish to read further on this topic in *Modern Methods of Valuation* in which analysis is used to determine appropriate rates.

That a different answer results using this approach should not necessarily be of concern, because valuation is in any event inevitably imprecise incorporating as it does so much that is opinion, albeit, it is hoped, informed opinion. The use of a different method in any case infers that a valuer is not satisfied with the traditional method, while if expectations are that the different methods should produce the same results, then there would be no advantage in using one method as opposed to the other.

Which method a valuer chooses to use in practice is therefore another instance where individual judgement and experience has to be used, and ultimately experience in use is likely to determine the final choice.

It should perhaps be noted that the use of the Layer Method can easily be incorporated into a DCF valuation format, although that will not in itself make the eventual choice of the appropriate rates of interest for the valuation of the layers of income any easier.

Chapter 17

The Valuation of Terminable Incomes

Income flows do not always continue in perpetuity and any income may be receivable for a limited number of years only, after which no income whatsoever is received. Such a situation arises when a leasehold interest in property is owned. Any income which it produces will end with the termination of the lease.

Many leasehold properties are occupied by the leaseholders and do not produce an actual income flow. If, however, the full rental value is greater than the rent the leaseholder actually pays under the lease, a "profit rent" is enjoyed, as annual profit could be made if the property were sublet at the full market value. If the lessee continues in occupation there will be a direct benefit from occupation at a cheap rent, which benefit could be sold to another would-be occupier.

Only if a leasehold property produces or is occupied at a profit rent, will it have a value that can be calculated. The basic calculation for the valuation of a leasehold interest is

	Full rental value
Less	Rent paid under the lease
	Profit rent
×	Multiplier
	Capital value

Profit rents may occur for a variety of reasons. They may arise if rental values have increased since a lease began, or if a rent agreed under a lease is below the full market rent at the date of agreement. The latter may happen if a lessee pays a capital sum (a premium) at the commencement of a lease, representing rent or partial rent paid in advance. A lessee who pays a premium will only do so if a reduction in rent is received to compensate for the payment of the premium. Leaseholders sometimes pay all the rent for the duration of their lease in advance, and they thereafter occupy the property at a "peppercorn rent", which amounts to paying no rent at all.

Whatever the reason for its existence, if a profit rent does arise on a leasehold interest, an annual income flow will exist in the form of that profit rent, and this could be attractive to investors.

When a freehold property is owned, the owner of the interest can expect the receipt of an income flow in perpetuity, as the interest from which it arises will always be owned. There will therefore always be a capital asset, and if the income flow and the yield do not vary, then the capital value will always remain the same. Even if the interest is sold the capital will be recouped from the sale proceeds, and the asset will remain intact in money terms.

In reality it is unlikely that the value of any freehold interest will remain exactly the same over a long period. In inflationary periods rental values may increase quite rapidly over time, while if buildings deteriorate or a location becomes less popular, rental values may decrease. Likewise, the underlying conditions of the market change frequently, with resultant changes in yields. However, the proposition that freeholders always retain their assets and that their capital invested remains intact is basically true, and contrasts with the position of a leaseholder.

In the case of a leasehold interest, the leaseholder has the right to use and occupy the property only during the term of the lease. Once it expires there is no common law right to retain the use of the property, which returns to the control of the freeholder. As the unexpired term of a lease gets progressively shorter with the passage of time, so will the value of the leasehold interest decrease, until at the end of the lease no value remains.

If two similar quality investment properties produce equal income flows, one being a freehold interest and the other a leasehold interest, an investor would obviously find the freehold a more attractive proposition. He or she would be prepared to pay less for the leasehold, and a valuation problem arises in determining what the difference in the value of two such properties should be.

Not only does the income flow from a leasehold investment cease when the lease expires, but there are also other disadvantages. Freeholders, within the limits of common law and statutory law, can do what they like with their properties and are virtually "kings of their castles". Leaseholders do not have unlimited freedom in the use of their properties, being restricted by the terms of their leases. They may find that even if they have assigned or sublet their properties on the understanding that the new occupiers will be responsible for complying with all the covenants and conditions of the leases, a freeholder may still be able to take action against them if an assignee or sublessee defaults. As the unexpired term of a lease grows shorter, it may prove very difficult to assign or sublet a property, as most would-be occupiers require a reasonably long period of undisturbed occupation. What is often referred to as the "fag-end" of a lease may therefore be virtually unsaleable, and leaseholders may be left with properties which neither they nor anybody else wishes to occupy for such short periods, but for which they nevertheless have to pay rent.

The fact that a leaseholder has to pay rent at regular intervals, although not a particularly onerous task, involves management work, which does not arise with a freehold interest. A serious disadvantage with a leasehold property is that its value will eventually decrease even in inflationary times, and this lack of security in real terms is a most unattractive feature to investors. For these and other reasons a leasehold interest is generally considered less attractive than a freehold interest in a similar property, and investors therefore require higher yields from leasehold investments. The differences between freehold and leasehold yields may be as low as 0.5% on good properties which are held on long leases and let to good tenants, increasing to 2% or more as the quality of the property becomes poorer and as the lease term decreases.

The extra yield is not, however, sufficient to compensate a leaseholder for the eventual disappearance of the asset. If one investor spends £200,000 on a freehold investment, and another spends £200,000 on a leasehold investment, the latter will eventually lose the £200,000 invested whereas the freehold investor will always own the property on which money has been spent. To ensure that they will not end up worse off financially through buying leasehold interests, wise leaseholders can set aside part of the annual income which their properties produce, and can invest it to ensure that, by the time a leasehold interest terminates, the accumulated savings are sufficient to enable them to purchase another property or another investment which will produce a similar income flow. If this is done a leaseholder will be no worse off than if a freehold interest had been purchased in the first instance.

The fund which the leaseholder could establish is called an Annual Sinking Fund, because payments are made into it and compound interest is earned by it generally on an annual basis, and because it grows in size as the leasehold interest "sinks" in value.

The objective of a sinking fund is to recoup the capital invested in a wasting asset and in theory this can only be ensured if there is a guarantee that the money will be replaced by a specific date. Sinking fund theory is therefore based on the assumption that the annual instalments are placed in investments which are virtually risk free. It is possible for a sinking fund policy to be taken out with an insurance company which will guarantee to pay a specified sum to the saver on a definite future date, in return for a series of annual premiums paid over the life of the policy. Because the insurance company guarantees payment of the future sum, the rate at which it pays interest on the annual instalments is low. The low rate reflects the low risk nature of the investment to the saver, and also leaves the insurance company a margin which it can earn on reinvestment of the premiums. This will cover its overheads, and allow a profit to be made as payment for its expertise and its risk taking. The insurance company will also discharge any tax liability and will pay interest on the sinking fund policy net of tax. For these reasons the interest earned on sinking fund policies generally ranges between about 2% and 4% net of tax. These rates are not quite as low as they at first appear as if income tax is payable at, for instance, 40%, 2% and 4% net of tax rates are in fact 3.34% and 6.67% gross of tax.

A sinking fund policy is in fact not really different from an endowment policy under which an insurance company agrees to pay an insured person a fixed sum at a stated future date in return for the payment by the insured of a regular series of monthly, quarterly, or annual premiums.

The leasehold investor faces a slightly modified calculation when valuing his or her interest, as follows:

	Net profit rent from property
Less	Annual sinking fund contribution
	Net spendable income
×	YP
	Capital value of leasehold interest

The logic of this is not difficult to understand, but the mathematics provides a problem, as until the capital value is known, it is impossible to calculate the size of the required annual sinking fund instalment. The way in which the problem is overcome is by use of a modified Years' Purchase. When valuing a perpetual income a YP is found by use of the formula $1/i$. The 1 represents £1 of capital and i represents the annual earnings of that £1. The YP which results is the multiple that capital is of annual income. The multiplier theory is exactly the same with a terminable income, capital value being related to annual income. However, as part of the annual income of a leasehold property should be devoted to a sinking fund, the YP formula is modified to include a sinking fund allowance, becoming

$$YP = \frac{1}{i + SF}$$

The YP which results shows the multiple that capital value is of annual income after it has been separated into the annual spendable income and the sinking fund instalment.

The formula works in the following way, as illustrated by reference to an annual income of £10,000 arising from a leasehold interest which expires in 20 years' time. The potential investor can take out a sinking fund policy at 2.5% and requires a 10% yield, and wishes to know how much to pay for the investment.

The YP formula is
$$\frac{1}{i + SF}$$

The annual sinking fund required to replace £1 of capital over 20 years @ 2.5% is 0.0391471, and i is 0.1.

$$YP = \frac{1}{i + SF} = \frac{1}{.1 + .0391471} = \frac{1}{.1391471} = 7.1866391$$

This can be found on p 4 of *Parry's Tables* as 7.1866.

The YP is then used in the normal way and applied directly to the total net income flow.

Net income	£10,000 pa
YP for 20 years @ 10% and 2.5%	7.1866391
Capital value	£71,866.391

The YP is stated to be at 10% and 2.5% and this is known as a *Dual Rate YP* as two different rates of interest are used in the calculation. The rate of 10% depicts the yield the investor requires (or the rate at which remuneration from the investment is required) and it is referred to as the *"remunerative rate of interest"*. The rate of 2.5% shows the earning power of the sinking fund or the rate at which interest accumulates. It is referred to as the *"accumulative rate of interest"*.

An analysis of the above valuation will reveal the effectiveness of the YP.

Annual income	£10,000
Yield required on capital invested	
$\dfrac{71866.391 \times 10}{100} =$	7186.6391
Balance available for annual sinking fund	£2813.361
Annual Sinking Fund necessary to redeem	
£1 in 20 years @ 2.5% (*Parry's Tables*, p 84)	0.0391471
Capital to be redeemed	£71,866.391
Annual Sinking Fund required to accumulate	
to £71,866.391 over 20 years at 2.5%	£2,813.3607

These calculations show that the YP has capitalised the income flow in a manner which enables the correct yield to be obtained from the capital invested, and which also enables a sinking fund to be established out of income to recoup the original capital invested by the end of the lease. A final check proves that the sinking fund can achieve its objective.

Annual Sinking Fund	£2813.3607
Amount of £1 pa in 20 years @ 2.5%	25.5447
Capital recouped after 20 years	£71,866.455

The figures in these calculations can be related to the layout shown on p156.

Net Income	£10,000 pa
Less Annual Sinking Fund	2813.3607
Spendable Income	£7186.6393
× YP in perpetuity @ 10%	10
Capital Value	£71,866.393

This illustrates the fact that once an allowance has been made out of income to keep the original capital intact, the remaining income can be treated as if it is receivable in perpetuity. This is logical, as at the end of the lease term the sinking fund of £71,866 can be used to purchase another investment which produces a 10% yield in perpetuity, or £7186.6 pa.

Sinking Fund theory as applied to the valuation of terminable incomes is often criticised on the ground that it is unrealistic, and that no one would invest money to earn interest at only about 2.5%. The fact that some investors do take out sinking fund policies (and endowment policies which earn returns at similar rates) with insurance companies rebuts this point to a certain extent. It is also often said that those who do so are unwise, as higher rates of interest could be earned from other sources, but if it is a basic requirement of an investor that the receipt of the future sum should be guaranteed, the requirement will not be achieved if acceptance of a higher yield reduces the security of capital. The choice between higher yields and absolute security must be the decision of individual investors, but whatever the criticisms of sinking fund theory may be, the resultant YP does ensure that an investor's original capital can be kept intact, and that the purchase of a leasehold investment leaves them no worse off than if a freehold investment had been purchased.

Another criticism of sinking fund theory is that in inflationary times it is unwise to base a decision on the recoupment of historic costs, and that provision should be made for the recoupment of a larger sum to allow for the ravages of inflation. If calculations were made on this basis, an investor would be allowing for part of the risk of the investment by the allowance made in the sinking fund. If by creating a larger sinking fund the risk to capital caused by inflationary trends was reduced or eliminated, the investor would presumably be prepared to accept a lower remunerative rate of interest, as some of the risk normally reflected in this rate would have been removed. The use of a lower yield would then tend to cancel out the effects of creating a larger sinking fund. Instead of using the conventional dual rate YP approach, valuations could quite easily be done using these other approaches and a discounted cash flow format as described in a later chapter.

Individual investors may obviously choose different ways of dealing with the sinking fund problem. Some may actually take out sinking fund policies with insurance companies at low rates of interest. Others may decide that rates offered by insurance companies are too low and that they will therefore put their sinking fund allowances into investments which, although not so risk free, pay a higher rate of interest. Such a course of action may be quite acceptable, particularly if the period over which a sinking fund must accumulate is reasonably short or if individual investors are sufficiently well off to be able to carry some of the risk themselves. Clearly, if investors earn money on their sinking funds at higher rates of interest than conventional theory suggests is appropriate, the annual sinking fund allowances need not be so large (that is if the recoupment of the historic costs of investments is all that is required). They could then bid more for a particular leasehold investment and still obtain the yield desired. Whether in such circumstances they actually will bid more will depend on whether competition in the market forces them into a position of having to do so, but clearly the advantageous sinking fund position would enable such action to be taken if it was thought to be appropriate.

Investors with considerable property holdings may decide not to take out sinking fund policies, working on the principle that they can reinvest earnings themselves in other property purchases to earn a far higher rate of interest. This would be the likely attitude of a property investment company or a large institution. They could even organise their purchases such that a purchase of a wasting leasehold interest was balanced by a purchase of another asset which should appreciate over the same time period, the loss on one asset being compensated by the gain on the other.

In practice there are probably few purchasers of leasehold interests who actually take out sinking fund policies with insurance companies. Attitudes will vary depending upon the circumstances and outlooks of the individual investors, but those who reinvest what would be regarded as the sinking fund element in high yielding investments must appreciate that if they use this higher accumulative rate of interest in the valuation process they will, all other things being equal, increase their bid for a leasehold property. Such an action would narrow the gap in market values between the leasehold interest and freehold interests, and individual investors would have to make their own subjective decisions as to just how small they are prepared to see the differential become.

Valuation is in any event an imprecise science because of the very wide range of variables involved. It can be argued that the results obtained by valuing using conventional sinking fund theory are equally as acceptable as those to be obtained by using any of the alternatives. It might be argued that the use of a particularly low accumulative rate of interest in the valuation process would result in an unreasonably large sinking fund allowance. If that large sinking fund was in fact invested at a higher accumulative rate, then a sum larger than the historic cost of the asset would be recouped. Such a result, although occurring by a somewhat unsophisticated process, would in fact go some way towards answering the critics who suggest it is wrong only to allow for the recoupment of historic purchase costs.

Obviously valuers should attempt to be as accurate as possible in all their calculations, and it is possible to dispense with the conventional approach and to allow for the recoupment of a future capital sum which attempts to allow for the change in values over the ownership of an asset. It is also possible to allow for this recoupment of capital using a higher accumulative rate of interest. As long as valuers are aware of exactly what they are doing and of the problems in predicting the eventual capital sum required and the correct long-term accumulative rate of interest, such a course of action is acceptable.

As a defence for conventional sinking fund theory, it is a fact that its use in the calculation of Dual Rate YPs results in the YPs for the valuation of terminable incomes being reduced substantially to reflect the unattractiveness of such incomes when compared with perpetual incomes.

The following examples illustrate the traditional approach to the valuation of varying incomes.

Example 1

A shop is held on a 21-year lease of which 14 years are unexpired. The full rental value is £20,000 pa on full repairing and insuring terms, and the rent payable under the lease is £16,000 pa on the same terms. The freehold interest in the property is considered to be an 8% risk.

Full rental value	£20,000 pa
Less Rent paid under lease	£16,000 pa
Profit rent	£ 4,000 pa
YP for 14 years @ 9%/2.5%	6.6429
Capital value of leasehold interest	£26,571.60

Note. If there was no profit rent, the capital value would be nil. The leasehold rate of 9% is 1% above the freehold rate, to reflect the disadvantages of leasehold interests. The sinking fund is calculated at 2.5%.

"**Full repairing and insuring terms**" refers to the agreement in the lease regarding repairing and insuring liabilities. When a property is let on FRI terms, as they are abbreviated, the lessee is responsible for all repairs and for insuring the property.

If the lessee could sublet the property on the same terms, the liability for these items could be passed to the sublessee, so no deduction for them is necessary in the valuation of the lessee's interest.

Example 2

A shop is occupied on a lease with 12 years unexpired at a rent of £12,000 pa for the next 5 years and £14,000 pa thereafter. It is held on FRI terms and the full rental value on the same terms is £18,000. The freehold risk rate is 7%.

Next 5 years			
FRV		£18,000 pa	
Less Rent paid under lease		£12,000 pa	
Profit rent		£6,000 pa	
YP 5 years @ 8%/2.5%		3.7003	
			£22,201
Last 7 years			
FRV		£18,000 pa	
Less Rent paid under lease		£14,000 pa	
Profit rent		£4,000 pa	
YP 7 years @ 8%/2.5%	4.7060		
× PV £1 in 5 years @ 8%	0.6805832	3.2028	
			£12,811
Capital value of leasehold interest			£35,012
Say			£35,000

Note. The method used to defer the income which is not received until 5 years have elapsed is the same as that used in chapter 16.

There is no variation in the rate of interest used for the two periods, as it is assumed that in both periods the property could and would be sublet at the full rental value. The receipt of rent would therefore always be at full risk.

Single rate years' purchase

In chapter 16 the same formula as that used in this chapter was used to find the Years' Purchase appropriate for the valuation of varying incomes. In that chapter, however, it was used to calculate what is known as the Single Rate Years' Purchase, which is a YP calculated using the same rate of interest for the remunerative rate and for sinking fund calculations. A comparison between Single Rate and Dual Rate YPs is shown below.

Single Rate YP for 10 years @ 10%	*Dual Rate YP for 10 years @ 10% and 2.5%*
$$YP = \frac{1}{i + SF}$$	$$YP = \frac{1}{i + SF}$$
i = .1 ASF to redeem £1 in 10 years @ 10% = .0627454	i = .1 ASF to redeem £1 in 10 years @ 2.5% = .0892588
$$YP = \frac{1}{.1 + .0627454}$$	$$YP = \frac{1}{.1 + .0892588}$$
$$YP = \frac{1}{.1627454}$$	$$YP = \frac{1}{.1892588}$$
$$YP = 6.1445669$$	$$YP = 5.2837701$$

It can be seen that the only difference in the calculations results from the use of different rates for the sinking funds. The YPs are, however, used in different circumstances, and the Single Rate YP is used in the valuation of incomes produced by freehold interests, when the role of the sinking fund in the formula is to make an arithmetical adjustment to the YP merely to permit valuations to be made for a limited period. In freehold valuations there is no suggestion that a sinking fund should actually be established, as there is no need to recoup capital, none being lost with the passage of time. The ASF is a reducing agent in the formula to "devalue" the YP so that no value is attached to the period during which an income does not flow.

To illustrate that the single rate YP formula does effectively adjust YPs to allow for time differentials, an income of £10,000 pa receivable in perpetuity will be valued at a 10% rate of interest.

Annual income	£10,000
YP in perpetuity @ 10%	10
Capital value	£100,000

This valuation could be done in two stages, and, unless the technique used is faulty, the valuation of an income of £10,000 pa for 10 years followed by the valuation of a reversion to £10,000 pa in perpetuity, should give the same total as the calculation above.

Next 10 years

Annual income		£10,000	
YP for 10 years @ 10%		6.1446	
			£61,446
Reversion			
Annual income		£10,000	
YP in perpetuity @ 10%	10		
× PV £1 in 10 years @ 10%	0.38554	3.8554	
			£38,554
Capital value			£100,000

The arithmetical adjustment in the YP formula effected by the sinking fund element of the denominator has allowed for the correct apportionment of the income over time.

The use of a *Dual Rate YP* envisages the need actually to establish a fund to accumulate money in circumstances in which it can only be invested to earn interest at a rate which differs from that earned by the property investment.

Chapter 18

Taxation and Valuation

Taxation has only been mentioned in passing, but, as most people are only too well aware, it plays a very important and prominent role in modern life, to the extent that income is frequently very much reduced after tax liabilities have been met. Almost all income is assessable for tax purposes, although, depending upon personal circumstances and the detailed provisions of particular tax systems, taxpayers may get various allowances and expenses against their income to reduce their liability. In the United Kingdom, earnings from certain sources, such as National Savings Certificates, are not taxed, while some earnings, such as interest earned on building society accounts, have been paid net of tax. The general rule, however, is that most income is taxable, and because there is often a wide discrepancy between earnings gross of tax and the same earnings net of tax, the majority of people are more interested in knowing the size of the latter. Wage-earners look at the last figure on their pay-slips to discover what is left for them to spend after the "taxman" has had "his share", and the majority of investors will also be principally interested in knowing what is left for them at the end of the day.

Because there is a wide variation in incomes, depending upon many factors such as the type of job or profession in which they are earned, whether they are earned in a high wage area or a low wage area, and the age of the earner, and because different earners have different allowances to set against tax, resulting from differing personal circumstances, there is a very wide range of average rates of liability to tax. Although the basic United Kingdom rate of income tax is currently 22% (with a lower rate of 10% on the first £1,520 of taxable income), probably very few people have that rate as the average rate of tax for their total income.

During the 1980s in the United Kingdom, there were considerable reductions in income tax rates. Prior to these reductions, there were periods when total tax on income was in some circumstances, as high as 98p in the £ on large incomes. There were also periods during which investment income was subject to a higher tax than other income, to the extent that although income from employment might have been subject to a maximum tax of 60p in the £, investment income was subject to an additional investment income surcharge of 15% giving a total liability of 75p in the £. Average liabilities may (at the time of writing) range from nil if total allowances exceed total income, to a figure approaching 40p in the £ if an income is large.

Because of the very wide range of possible tax liabilities, yields from investments are normally quoted gross of tax, as it would be difficult to decide what adjustment would be appropriate to the majority of investors if yields were to be quoted net of tax. Also, if yields were quoted net of tax it would be difficult to compare yields over extended periods of time, as any tax changes would result in revisions of net of tax yields, making immediate comparison impossible. Consequently, yields from stocks and shares, loans and property are conventionally quoted as the yield before income tax has been deducted, and investors must allow for their own personal tax liability if they wish to find out the net of tax yield.

There is a school of thought which considers that valuers should use net of tax incomes and yields in their calculations, as, in a similar way to sinking fund allowances, tax liabilities cannot be spent by an investor. To include them in calculations amounts to valuing an element of income which the investor does not in fact receive. This is a perfectly sound argument, but the valuer should always give consideration to the purpose of a valuation before deciding on appropriate tactics. When a valuation is being made to find the market value of a property, if all the potential purchasers are valuing on a gross of tax basis, that must be the appropriate approach to use in order to determine the market value their competition would create. On the other hand, the valuer would be perfectly justified in valuing on a net basis if his or her instructions were to determine the value of a property to a particular investor, in which case allowance could be made for that individual's tax liability. The market value and the value to the individual might in fact be quite different.

Perpetual income and net of tax valuations

When an income is receivable in perpetuity, a valuation will give exactly the same answer whether calculations are done gross or net of tax. To illustrate the fact, an income of £10,000 pa receivable in perpetuity at a gross of tax yield of 10% will be considered.

Valuation

Annual income	£10,000
YP in perpetuity @ 10%	10
Capital value	£100,000

The same income will now be considered on a net basis from the point of view of a person paying tax at 40p in the £.

For every £1 of income he or she receives the tax payment will be 40p and the net of tax income will be 60p. Therefore, if £100 is invested at 10% gross, the gross of tax income will be £10, while the net of tax income will be £6, giving a net of tax yield of 6%. In valuing a net of tax income, the net of tax yield must obviously be used for consistency and accuracy, and the value to the taxpayer is found as follows:

Annual income	£10,000
Less Income tax @ 40p	4,000
net of tax income	£6,000
YP in perpetuity at net of tax yield of 6%	16.666
Capital value	£100,000

If tax rates were such that an investor had a tax liability of 80p in the £ the calculation would be:

Annual income	£10,000
Less Income tax at 80p	£8,000
net of tax income	£2,000
YP in perpetuity at net of tax yield of 2%	50
Capital value	£100,000

With an unvarying income receivable in perpetuity the gross of tax and net of tax approach will always give exactly the same answer, irrespective of differing tax liabilities.

Varying incomes and net of tax valuations

An income of £10,000 pa receivable for 10 years with a reversion to a perpetual income of £20,000 pa will be considered. For illustration purposes a gross of tax yield of 10% will be used throughout.

Gross of Tax Valuation

Annual income	£10,000	
YP for 10 years @ 10%	6.1446	
		£61,446.00
Reversion to income in perpetuity	£20,000	
YP in perpetuity @ 10% deferred 10 years	3.85543	£77,108.60
Capital value		£138,554.60

The same income will be considered on a net of tax basis with a tax liability of 40p in the £, in which case the net of tax yield is 6%.

Net of tax @ 40p valuation

Annual income	£10,000	
Less Tax @ 40p	£4,000	
Net of tax income	£6,000	
YP for 10 years @ 6%	7.3601	
		£44,160.60
Reversion to income in perpetuity	£20,000	
Less Tax @ 40p	£8,000	
Net of tax income	£12,000	
YP in perpetuity @ 6% deferred 10 years	9.30658	
		£111,678.96
Capital value		£155,839.56

This valuation gives a much higher result than the gross of tax valuation, and a valuation for an 80p tax liability will also differ.

Net of Tax @ 80p valuation

Annual income	£10,000	
Less Tax @ 80p	£8,000	
Net of tax income	£2,000	
YP for 10 years @ 2%	8.9826	
		£17,965.20
Reversion to income in perpetuity	£20,000	
Less Tax @ 80p	£16,000	
Net of tax income	£4,000	
YP in perpetuity @ 2% deferred 10 years	41.01741	
		£164,069.64
Capital value		£182,034.84

Yet again a higher value results than that found by use of the gross of tax method. There is nothing magical about these differences. They arise purely as a result of discounting the future at different rates of interest, and if reference is made to p113 it will be recalled that the lower the rate of interest that is used, the higher the future is valued in comparison to the present. Consequently, if there is an increase in income at a future date, the lower the rate of interest used, the higher will be the value placed on the future rental increase, and the lower the relative value placed on the immediate income.

In terms of YPs this fact can be illustrated as follows by reference to the three valuations.

	Gross basis (10%)	Net of tax @ 40p (6%)	Net of tax @ 80p (2%)
YP for first 10 years	6.14460	7.36010	8.98260
YP for reversion	3.85543	9.30658	41.01741
Total YPs used	10.00003	16.66668	50.00001
% of YPs allocated to first 10 years	61.45%	44.16%	17.97%
% of YPs allocated to reversion	38.55%	55.84%	82.03%

The last two lines show quite clearly the change in the apportionment of values between the term and reversion as the rate of interest decreases. As the use of a net rate of interest inevitably involves a reduction in the rate, net of tax valuations will throw up this pattern of values whenever there is a future increase in rent.

Although the Gross of Tax and Net of Tax approaches can lead to considerable differences in the results obtained in a valuation, the student should remember that in the market most valuations are in fact done on a gross of tax basis. If the market values in this way, then the gross of tax approach must be used to find market value, but the significance of the net of tax approach to the individual investor must always be remembered.

Tax and the valuation of leasehold properties

Liability to income tax creates a problem in the valuation of leasehold interests that does not exist with freehold interests. A major difference between the two types of valuation is that in leasehold valuations the rental income is allocated partly to an annual sinking fund to recoup the capital invested, the remainder being spendable income.

In order to examine the problem caused by the burden of income tax, a leasehold property which produces a profit rent of £10,000 per annum for the next 12 years will be considered. A yield of 10% is required and a sinking fund can be established to accumulate at 3%.

Valuation
Profit rent   £10,000
YP for 12 years @ 10% and 3%   5.8664
Capital value   £58,664

A yield of 10% on £58,664 is £5,866.4 pa. The ASF contribution is therefore £10,000-£5,866.4 or £4,133.6 (or ASF to recoup £1 in 12 years @ 3% = 0.0704621 × Capital £58,664 = £4,133.6).

If the owner of the leasehold interest has to pay income tax at 40p in the £ the following overall pattern results.

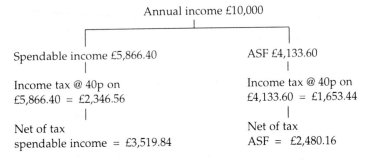

Annual income £10,000

Spendable income £5,866.40   ASF £4,133.60

Income tax @ 40p on   Income tax @ 40p on
£5,866.40 = £2,346.56   £4,133.60 = £1,653.44

Net of tax   Net of tax
spendable income = £3,519.84   ASF = £2,480.16

The total tax payment is £2,346.56 + £1,653.44 which is £4,000 or 40p in £ on £10,000.

It is quite apparent that if an ASF contribution of £4,133.6 is necessary to recoup the capital of £58,664 over 12 years, annual contributions which have been reduced by tax to £2,480.16 cannot possibly achieve the same objective. The ASF contributions must therefore be increased to the original £4,133.60 but if extra money is taken from the spendable income the correct yield will not be obtained from the investment.

The solution which is adopted is to increase the original ASF contributions so that after income tax has been paid on them they are still large enough to recoup the original capital. As the ASF contribution is increased, the spendable income must be reduced, but if the bid for the property interest is at the same time reduced so that the spendable income gives the correct yield on capital, the investor will achieve all his or her objectives.

The ASF contributions are increased in the ratio of income before tax to income after tax by multiplying the ASF by the equation

$$\frac{100}{100 - \text{Tax}}$$

This process is done within the YP formula, which becomes

$$YP = \frac{1}{i + SF \times \dfrac{100}{100 - T}}$$

If this formula is used to obtain the YP for 12 years at 10% and 3% allowing for income tax at 40p in the £, the calculation becomes:

$$YP = \frac{1}{.1 + (\text{ASF to recoup £1 in 12 years @ 3\%}) \dfrac{100}{100 - 40}}$$

$$= \frac{1}{.1 + (.0704621)(100/60)}$$

$$= \frac{1}{.1 + .1174368} = \frac{1}{.2174368} = 4.5990375$$

This is known as the Years' Purchase for 12 years at 10% and 3% adjusted for tax at 40p in the £. The effect of the adjustment for tax has been to reduce the size of the YP from 5.8664, and to increase the ASF allowance for the recoupment of each £1 of capital invested from £0.0704621 to £0.1174368.

The interest will now be valued, using the new YP, and the result will then be analysed.

Profit rent	£10,000
YP for 12 years @ 10% and 3% adjusted for tax @ 40p in £	4.5990375
Capital value	£45,990.375

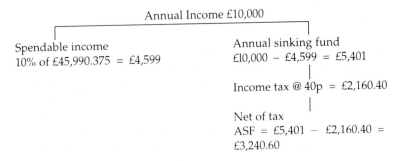

Annual Income £10,000

Spendable income
10% of £45,990.375 = £4,599

Annual sinking fund
£10,000 – £4,599 = £5,401

Income tax @ 40p = £2,160.40

Net of tax
ASF = £5,401 – £2,160.40 = £3,240.60

The ASF can be checked for effectiveness

ASF net of tax	£3,240.60
× Amount of £1 pa for 12 years @ 3%	14.1920
Capital recouped	£45,990.595

It can be seen that the adjusted YP has reduced the purchase price sufficiently to enable both the correct yield to be obtained from the investment and the capital to be recouped, despite the tax burden on the sinking fund element of the income. If the tax liability at 40p on the spendable income is checked, it will be found to be £1,839.6 giving a net of tax spendable income of £2,759.4 and a total tax liability of £4,000.

A similar adjustment to the sinking fund can be made to allow for any rate of tax, and the valuation could be done to reflect the value to an investor paying tax at 60p in £.

Profit rent	£10,000

YP for 12 years @ 10% and 3% adjusted for tax at 60p

$$\cfrac{1}{i + (.0704621)\cfrac{100}{100 - 60}} = \cfrac{1}{.1 + (.0704621)\cfrac{100}{40}}$$

$$= \cfrac{1}{.1 + .1761552} = \cfrac{1}{.2761552} = \qquad\qquad \underline{3.6211521}$$

Capital value	£36,211.521

As income tax liability increases, so does the value decrease, simply because a greater sinking fund allocation becomes necessary. This reduces the spendable income, and therefore the capital value to the taxpayer also falls.

Sinking fund earnings and income tax

Earlier in the book it was mentioned that the interest rate earned on sinking fund policies was usually quoted on a net of tax basis, that is on the assumption that tax was paid by the insurance company prior to payment to the policy holder. The YP calculations above were made on this assumption and the Dual Rate YP figures in Parry's Tables are compiled on the same assumption. There is therefore no need to make any adjustment in the YP formula for this tax burden.

It should be remembered that the rate at which a sinking fund earns money is not quite as low as the net of tax rate of interest makes it appear. If income tax is paid at 40p in £, 3% net of tax is the equivalent of 5% before tax, and in the examples above this would be the gross of tax rate of interest on the sinking fund.

To adjust rates of interest to allow for tax, gross rates can be reduced to a net of tax rate by using the formula:

$$\text{Gross rate} \times \frac{100 - T}{100} = \text{Net rate}$$

Net rates can be increased to a gross of tax rate by using the formula:

$$\text{Net rate} \times \frac{100}{100 - T} = \text{Gross rate}$$

In each case T represents the rate of tax quoted as a percentage or as pence in the £.

The remunerative rate of interest and income tax

In none of the calculations made above was any adjustment made to the remunerative rate of interest to allow for tax payments on the spendable income. This is because the convention is to quote all yields on a gross of tax basis for comparison purposes, and unless a valuer has a specific reason for departing from the conventional approach, no adjustment is made to the yield in respect of income tax liability.

The use of tax adjusted dual rate years' purchase tables

The valuer should use tax adjusted tables whenever a valuation is made of a leasehold property for which the potential purchasers are tax payers. If circumstances are unusual enough for all the would-be purchasers to be exempt income tax, there is no need for tax adjusted YPs. When tax adjusted tables are used it is normal for them to be adjusted at the standard rate of income tax. For correct usage the rate of tax adopted should really be that which is likely to apply to the potential purchasers of the interest concerned, although this may not necessarily be easy to determine, or the rate applicable to the client for whom a valuation is being made.

Chapter 19

Discounted Cash Flow Techniques

The use of discounted cash flow techniques is said by some to have been introduced into property valuation in the last twenty years or so, and in the relatively recent past one heard discussions as to whether this was a welcome development and whether such techniques were proper and acceptable methods of valuation. Although it would be inappropriate to go too deeply into this subject area in this book, in view of the increasing prominence of this topic it is essential to consider it. There is no intention of providing anything other than an introduction to the topic, and the reader will be able to follow this up with further study in more specialised books.

It is appropriate to recall some of the fundamental points made previously in this book. Valuation is a method of assessing the various advantages and disadvantages of an interest in property and expressing them in money terms. When we place valuations on them we enable different properties with different advantages and disadvantages to be compared with each other in money terms. Discounted cash flow techniques facilitate such comparisons, and it is worth considering why they have become popular in recent years, and, conversely, why they were not used more widely previously. In posing these questions it must be stated that it is the author's opinion that such techniques have in fact been used by valuers for a very long time, although not in the same format as today. It is hoped the reasons for this opinion will shortly become apparent.

The traditional approach to the valuation of an income flow from a property has already been considered and is as below.

Example 1

What is the value of an income flow for the next 4 years of £10,000 per annum payable in arrears from a freehold property if the property is considered to be an 8% risk?

Net income	£10,000 pa
YP for 4 years @ 8%	3.3121
Capital value	£33,121

As considered in chapter 14 the Present Value of £1 Per Annum (or the YP) is nothing more than an addition of the present values of £1 for each of the years under consideration, so the above calculation could have been achieved by the following alternative method.

Example 2

First year's income:	£10,000 × PV £1 in 1 year @ 8% = £10,000 × 0.9259259	= £9,259.259
Second year's income:	£10,000 × PV £1 in 2 years @ 8% = £10,000 × 0.8573388	= £8,573.388
Third year's income:	£10,000 × PV £1 in 3 years @ 8% = £10,000 × 0.7938322	= £7,938.322
Fourth year's income:	£10,000 × PV £1 in 4 years @ 8% = £10,000 × 0.7350299	= £7,350.299
Total capital value		£33,121.268

The addition of the present values of £1 for each of the four years gives a figure of 3.3121, or the Present Value of £1 Per Annum for 4 years at 8% as used in example 1.

What is made clear by the layout in example 2 is that we have taken a series of cash flows and have discounted them to the present. The abbreviated layout of example 1 does not make this so apparent, but from this we can see that the traditional methods of valuation are in fact techniques of discounting cash flows.

It is therefore not true to say that the use of discounted cash flow techniques is new in property valuation. What has happened in recent years is that more sophisticated and more complex discounted cash flow calculations have been done, and that more variables and different layouts have been utilised for valuations. The two are very much related to each other in that the building into a valuation of more variables entails more calculations, and a new format to take account of the more complex nature of the valuation is a logical development.

There are numerous reasons why these more complex valuations have been adopted in recent years, but in general they derive from two main factors: first, there is probably a need for more complex valuations, and, second, the machinery to perform complex calculations is now cheaply available to all valuers, making such calculations a practical proposition. The need for more complex valuations arises because there are now many properties available for purchase which cost many millions of pounds. When such large sums of money are being committed for many years to come it is essential that the purchasers look very carefully at their commitments, taking into account in their valuations every conceivable variable which could affect the ultimate price at which they purchase, or, looked at from a vendor's viewpoint, at which a property sells. The need for care and precision in valuation is even greater if the vendors or purchasers are trustees of funds for others, as will often be the case with pension funds, insurance companies and other institutions which are active in the property market.

The need for complex valuations also arises from background factors which were perhaps not so important in past years. In the past it was often possible to borrow large sums of money for long periods at fixed rates of interest. Nowadays, it is likely that the cost of borrowing will be related to other variable factors, such as the base lending rates of banks, and there may therefore be a need to build into a valuation calculation variations to allow for anticipated changes in the cost of finance. Taxation is likewise more complex than in the past, and it may also be necessary to build in variables to allow for known or anticipated future changes in tax commitments. Similarly, if an important factor in the appraisal process is an element of tax relief on interest charges on borrowed funds, it may also be necessary to take account in the appraisal of variations in the amount of interest payable in the future and consequent variations in anticipated tax relief.

These, and other factors, give rise to the need for complex appraisal calculations which only a few years ago would have proved exceedingly difficult for all but a few large and well equipped firms. Not only that, but such calculations would have been too time-consuming in most instances, as valuations are regularly required at short notice. The electronic calculator has changed that situation and, as long as they understand what they are attempting to do, any valuer can now perform involved and difficult calculations with speed and ease. The development of computing technology has assisted the development of investment valuation and analysis techniques.

There is now no excuse for avoiding complex valuation calculations. Equally, though, there is no point in using them unless there is a need for their use, and unless they provide as good a result, and preferably a better result than the simpler, traditional approach would provide. If there is no positive benefit to be obtained from using the more elaborate approach, then there is no point in using it.

Whenever modern discounted cash flow techniques are used it is well worth remembering that they are nothing more than variations on the basic compound interest theme which is the basis of all valuations. All future receipts and liabilities are discounted to a common date to enable an accurate appraisal to be made and, quite frequently, to enable comparison to be made with other investments for which the same exercise has been carried out.

There are several kinds of discounted cash flow appraisal but consideration will be restricted to the two most commonly used, the Net Present Value Method and the Internal Rate of Return Method.

Net Present Value Method

This is just like the Investment Method of valuation in that all future items are discounted to a net present value by using compound interest calculations. However, whereas in an investment valuation it is conventional only to consider the net income of each year, the use of a tabular format for the Net Present Value Method facilitates the inclusion as separate items in the appraisal of all items of income and expenditure for each year. The valuer can decide whether he or she will then in fact simply discount the net income or the net deficit for each year to the present, or whether, if there is sound reason for so doing, items of income and expenditure will be discounted separately using different rates of interest for the purpose as appropriate. Not only can different rates of interest be utilised but calculations can be made for individual items at more precise dates rather than at the end of each year, which the conventional investment valuation usually presupposes.

The precise layout can be varied to suit the requirements of an appraisal or the preferences of a valuer, but a typical layout is illustrated below.

End of Year	Particulars	Outflow	Inflow	Net Flow + or −	PV of £1	Net Outflow	Net Inflow

The use of the method can best be illustrated by consideration of examples. In each case it will be assumed for simplicity that calculations are needed at twelve-monthly intervals, although different time periods could be utilised as required.

Example 3

Property A is available at a purchase price of £100,000 and the total costs of purchase would be £8,000. It is let at a rent of £11,000 pa receivable in arrears on full repairing and insuring terms for the next four years. The potential purchaser expects that the property will be sold after four years for £140,000, the total costs on the sale being £6,000 (see Table 19.1 on p180).

A rate of 12% has been chosen (known as the "target rate") and at this rate the calculations reveal that in net present value terms the investment shows a positive balance of £10,570 representing an excess of inflows over outflows. A positive balance at the end of net present value calculations suggests that an investment will be worth entering into as it should be viable at the target rate. Conversely, if there is a deficit at the end of the calculations the inference is that it will prove an unacceptable investment. If net outflows and net inflows are exactly equal the investment will earn money at the same rate as that used in the calculation and will be an acceptable investment as it should not entail loss.

In choosing a rate of interest of 12% it may be that the investor is influenced by the fact that this is the cost of money: 12% pa interest has to be paid on money borrowed. It may be that there is no need to borrow money, and 12% is the opportunity cost of money to the investor, who is prepared to invest if that rate of return can be obtained on money invested. Again, the choice of 12% may be made because the property which produces the income is regarded as a 12% risk when compared with all the other investments in which money could be placed.

Whatever influences the choice of rate the calculations show that the "target rate" of 12% is achieved with a bonus of £10,570. In effect the rate of return on the investment is in excess of 12% pa. It should be noted that this will be the case as long as the predictions of inflows and outflows are correct, and as long as the appropriate rate remains at 12%.

Table 19.1 Example 3

End of Year	Particulars	Outflow	Inflow	Net Flow + or −	PV of £1 @ 12%	Net Outflow	Net Inflow
0	Purchase Price	100,000		−108,000	1	108,000	
	Purchase Costs	8,000					
1	Rent		11,000	+11,000	.8928571		9,821
2	Rent		11,000	+11,000	.7971939		8,769
3	Rent		11,000	+11,000	.7117802		7,830
4	Rent		11,000	+145,000	.6355181		92,150
	Sale Proceeds		140,000				
	Sale Costs	6,000					
						108,000	118,570
Net Present Value							+10,570

Note: Outgoings on management have been ignored in this example for simplicity.

It is worth considering what happens when the target rate is altered.

Example 4

The same facts as in example 3, but a rate of 16% is used instead of 12% (see Table 19.2 on p182).

The calculations show that if the target rate is 16% the investment will not be acceptable if the predictions of inflows and outflows are correct. The investment shows a deficit over the four years of £3,213 when all future costs and returns are translated into terms of present value.

What was an acceptable investment at a target rate of 12% is not so at 16%; so somewhere between these two figures will be a rate of interest at which net outflows and net inflows are exactly equal, and this figure will be the rate of return on the investment. The investor must then decide whether this rate of return is acceptable.

The rate of discount is critical as it determines whether an investment will be regarded as acceptable. It can be seen that an investment may be attractive to a person who can borrow money at 12%, but unattractive to someone who has less favourable borrowing terms.

The Net Present Value approach is a very useful method of analysis and enables investors to determine whether investments are likely to prove acceptable at a given target rate.

It also enables alternative investments to be compared to show which will give the biggest profit (or the biggest loss). There may be occasions on which expenditure cannot be avoided, and if a loss of some sort is inevitable it may be helpful to know which option will give rise to the smallest loss. Its use as a method of comparison will be considered in the following example and by subsequent comparison with example 3.

Example 5

Another property, Property B, is on offer at £85,000 for which the expenses of purchase would be £4,000. It is let for four years on internal repairing terms at a rent of £10,500 pa receivable in arrears. It is anticipated that it could be sold after four years for £112,000 the estimated costs of sale being £6,000 (see Table 19.3 on p183).

Table 19.2 Example 4

End of Year	Particulars	Outflow	Inflow	Net Flow + or –	PV of £1 @ 16%	Net Outflow	Net Inflow
0	Purchase Price	100,000		–108,000	1	108,000	
	Purchase Costs	8,000					
1	Rent		11,000	+11,000	.8620690		9,483
2	Rent		11,000	+11,000	.7431629		8,175
3	Rent		11,000	+11,000	.6406577		7,047
4	Rent		11,000	+145,000	.5522911		80,082
	Sale Proceeds		140,000				
	Sale Costs	6,000					
						108,000	104,787
Net Present Value							–3,213

Note: Outgoings on management have been ignored in this example for simplicity.

Table 19.3 Example 5

End of Year	Particulars	Outflow	Inflow	Net Flow + or –	PV of £1 @ 12%	Net Outflow	Net Inflow
0	Purchase Price	85,000		−89,000	1	89,000	
	Purchase Costs	4,000					
1	Rent		10,500				
	Repairs	700					
	Insurance	220					
	Management	525		+9,055	.8928571		8,085
2	Rent		10,500				
	Repairs	770					
	Insurance	242					
	Management	525		+8,963	.7971939		7,145
3	Rent		10,500				
	Repairs	847					
	Insurance	266					
	Management	525		+8,862	.7117802		6,308
4	Rent		10,500				
	Repairs	932					
	Insurance	293					
	Management	525					
	Sale Proceeds		112,000				
	Sale Costs	6,000		+114,750	.6355181		72,925
Net Present Value						89,000	94,463
							+5,463

Note: Management costs are calculated as 5% of rent received.

In this case there is a predicted profit in net present value terms of £5,463 using a target rate of 12%. At 12% Property A produced £10,570 above the target return which is 93.5% greater but on an initial investment which required 21.35% more capital. It therefore appears that Property A would be more attractive, although much may depend on whether there are problems in raising the extra £19,000 capital.

Whatever the circumstances, if our investor needs a steady income for each of the next four years Property B will not be attractive. The calculations have been made on the not unreasonable assumption that the costs of repair and insurance will rise by 10% pa, and this would result in a decreasing annual income. This may make investment B less attractive to the investor especially if an income flow is required to meet commitments which may increase with inflation. The arrangements the investor could make with the balance of £19,000, which would be saved on the purchase of Property B, would be an important factor in determining the eventual investment decision, although it would appear unlikely that arrangements could be made which would tip the balance in favour of Property B.

Example 6

To complete the picture the same facts are used as in example 5 but with a target rate of 16% instead of 12% (see Table 19.4 on p185).

The effect of the higher target rate is again to make the investment unacceptable. Whereas with Property A the deficit at the higher rate was £3,213, in this case there is a deficit of £5,479.

These examples illustrate why discounted cash flow methods have become popular as with a relatively simple method of layout a great number of variables can be built into calculations without too much drudgery, an electronic calculator or a computer making the calculations easy. Other variables can be included as appropriate and these can include allowances for sinking funds and for tax liabilities.

Discounted cash flow calculations more easily permit the cost of outgoings to be calculated at the precise time they are likely to occur, rather than assuming, as is generally the case in the conventional investment valuation, that all outgoings will occur at the ends of the relevant years. Tax or rate liabilities in particular may be phased over a period of time rather than being payable at

Table 19.4 Example 6

End of Year	Particulars	Outflow	Inflow	Net Flow + or –	PV of £1 @16%	Net Outflow	Net Inflow
0	Purchase Price	85,000		–89,000	1	89,000	
	Purchase Costs	4,000					
1	Rent		10,500	+9,055	.8620690		7,806
	Repairs	700					
	Insurance	220					
	Management	525					
2	Rent		10,500	+8,963	.7431629		6,661
	Repairs	770					
	Insurance	242					
	Management	525					
3	Rent		10,500	+8,862	.6406577		5,678
	Repairs	847					
	Insurance	266					
	Management	525					
4	Rent		10,500	+114,750	.5522911		63,375
	Repairs	932					
	Insurance	293					
	Management	525					
	Sale Proceeds		112,000				
	Sale Costs	6,000					
						89,000	83,520
							–5,480

Net Present Value

Note: Management costs are calculated as 5% of rent received.

one specific date, and such phasing can be incorporated in a discounted cash flow calculation so that more precise estimates of the discounted burden of future liabilities are made.

The approach sets out clearly on paper the various outflows and inflows and the discounting calculations, and it enables easy comparison to be made between alternatives. All of these variables could be incorporated into the conventional valuation format, although probably not with such ease or such clarity for comparison purposes.

Valuation using the Net Present Value Method

The above examples have illustrated the use of the method for analysis. The purchase price and costs have been known and have been incorporated in the overall analysis of the investment.

When a valuation is made the purchase price and costs are not known; to find them is the object of the exercise. The Net Present Value Method can be used just as easily as a method of valuation. This is done by omitting the purchase price and costs of acquisition and then taking the balance of the discounted net inflow and net outflow figures for the years concerned. This balance will give the figure which one can afford to pay for a property at a given rate of interest. Considering examples 3 to 6 as methods of valuation:

In example 3 Property A at 12% is valued at £118,570,
In example 4 Property A at 16% is valued at £104,787,
In example 5 Property B at 12% is valued at £94,463, and
In example 6 Property B at 16% is valued at £83,520.

These figures represent the values placed on the expected future net returns from the properties at the given target rates. They therefore represent the maximum figures (inclusive of purchase costs) which potential purchasers should be prepared to bid unless they revise their valuations for some valid reason. Potential purchasers would only bid up to these maximum figures if the pressures of the market forced them to do so, and they would purchase at lower figures if possible.

General comments on Net Present Value Method

The method permits a more sophisticated treatment of investment valuations than does the traditional approach, but it is only as good as the person using it. It can be very accurate, but it can also be very inaccurate, and the degree of accuracy will depend upon the accuracy of the various inputs.

It has the virtue of enabling many relevant items to be included in the calculations in a format which allows ease of understanding, and the same calculation can be used both for valuation and comparison. It also forces the valuer to give individual consideration to a number of factors which are considered collectively when using the traditional method of valuation.

The net present value approach shows whether an investment is expected to be viable or not, and it predicts the size of the profit or loss, but it does not give a precise indication of the rate at which an investment will earn money.

The internal rate of return

This is the other discounted cash flow method which will be considered and its objective is to show the rate at which an investment earns money.

The only indication the Net Present Value Method gave of the rate at which money was earned was to show whether the rate of earning was below or above the rate chosen for the calculations (that is the target rate). To know the size of a profit (or a loss) may be helpful, but if the highest return also requires the highest capital outlay the information may be of limited use. An investor's real wish will be to know the rate at which capital will earn money. The Internal Rate of Return Method therefore seeks to find the precise rate at which the capital invested will earn money thus enabling a more precise comparison to be made between competing investment opportunities.

The rate of interest at which the net present value of the outflows and the net present value of the inflows are equal will be the rate at which an investment earns money. Because the Internal Rate of Return reveals the earning power of money it not only permits comparison between different property investments, but it also allows comparison to be made with other types of investment, such as gilt edge stock and equities.

The Internal Rate of Return can in fact be found by trial and error, or by use of a programmed financial calculator or a suitable computer program.

In example 3 at a target rate of 12% a surplus of £10,570 could be expected, while in example 4 the use of the target rate of 16% for the same investment revealed an expected deficit of £3,213. Somewhere between 12% and 16% there must therefore be a point at which discounted outflows and inflows are exactly equal and that rate of interest will be the Internal Rate of Return.

By interpolation we can anticipate that the Internal Rate of Return will be about 15%, and we can then make calculations to check the precise rate.

Example 7

The facts are as in examples 3 and 4 but the objective is to find the Internal Rate of Return (see Table 19.5 on p189).

At a discount rate of 15% the discounted net outflows and net inflows are almost equal, and it can be seen that the Internal Rate of Return of the investment is therefore 15%. It would be possible to work out the precise rate at which the two columns are equal, but for most purposes the rate of 15% would be sufficiently accurate.

A Comparison of the Net Present Value and Internal Rate of Return Methods

In the Net Present Value Method one rate of interest is chosen (the target rate) to find whether an investment is profitable at that rate. Because of this the method is simple involving no real mathematical difficulties. The method is particularly suitable for valuation, and the only real problem is the choice of the target rate. This could be determined by the cost of borrowing, the opportunity cost of money or the risk rating of the particular investment.

In the Internal Rate of Return Method calculations have to be made at a number of different rates of interest to determine the actual earning power of an investment. More calculations are therefore involved but with modern programmable calculators this really poses no problem. Whereas the Net Present Value Method only gives the size of an excess or a deficit, the Internal Rate of Return Method reveals the earning power of an investment and this makes the method more suitable for the analysis of investments and for comparison purposes.

General

Discounted cash flow techniques do not conflict with traditional investment valuation methods; rather do they represent a

Table 19.5 Example 7

End of Year	Particulars	Outflow	Inflow	Net Flow + or –	PV of £1 @ 15%	Net Outflow	Net Inflow
0	Purchase Price	100,000		–108,000	1	108,000	
	Purchase Costs	8,000					
1	Rent		11,000	+11,000	.8695652		9,565
2	Rent		11,000	+11,000	.7561437		8,318
3	Rent		11,000	+11,000	.6575162		7,233
4	Rent		11,000	+145,000	.5717532		82,904
	Sale Proceeds		140,000				
	Sale Costs	6,000					
						108,000	108,020
Net Present Value							20

development and refinement of the same basic principles used in the traditional approach. Their use will not necessarily give better or more accurate results, as much will depend upon the skill and judgement of the person doing the calculations, the results obtained from the calculations being very much dependent upon the various assumptions made by the user.

However, it is probably true to say that where a calculation or valuation is dependent upon a considerable number of variables and a large number or varying pattern of time periods, the use of discounted cash flow techniques is more appropriate than the traditional investment valuation approach. It is also easy to incorporate in discounted cash flow calculations estimates to reflect the effect of inflation upon future income flows and outgoings. Likewise, whereas traditional methods cannot easily reflect the possibility of the future sale of the income producing asset at an enhanced value, such a calculation is easily incorporated in a discounted cash flow calculation. (The reader should not overlook the fact that it may in some instances be necessary to allow for future decreases in income or value.) In this respect discounted cash flow techniques are particularly useful as they facilitate the estimation of a true return over time allowing for changes in value and liabilities over the time period. The traditional valuation approach generally assumes a rate of return and utilises estimates of returns and liabilities at figures appropriate at the time the valuation is made and consequently is not likely to give an indication of the true return to an asset over the period of ownership.

Advocates of the use of the DCF approach for all valuations stress the ability to easily incorporate future variations in income, outgoings, or yields into the calculations, something which is not so easily done with the conventional investment (or capitalisation) approach. They point out that rather than placing most of the emphasis in the valuation process on what is actually the state of affairs at the time of valuation, it is easy in the DCF approach to make explicit in the valuation the valuer's expectations regarding future changes in a wide range of variables relevant to the valuation. It is argued that doing so will provide a more reliable valuation, and that indicating all the variables and the assumptions about them in the valuation makes it more easily usable and more useful to the end user.

When reasonable assumptions can be made about future events there is no doubt they should be incorporated in valuations, and there is also no doubt that the DCF format enables this to be done with ease and that it also makes much useful information clear to users of a valuation. As with any valuation, though, care has to be taken by valuers to thoroughly research and test the reliability of the various inputs used in a valuation. The ease with which computerised DCF programs can be used may too readily encourage valuers to include a wide range of variables, and it is not unknown for DCF valuations to include such things as rental growth predictions for periods 10 or 20 years into the future (which could, of course, also be done with the investment approach). There are many variables which are difficult to predict more than one or two years into the future, and the further ahead the valuer seeks to predict the more unreliable the predictions are likely to become.

Valuers can confidently include in valuations future variations about which there is a reasonable degree of certainty, such as rent increases agreed in leases, and, indeed, should do so if valuations are to be reliable. However, it has to be emphasised that valuers should be confident of the reliability of assumptions and inputs incorporated in valuations, and should not allow the ease with which variations can be made in the DCF approach to cause changes to be made for the sake of change.

As with any valuation approach, the quality of the results obtained from the use of the DCF approach will only be as good as the valuer using the approach and the quality and reliability of the inputs used in a valuation.

Chapter 20

The Effect of Statutes

Valuations so far have been discussed taking into account what might be termed the normal market factors which affect value. These include the physical features of a property, geographical features, the legal interests and liabilities, and finance and the economy generally. There may, however, be somewhat abnormal factors introduced, in that governments pass statutes, or Acts of Parliament as they are more often described, which may interfere with normal market conditions.

The passing of an Act may suddenly create completely new conditions from those which previously existed. It may introduce factors which could not previously have been foreseen, and the property market may be taken completely by surprise. In such circumstances the passing of an Act of Parliament may have an immediate effect on the market, resulting from the new conditions suddenly imposed. However, if the market has been awakened to the possibility of a particular Act of Parliament some time before it is actually drafted in Bill form, it may have anticipated the effects of such an Act, and the passing into law of the Bill may have a less marked effect on market values than would otherwise have been the case.

Indeed, the market may possibly have over reacted to the promise of a particular piece of legislation, and on the publication of a Bill it could be that values rise rather than fall, because its implications may not be as severe as the market had previously feared.

Whatever the effects may be, large or small,. immediate or gradual, it is the valuer's role to interpret the implications of statutes and to try to calculate in money terms the changes in value which may result.

It is not intended in this book to discuss any Acts of Parliament in detail. Such can only be done usefully and effectively after a student has attained a relatively high technical standard and has a reasonable legal grounding. At this stage all that will be done is to discuss in very general detail the effect on market values of the passing of Acts of Parliament, and the valuer's role in attempting to assess such effects.

Governments pass statutes for a wide variety of reasons; some are passed for tax-raising purposes and may have a direct impact on the values of investment properties if they reduce the net of tax income produced; some are passed to affect the legal relationships between different people or organisations, and may influence property values if they affect the bargaining powers of the parties to a transaction; some are passed to enable public bodies to compulsorily acquire property owned by others and such powers may have a considerable effect on both the use and value of property.

Statutes may relate to all properties, or they may relate to a specific class of property. Indeed, they may not necessarily relate to property at all but may nevertheless have an important impact on the property market if they have an effect on investors and their general approach to investment. An example of the type of legislation which affects all properties is planning legislation which often imposes quite rigid control on the development and use of land. Such legislation therefore has an effect on the value of any piece of land, in that the freedom to use it as the owner may wish may be withdrawn, and he or she may only be able to use it for a particular purpose if the planning authority deems it appropriate to grant permission for such use. The power of an authority to refuse permission for certain uses or, alternatively, to grant permission, can create situations in which a hectare of land could be worth perhaps £5,000 for agricultural purposes or, alternatively, £250,000 or more for development.

If a planning authority will not permit property to be put to its most valuable use, then its value must inevitably be lower than if the higher and more valuable use were allowed. A typical example of such a situation may arise with residential property which could easily be used for prestige office accommodation were planning permission to be given, but for which the local planning authority is unwilling to grant a permission. The property is thereby restricted to residential user, for which purposes it may be very much less valuable than if office use were permitted. The valuer will be able to assess this lower market value relatively easily in most cases by consulting office records and by using knowledge of the area.

Other statutes may affect specific types of property only, and there are numerous instances of such legislation in the United

Kingdom. Business properties have for many years been affected by the Landlord and Tenant Acts of 1927 and 1954 and related legislation, while the various Factory Acts are examples of a type of legislation which affects only a restricted class of business properties. Agricultural properties are directly affected by the Agricultural Holdings Act 1948, the Agricultural Act 1958 and several other more recent statutes. Residential properties have since the First World War been affected by a whole series of Rent Acts, Leasehold Reform legislation, various Housing Acts and other statutes.

There are many other statutes which affect property values, but mention of just these few will suffice to show the reader that the valuer has much legislation to contend with. Such is the case in most countries and is not confined to the United Kingdom only.

The passing of an Act of Parliament, or indeed the publication of a Bill, may have an immediate effect on the market, or possibly a more gradual effect. A gradual effect may occur if an Act provides for the introduction of new provisions affecting properties, and these provisions are phased to come into effect at a number of future dates. As the various provisions become operative, so the Act has a greater impact and in consequence more effect on values.

Alternatively the effect of a statute may be restricted in that its provisions may affect only a few properties in a particular category. A Housing Act may affect only certain types of residential property, and generally the types affected will be those less fit for human habitation. It might well be that a particular Housing Act referred only to properties of a certain number of storeys, or in a certain area. The scope of the provisions and the type of property which they affect will generally be self-evident on reading an Act.

It is the valuer's task to interpret the provisions of each statute, and the only way to do this correctly is to take up the statute and read it. This may seem an obvious suggestion but often it is all too easy to refer to a textbook written by a specialist on the subject, and to rely upon this source of information rather than referring directly to the Act. Any who adopt this course do so at their own peril, because in doing so they are placing their faith in the wisdom and skill of the writer. There is no doubt much to be gained from reading such textbooks and treatises, but this should be done not to the exclusion of an inspection of the statute but rather to broaden one's outlook and at the same time get another opinion on the interpretation of what is often far from clear-cut legislation.

Having read and interpreted the various provisions of a statute, the next step is to decide which provisions are important in the valuation process. The valuer may decide either that he or she must vary the method of valuation used or the figures used in a valuation. Either of these approaches, or possibly a combination of the two, may enable account to be taken of the provisions of a statute and to allow for their effect on value.

Statutes vary considerably in that they may have what can be described as either a direct effect on value or an indirect effect. The direct effect may arise in cases in which a statute states that a specific valuation approach is to be used in particular circumstances. An example of one which did this in the United Kingdom was the Land Commission Act 1967, which laid out in very detailed form the various valuation approaches to be used in calculating the development value of a property, on which betterment levy would be charged. Likewise, the Leasehold Reform Act 1967 introduced provisions relating to residential properties held on long leases at low rents. It specified the valuation approach to be used in calculating the sum to be paid by a long leaseholder who claimed the right to purchase the freehold interest in his or her house from the freeholder. Both these Acts had a direct effect on the valuation approach to be used.

An Act may have a direct effect on the market in that certain properties may be removed from the market because it has been passed, and this affects both the value of the properties so removed and the value of the properties remaining in the market, as they may well become more valuable because the total supply has been reduced by the passing of the statute.

An indirect effect of the passing of a statute may be that the underlying conditions of the market are altered by its introduction. Landlord and tenant relations have been altered in the past in a wide range of properties by the introduction of such Acts as the Agricultural Holdings Act 1948, and the Landlord and Tenant Act 1954, the latter Act affecting the landlord and tenant relations in both residential properties and business properties. An alteration in the rights of landlord and tenant has an effect on value in that there is a tendency for the value of one party's interest to be decreased while the value of the other party's interest increases. The extent of such increases and decreases will vary, depending upon how far reaching the provisions of an Act are, and how the normal bargaining positions of the parties are affected by it. In some cases the variation may be in a form which does not necessarily have a marked effect on the value of a property. However, the converse often seems to apply. The UK Rent Act 1957 is an example of an Act which had a profound effect on the value of certain types of property. Prior to its passing a very large proportion of unfurnished rented accommodation was controlled, and the tenants were enjoying very great security of tenure with rents fixed at levels which the landlord could not exceed without breaking the law. The 1957 Act removed certain types of residential property from these controls, but others remained within the system of control. Those that remained within control were subject to stringent conditions regarding the rent which could be charged. This was assessed by reference to the gross values for rating purposes of the property in 1956, and rough and ready as such a method of rent assessment was in 1957 when the Act was passed, with the passage of time the method and the level of rent charged became even more rough and ready, to the extent of being positively ludicrous. Rents on many properties which remained controlled stayed at basically the same level for periods in excess of 20 years with the result that the income to a landlord was often so low that it was quite impossible for properties to be maintained in a reasonable state of repair. Even where the income exceeded the outgoings, and there were many instances where the reverse was the case, the general level of net income arising from such properties was so low that they became more and more unattractive as investments and changed hands at figures which, in comparison with the vacant possession value of similar properties, were incredibly low. The 1957 Act therefore had a great effect on the value of those properties which remained controlled under its provisions, and their values related more to 1957 values than to the comparable values of residential properties which escaped control under the Act. The effect of this Act on residential property values was both obvious and considerable.

A valuer may often encounter a different type of problem arising from the existence of a statute, in that it may effectively destroy the open market for a particular type of property. The UK Leasehold Reform Act 1967 gave to certain long leaseholders of residential properties the right to buy the freeholds of their homes from the freeholders, and the very existence of these provisions virtually destroyed the market for long leaseholds in many sectors of the residential market. A valuer consequently has to attempt to assess value under the terms of the Act without in many cases there being any current evidence of the value of such interests in the open market. The valuer is in the position of having to estimate market value on a hypothetical basis, which creates many difficulties and leaves many doubts at the end of the exercise. In such circumstances it is impossible to do little more than make an honest assessment of what is considered to be value under the terms of the Act, but only rarely will it be possible to argue to uphold the value assessed in such conditions with as much conviction as could be done if value was being assessed in normal open market conditions.

A somewhat similar situation arises when property is compulsorily acquired in the United Kingdom. There are several Acts of Parliament which may affect the basis of valuation in such circumstances, but in most cases the main code of compensation will be that contained in the Land Compensation Act 1961. This Act details various provisions under which an acquiring authority has to assess the compensation payable to the owner of any interest in land which is compulsorily acquired by it. The basic figure which has to be paid is the value of an interest as if it were offered for sale in the open market by a willing seller. There are many other provisions contained in the Act, all intended to clarify the valuation approach to be used in any situation, but their general effect is that an owner whose interest has been compulsorily acquired shall be paid sufficient money to put him or her in an equivalent financial position after the acquisition to that which existed prior to the acquisition. Reimbursement has to cover all losses which are a direct result of a property being compulsorily acquired.

On paper and in theory this proposition is quite straightforward and very fair; but to put it into practice may be far from simple. The mere hint of a possible compulsory acquisition may immediately kill any potential market for the property involved. It may be that several properties are to be compulsorily acquired, or indeed a whole area of a town. No matter how many properties are involved, it is almost inevitable that the time lapse from the first mention of a scheme until its final implementation, when the properties are compulsorily acquired, will be several years, possibly even many years. Because of the long time between the inception of the scheme and its completion, it may be that there are no market transactions whatsoever in a particular area for a period of several years, as, whenever an affected property is offered for sale on the market, would-be purchasers are scared away on learning of the impending compulsory acquisition. Valuers may therefore find themselves in the position of having to assess values for open market purposes in an area where there have been no open market transactions for several years. Even if there have been some transactions in the area, they may have taken place at considerably lower figures than would have been the case but for the threat of compulsory acquisition.

It may be that a landowner has been put into the position of having to dispose of a property and yet has been unable to force the acquiring authority to purchase the interest. Such a situation might arise if property has to be sold to pay taxes due following a death and the inheritor of the property is left with the choice of either holding on to it and borrowing money to pay the tax or selling the blighted property at a figure well below the normal market value. The latter may be the most expeditious course of action open to the inheritor, but the sale on the open market of a blighted property may create a low tone of values, and an unrealistically low base for assessing the compensation payable on other affected properties.

There may be several courses open to valuers in their attempts to find the true market value, that is the value ignoring depreciation caused by the threat of compulsory acquisition. They may attempt to relate open market sales in other areas which are not affected by the threat to the area with which they are dealing. They may look at the evidence of such areas and calculate that values have risen by certain percentages since the compulsory acquisition was first suggested. They may then attempt to update the values in the affected area by using graphs projecting the values in existence at the date the scheme was first announced in an effort to relate them to the current market.

Such projections would assume that the values of the property being dealt with would have increased at the same rate as properties in other areas over the same period. This is a valid approach to the problem but it does not necessarily prove anything – it merely suggests what might have been.

In attempting such an exercise a valuer must pay great attention to all the other factors which could have affected the market in the area, for it may be that, even without the threat of compulsory acquisition, values in a certain area might have decreased. In circumstances in which properties are to be acquired for the construction of a dual carriageway road, traffic in the area may have become so bad that the existing single carriageway road is quite inadequate to cope with it, and the area is being harmed considerably by the excess traffic. Had there been no road scheme the area would have deteriorated because of the adverse traffic conditions. If this were the case a valuer would have to isolate the depreciation which would have occurred in any event, from the total depreciation, the balance being the depreciation in value which results from the threat of compulsory acquisition. It is not difficult to foresee the problems in such an exercise. A valuer can at best believe that the figure put forward is reasonable, rather than believe it is the only acceptable figure.

Another way of trying to assess what open market value would have been but for the threat of compulsory acquisition might be to look at the market trend in the area prior to the first threat of acquisition, and to project the previous trends in value forward from that date on the assumption that they would have continued along the same pattern as before. Subject to the type of reservation outlined in the previous paragraph, this might be an acceptable solution.

Yet another attempt to solve the same problem might be to look at any properties within the area concerned which have not been adversely affected by the threat of compulsory acquisition, in cases where only certain classes of property are affected. This may be an easier situation for the valuer to deal with, and it may be perfectly reasonable to assume that values in the properties affected by acquisition would have moved similarly to those of properties not affected. Here again, though, there are serious reservations, as it may be that the different types of property are so different that their markets are completely dissimilar and subject to different factors where value is concerned.

It is not difficult to see the problems facing valuers when legislation effectively destroys the market, and much will depend on their judgment and interpretation of the various statutes if reasonably correct answers are to be obtained. Two equally conscientious and skilled valuers, one working for an acquiring authority, the other working for a claimant, may arrive at quite different answers concerning the same piece of property. Although a layman might consider this to be unreasonable, it is only fair to point out that in any situation which involves judgment a difference of opinion of this nature is likely to arise. Two barristers may consider the same case and each reach a different conclusion, just as two doctors may each inspect a patient and recommend different cures for the same ailment. Where facts are uncertain it is almost inevitable that there will be differences of opinion between people, but the role of valuers in this type of situation will be to use their technical knowledge, their skill at interpreting statutes, and their experience and judgment in an attempt to reach as nearly as possible the correct valuation.

A more general type of statute which may affect values is the various Finance Acts. Indeed a Budget itself may do so, the Finance Act coming later and confirming the promises made in the Budget. The availability of credit and the cost of credit is as important with property investment as with any other investment. Promises or threats contained within a Budget may affect the thinking of investors and influence their subsequent actions. The Budget may encourage investment, or it may equally well discourage it. If it encourages investment, it may encourage it in particular forms while discouraging it in others. It may cause a particular form of investment to be relatively more attractive than others, even when the general effect of the Budget has been to encourage all types of investment.

If new taxes are promised or threatened such as taxes on the development value of land, wealth taxes or capital gains taxes, the reaction of investors may be to wait and see the precise content of the detailed legislation before embarking on any new projects. There may be many reasons why investors will wait and see. It may be that they merely wish to see which investments will be most attractive after the new measures are put into effect, and then to invest in those particular investments, or alternatively, it may be that they wait to see whether any form of investment is likely to be worthwhile, or whether immediate consumption will be more beneficial to them than saving and investment.

There are many possibilities and valuers have to look at budgetary announcements and endeavour to assess what effect they will have on the investment market generally. Here again much skill will be required, but, as professional people, valuers will have to commit themselves to a certain extent and make predictions of what they think will happen and what the effect will be on property values.

Many hours of study are required to understand fully the effect of statutes and government action on property values, but it is hoped that this brief look at the problem will have alerted would-be valuers to the problems they will have to face. This should not deter them, as these are the very factors which help to make the whole science or art of valuation intriguing and interesting. To summarise, valuers have to be adept at understanding Acts of Parliament, and have to be able to isolate the important factors which affect property values. They have to understand their effect and then be sufficiently competent in the mathematical process to vary their normal valuation procedure to take account of the new circumstances resulting from an Act. They have to be analytical, practical and methodical, and, needless to say, experience will be very useful to them in this task. A good knowledge of people and an understanding of the way they think and the way they react to various situations are also very useful, as the market for property, like any other market, is made up of people. To understand people is to a large extent to understand values.

Chapter 21

Taxation and Property Investment

Earlier in this book tax was mentioned and its effect on sinking funds was discussed. Calculations were made and it was shown how simple adjustments could be made to allow for the payment of income tax on sinking fund allotments. The question whether yields should be considered gross of tax or net of tax was also examined, and it was pointed out that as yields on most other investments were normally considered on a gross basis and different taxpayers had different tax liabilities it was convenient to consider yields from property on a gross basis. Because of the multiplicity of tax liabilities, no single adjustment for tax would be appropriate to all potential investors and, indeed, an adjustment at the standard rate might, in reality, be applicable to no one.

However, the effect of tax is very important when investment is considered, and this importance tends to grow if tax systems become more complex, as is often the case, and if the number of different types of taxes increases. Although income earners normally compare gross salaries when comparing the merits of different jobs, it is nevertheless probably true that once employed in a particular job the majority are more interested in their net salaries, as it is the amount of money available to them after payment of tax which dictates their standard of living. The net income available from property investment must inevitably be very important to an investor as it is pointless considering the gross income if, after tax has been deducted, the remaining sum bears no relationship to the original income flow. Investors cannot spend the tax element of their income, and it seems logical that their real interest should lie in considering the income available for personal use.

This concept may appear contrary to the conventions of property valuation, but if attention is given only to the gross yield before payment of tax investors will in fact be considering the combined yield which properties give to them and the Revenue. The amount that the Revenue receives is hardly beneficial to investors, and it therefore seems more appropriate to consider the net of tax value of investments to them personally. As yields in the market are normally quoted gross of tax, it would appear that two valuations at least should be done for almost every market appraisal. A valuation at a gross of tax rate may be appropriate to discover what the market will be prepared to pay for a property, and a net of tax valuation will endeavour to find the figure which a specific investor can afford to pay. The two figures may well be quite different, and if an investor's personal net of tax valuation proves to be lower than the normal market valuation then it is probably because he or she is in a less advantageous position after tax than others in the market.

As there is a multiplicity of tax rates applicable to different investors depending upon their personal circumstances, it may be that income tax liability for different investors will vary considerably, UK tax rates on income, for example, ranging in 2000–2001 from 10% to 40%. The UK tax liability of companies is on a different basis, and corporation tax is chargeable to small companies from a rate as low as 10% on profits up to £10,000 to a top rate of 30% on large companies, all these rates being considerably lower than in many earlier years.

Tax structures often reveal considerable differences between individuals and, indeed, between individuals and companies, and people might decide that activities would be better carried on by companies than by them as individuals as a lower tax liability might be incurred by a company than an individual would incur. There would be many other considerations to take into account, however, before reaching such a decision.

For much of the time in the United Kingdom following the introduction of Capital Gains Tax (CGT) in the 1960s, CGT rates were much lower than the higher rates of income tax with the result that for many people it was advantageous to try to make gains which would be subject to taxation under CGT rules rather than them being taxed as income. Such a consideration played an important part in many investment decisions and making the right decision could result in considerable tax savings. Since the early 1990s in the UK the rate at which tax on capital gains is charged has been the same as the income tax rate for the respective taxpayer and much of the motivation for tax planning in this area has therefore been removed.

However, the existence of a £7,200 (as at 2000–2001) annual exempt amount for each individual could still be an important factor for some people, as being able to utilise such an exempt amount could still significantly reduce tax liability for individuals, the tax saving for someone paying tax at 40% being £2,880 if the full exempt amount can be used. If a capital gain is made there could therefore be a reduced tax liability on that gain so that an individual paying tax at 40% may well be better off attempting to make capital gains each year to enable advantage to be taken of that annual exemption. The elimination of the large differential that used to exist between the top income tax rate and the CGT rate has, however, made such considerations less significant than previously, and they are now only likely to be important to a limited number of individuals rather than playing a very important part in the decision making process for many as was the case in earlier years.

There have been a number of different taxes on the development value of land in the UK since the Second World War, the most recent having been Development Land Tax which was introduced by the Development Land Tax Act 1976.

There were some exclusions from this tax, including the first element of development value realised in any financial year, the amount exempt having varied from time to time. An individual who was in a position to realise development value in excess of the exempt amount should therefore have considered whether it might not have been advisable to sell properties in different years in order to restrict the development value realised in any one year to the exempt sum. In subsequent years advantage could again have been taken of the exemption. By adopting such a policy a vendor might well have escaped tax liability completely, and the consequent tax saving may well have more than repaid the waiting. Tax planning of this nature can be a very important activity for the property owner or property adviser.

Taxes which become due at the transfer of property on death have varied considerably in the United Kingdom since the early 1970s, "death duties" having existed at various times in the form of Estate Duty, Capital Transfer Tax and now Inheritance Tax, each with a range of tax levels which varied or vary with the total value of the property transferred. In addition, variations in the details of each tax from time to time have resulted in constantly changing liabilities for a given value depending upon the precise date of death.

Some other countries have tax systems which levy tax on transfer at death, and where there is such a tax the valuer may often have to assess the tax due on the transfer of property interests both freehold and leasehold. In some cases it will be necessary to make such assessments even where death has not occurred in order to assess the contingent tax liability which would arise should the property owner die. Such a valuation could, for instance, be needed for insurance purposes to provide for an insurance payment to cover the cost of Inheritance Tax (in the UK) when it in due course becomes payable.

Such taxes are often quite complex and will not be discussed in detail here, but it is worth noting that following the UK 1987 Budget Inheritance Tax on transfers after 18 March 1987 ranged from Nil, to a rate of 30% (or 15% for transfers during a lifetime) to a top rate of 60%. Because of the "banding" of liability and the fact that some elements of any estate would be taxed at less than 60%, the maximum average rate of Inheritance Tax at that time would in fact always have been something less than 60%. A similar change has occurred with Inheritance Tax as with CGT and the top rate at which it was levied in 1993–1994 was 40%. This remains the top rate at the time of writing, although with the passage of time the amount of chargeable transfers to which the Nil rate of inheritance tax applies has been increased to £234,000.

The exemption from Inheritance Tax liability for lifetime gifts made more than seven years prior to death, and the existence of lower rates of Inheritance Tax for transfers within seven years of death, create a situation in which tax planning considerations may dictate that the gift of property to one's children during one's lifetime is preferable to the transfer of the property following one's death, the lifetime transfer resulting in either complete exemption or a lower tax charge.

Simplification of tax systems, which is very desirable, and the move towards similar rates of tax for different types of tax, not only make decision making easier for individuals but they also reduce the distortion to investment decisions that results from the need for careful tax planning caused by the existence of different bases of taxation, variations between tax rates, and a range of exempt allowances.

When new taxes are introduced or changes are made to existing taxes, wise investors consider their effect on their investments and when the full implications are known investment policies will probably be adjusted to give the most advantageous net of tax position. This is yet another consideration for the taxpayer or the investor to take into account, and there is little doubt that for many years past policy decisions by investors have been very much influenced by tax considerations.

A number of relatively simple examples follow to illustrate some considerations which an investor might have to take into account in formulating policy. The first example concerns an investor who is faced with the decision as to whether an asset should be sold and the money reinvested in an alternative investment. In many cases if the tax liability is ignored there will be little doubt than investors would be well advised to sell current investments and divert funds elsewhere, but because the liability to capital gains tax on any gain made will reduce the net amount available for reinvestment, they may well be encouraged to retain the current but inferior investment rather than sell it.

Readers should note that these examples are given to illustrate the effect that tax considerations may have on investment decisions. Rules and rates of tax used are those applicable in the UK at the time of writing (which precedes the date of publication) and both may vary with the passage of time. The rules and rates should not therefore be used by readers for tax calculations but those current at the relevant time and for the appropriate country should be used, it being advisable to check them by reference to an up-to-date tax encyclopaedia.

Example 1

An investor owns a property which he bought in 1968 for £15,000 and which gives a net annual income before tax of £18,000. He has been offered £180,000 by the lessee, who would like to purchase the freehold. The investor would like to accept the offer as he is of advancing years and the management of the property is proving troublesome to him, it being situated 200 miles from where he lives. He would like to acquire another investment nearer his home and has asked you to advise him. You are aware that he pays income tax at 40%.

The investor's present income shows a 10% gross of tax yield on the capital value as evidenced by the offer of £180,000.

Income net of outgoings	£18,000 pa
Less Income tax at 40%	£7,200 pa
Income available for spending	£10,800 pa

Disposal after 5 April 1988 of an asset owned on 31 March 1982 now creates a situation in which the base value of the asset for CGT calculations has to be taken as its value on 31 March 1982 provided this value exceeds the original cost of acquisition. In addition, a capital gain indexation allowance could be deducted prior to 6 April 1998 but from that date a new "taper" relief was introduced. Fuller details of provisions such as these can be studied in *Modern Methods of Valuation* or in tax books.

In this case for the purpose of simplicity it will be assumed that the appropriate value at 31 March 1982 plus allowances totals £65,000.

Proceeds of Sale on suggested terms:

Sale proceeds net of expenses, say	£174,000
Less Base value, say	£65,000
Capital gain	£109,000

Capital gains tax liability = £109,000 × 40% = £43,600.

The sum available for reinvestment amounts to the net sale proceeds of £174,000 less the capital gains tax payable of £43,600 or £130,400. If the freeholder could reinvest all this money in a similar risk property he would again get a 10% yield or an annual income of £13,040 before the payment of income tax. The tax payable on this income at 40% would be £5,216 and his annual income after tax would therefore be £7,824 or £2,976 pa less than the current income.

As a result, although both the landlord and the lessee would otherwise like to do a deal, the potential liability to capital gains tax means that the investor cannot afford to proceed with a sale unless it is possible to reinvest in a property with an anticipated early growth in income. The incidence of capital gains tax may therefore result in the rejection of a course of action which in all other respects would be advisable.

Alternatively, where on property management and investment management grounds, the retention and leasing of a property might be advisable, a high liability to tax may make retention so unprofitable after the payment of tax that the investor may decide on tax grounds alone to sell the property.

The removal by the British Government of the difference which previously existed between the rates of tax levied on income and capital gains was sensible in that it removed the distorting effect that the existence of two different rates of liability could have on decision making. Where there are different potential liabilities, property management and investment decisions (which in turn can have considerable economic consequences) may be made not for the best policy reasons but primarily for tax minimisation reasons.

Consideration of the type of problems which can arise for taxpayers when differing tax burdens arise under different types of taxes can be made by reference to the situation which existed in the UK in the 1980s.

Example 2

A businessman purchased a property in 1966 for £20,000 which in 1987 had a market value of £160,000. At that time he wanted to dispose of his business and he offered his manager the property at market value. The manager was unable to raise the necessary capital and offered to rent it at £16,000 pa. The businessman paid tax at a rate of 60% and the rate applicable at that time for capital gains was 30%.

Position if the property had been let to the manager

Annual income	£16,000
Tax at 60%	£9,600
Balance available after tax	£6,400

This was equivalent to a return of 4% net of tax on the market value.

Position if the property had been sold to a third party

Net sale proceeds, say	£155,000
Less Total purchase costs	£21,000
Capital gain	£134,000

A capital gain indexation allowance of £32,350 was calculated as being allowable against this gain leaving a gain chargeable to CGT of £101,650 and a tax liability of £101,650 × 30%, or £30,495 tax payable.

The net sum available to the businessman would have equalled:

Net sale proceeds	£155,000
Less Tax payable	£30,495
	£124,505

Analysis of net gain on original purchase:

Net of tax proceeds of sale	£124,505
Original expenditure	£21,000
Net gain	£103,505

This would have been equivalent to a net of tax yield of almost 8.5% pa over a period of ownership of 22 years.

Rather than accept a net of tax rent of £6,400 which was a net yield of about 4%, even though he would have liked to have assisted his manager, the businessman would have been better advised to sell the property and to have reinvested his money in a way in which he would have stood a chance of making another capital gain which would have shown him the equivalent of an 8.5% pa net of tax yield, in addition to the annual income it would have provided during the period of ownership.

If his liability to income tax had been only 45% the position would have been as follows:

Annual income	£16,000
Income tax at 45%	£7,200
Income net of tax	£8,800

This would have given a yield of 5.5% on the market value, which would have narrowed the gap between the alternatives, while if the income tax liability had been at only 30% the net of tax yield on letting would have been 7% and the net of tax gap between the alternatives would have been further reduced.

This example illustrates the distortions that were introduced to decision making with the existence of widely different tax rates under different types of taxes even when only one individual was concerned, and indicates why it is desirable for there to be some uniformity in tax rates between the different types of taxes which exist at any one time.

The previous two examples considered one individual and his decisions regarding future investment policy, but there will also be situations in which two individuals looking at the same investment will reach completely different decisions simply because they have differing liabilities to tax. An investment which is attractive to a person paying income tax at 20% may not be attractive to one whose tax liability on income is 40%, and the contrasts are likely to be even greater if the top tax rates are even higher.

Example 3

A leasehold property yields a gross of tax income of £20,000 pa and to provide for the recoupment of capital an annual sinking fund of £4,000 pa is required.

An investor with a tax liability of 25% will be in the following position:

Gross of tax income	£20,000
Less Income tax at 25%	£5,000
Net of tax income	£15,000

The net of tax income will comprise:

(1)	Spendable income	£11,000
(2)	Annual sinking fund	£4,000

An alternative way of looking at these calculations is:

Gross of tax income	£20,000
ASF required after tax	£4,000

If the sinking fund is enlarged to allow for income tax liability on it, it becomes:

$$\text{Net ASF} \times \frac{100}{100-25} = £4,000 \times \frac{100}{75} = £5,333.33$$

The separate elements of the total income flow would be:

Gross spendable income	£14,666.66
Less Income tax at 25%	£3,666.66
Net of tax spendable income	£11,000.00
Gross sinking fund allocations	£5,333.33
Less Income tax at 25%	£1,333.33
Net of tax sinking fund	£4,000.00

The net of tax spendable income represents 55% of the total gross of tax income.

From the point of view of a taxpayer paying tax at 40% the alternative calculations would be as follow:

Annual sinking fund after tax   £4,000.00

Grossed-up annual sinking fund $= 4{,}000 \times \dfrac{100}{100-40} = \dfrac{4{,}000 \times 100}{60}$

$$= £6{,}666.66$$

This taxpayer has to devote a larger part of the total income to establish an annual sinking fund to the extent that the gross of tax spendable income will be £13,333.33, the tax liability on it £5,333.33, and the net of tax spendable income will be £8,000, or only 40% of the original gross income. Clearly the second taxpayer will be able to bid less for the property than the first taxpayer unless he or she is prepared to accept a lower net of tax return from the property.

In the past the total tax liability on such investments could have been as high as 98% in which case the calculations would have been:

Grossed-up annual sinking fund $= 4{,}000 \times \dfrac{100}{100-98} = \dfrac{4{,}000 \times 100}{2}$

$$= £200{,}000$$

Obviously investors who had tax liabilities of 98% could not even have contemplated the purchase of an interest on which the tax liability resulted in the grossed-up annual sinking fund requirement being 10 times the gross of tax annual income produced by the investment. Even when the liability to tax was 60% the tax-burden on the sinking fund instalments would have resulted in 50% of the total gross income being monopolised by those instalments, which would have left a net of tax spendable income of only £4,000, or 20% of the total gross of tax income.

Our original taxpayer under current tax rates, however, retained 55% of the gross income as spendable income after paying income tax and after providing for a sinking fund to recoup the capital.

The above example illustrates a scenario in which the individual situations varied widely with different tax liabilities, and there will be many instances in which a marginally more favourable tax position for one investor will result in him or her being able to outbid other potential investors who are more heavily taxed.

The examples have purposely been kept both few and simple as it is not the intention to become involved with teaching tax, nor to get involved in complicated calculations which are better left until the reader has a thorough grasp of basic valuation principles and is able to understand most of the implications of property investment. The main objective at this stage is simply to illustrate the fact that the existence of different rates of tax and the resultant different tax liabilities (which may vary dependent upon the decisions an individual investor takes), are factors which must inevitably have a great effect on policy decisions. Indeed, the importance of such considerations is so great that tax planning is an important science these days. People who fully understand all the implications of complicated tax systems, if there are such people, are worth their weight in gold to investors and, indeed, are able to command high fees if able to give quality advice.

Chapter 22

Finance and Gearing

Obviously, before an investor can contemplate purchasing a property interest, he or she must have the necessary money, and a relatively large sum will be required, for even the purchase of a small property will probably cost in the region of £50,000. Larger properties will cost many times more, and some cost many millions of pounds to purchase. Because of the need for such large sums, the majority of purchasers cannot provide all the necessary money from their own funds, and they have to borrow to obtain sufficient money to proceed with a purchase.

This raises the problem of finding where money can be borrowed, and as the type of property varies, so the source of finance may vary. Irrespective of the source of finance, the better the terms the borrower can arrange, the more beneficial the overall property investment is likely to prove. If an investor is able to borrow a substantial sum of money at a relatively low rate of interest and is able to ensure that the money will be available for a long period, the overall security of an investment is likely to be considerably stronger than if funds were only available at a high rate of interest for a short time. The terms on which an investor can borrow will depend not only on the type of investment, but also upon the general state of the economy (local, national and international), at the time of the loan. If overall interest rates are high then the cost of borrowing is likely to be high, whereas during times of low interest rates, there will naturally be a greater chance of borrowing money on attractive terms.

Other critical factors influencing the terms upon which money can be borrowed are the reputation, character and standing of the borrower. A person of proven character and ability is far more likely to be able to borrow on good terms than those of doubtful character who have no previous dealings in the field into which they wish to venture. There is nothing particularly surprising or unreasonable about such a state of affairs. It would be a foolish person who would offer as favourable terms to a completely unknown individual as they would to someone of proven ability.

The more reputable and established investors become in their particular lines of business, the more preferential will be the terms they are likely to obtain.

As far as the type of property is concerned, the borrowing sources can be split into two basic markets; the market for residential property finance, and that for commercial property finance.

For many years in the United Kingdom the chief source of finance for the purchase of residential property was the building society movement. Loans from this source were normally only obtainable for the purchase of residential properties for the borrower's own use, Societies rarely lending money to finance the purchase of residential properties for investment purposes or for the purchase of commercial properties. Funds may occasionally have been provided for the purchase of a property with a mixed residential and retail or commercial use if the property was being bought for the purchaser's own use. Building society loans in general, provided an incredibly cheap form of borrowing at rates of interest which were normally much lower than those charged on loans for the purchase of non-residential property. Additionally, in the United Kingdom borrowers of funds for the purchase of a house for their own use were for many years able to get income tax relief on interest payments which further reduced the cost of purchasing a home. Reduction in recent years of some of the concessions previously obtainable reduced the tax relief available, and the complete abolition of concessions in April 2000 will remove the significance of income tax saving in the purchase of a home in the United Kingdom.

During the 1980s British building societies expanded into other areas of activity such as the provision of banking services, and changes to the legislation under which they operate enabled them to become involved in many activities previously denied to them. Indeed, many of the larger building societies have so changed their roles as to be rivals to the major banks in many areas of activity and they are now different from the traditional building society of earlier years. Despite these developments, there is little reason to suspect they will not continue to be a major source of funds for house purchase in the future.

There are several reasons why the rate of interest charged for housing loans is low, a major one undoubtedly being the fact that traditionally building societies have not been profit-making bodies.

Their objective has been to provide money for house-purchase, and they do so by borrowing from savers who wish to deposit money in a safe investment from which it can be readily withdrawn if needed.

This money is then lent to others who wish to purchase houses. Societies borrow money at a lower rate of interest than the rate they charge on their loans, but the difference between these rates of interest is not intended to provide them with a profit, merely to cover their operating costs. Although there appear to have been some extravagances in the building societies movement since the 1960s, in that a multiplicity of offices owned by a multiplicity of competing societies seem to have been opened, it is a fact that their overall operating costs are in general very low, and this has helped to enable them to offer funds to the public at low rates of interest. Borrowers who consider that their rates of interest are not low merely need to investigate the commercial money market to persuade themselves otherwise.

In the 1990s in particular there has been a substantial change in the United Kingdom in that many of the major building societies have now converted into banking institutions offering most of the services provided by the traditional High Street banks, and competing with them. The result is that, with a few notable exceptions, it is now mainly what were once the smaller building societies only which continue to fulfil the traditional role of building societies.

There are also other reasons why the rates charged for housing loans are relatively low. Whatever may be the state of the economy, people will always require houses in which to live, and so, although a depressed house market may exist in times of depression in the economy, the market for houses is not likely to disappear completely. There is consequently nearly always a market for residential property, even though prices may at times fall. The record over the post Second World War period was for many years of steadily increasing house prices in spite of periodic depressions, with the security of a loan made by a building society generally increasing over time rather than decreasing or staying static.

This situation changed markedly in the late 1980s and early 1990s with the advent of a major economic recession which affected the economies of many countries. In previous years house prices had risen considerably, but this recession saw prices fall in many areas by as much as 30% to the extent that many housing loans were no longer adequately covered by the value of the properties on which they were lent. The facts that loans of as much as 90% of market value were not uncommon and that many borrowers became unemployed as a result of the recession exacerbated the situation, and in the early 1990s repossessions of mortgaged houses on the default of the mortgagors became relatively commonplace rather than the rarity they had previously been.

Sadly, many people learnt the hard way that the purchase of a property does not guarantee a capital gain and that if general economic circumstances are unfavourable property prices can and do fall. Too many people had, it is suggested, bought properties on the basis of capital gains they hoped to make, rather than for the benefit of the value-in-use they expected to get from them, which for the vast majority of property purchasers ought to be the major reason for purchase.

Despite this phase of events, over a long period of time and in most instances the risk entailed in lending funds for house purchase by owner-occupiers is a good one, particularly as it is only secured on one person or at the most on one couple. It is easier to gauge the overall risk when only one or two people are involved than if several people are involved, as may be the case with, for example, a newly-formed company.

The risks that exist with building society mortgages when they are originally made are reduced by the fact that borrowers are normally paying back part of the capital borrowed with their interest payments, and the original sums lent are therefore decreasing with time. Allied to the decrease in the size of the loans, if there also happens to be an increase in house prices over a period, then the two factors operate together to increase the security of the loans. As the size of the loan decreases, the value of the property on which it is secured increases.

There are other sources of finance for residential house purchase, and a substantial amount of finance for house purchase is now provided by banks. In the 1980s there was a change of policy amongst the High Street banks regarding offering loans for house purchase. In the past United Kingdom banks had not been active in this field, and such loans as they did make tended to be for relatively short terms and at higher rates of interest than those offered by building societies. However, the major banks entered the house-mortgage market in direct competition with the building societies and now offer mortgages at very competitive terms in a similar way to banks in other countries. The fact that many building societies have now become banks has, of course, also increased the proportion of housing finance provided by the banking sector.

Private mortgages can also be obtained, that is, a mortage from one individual to another, the house purchaser. These now tend to be few in number, possibly because there are few people with sufficient liquid capital to lend for such purposes, and because those who wish to lend on a private mortgage would be competing with the building society rate of interest, which would often be lower than that which they could obtain from alternative sources. Also, private lenders in most cases will wish to retain a degree of liquidity which might be unacceptable to the would-be borrower, who would not want to run the risk of having the borrowed money recalled at short notice. To negotiate a satisfactory private mortgage, the terms would have to be such as were acceptable to both the lender and the borrower.

When money is required for investment purposes or the purchase of commercial properties, the sum the borrower requires will probably be considerably larger than that which the average house purchaser requires, since units of commercial property are generally much more valuable than the average house. The would-be investor or purchaser will usually have to have recourse to the commercial money market and the larger lending institutions for funds. At times borrowers may be able to obtain finance from the commercial banks, but such lending is likely to be for a short term only, and therefore likely to be relatively unattractive unless it is reasonably certain that the loan will be renewed at regular intervals, thereby making it in effect a long-term loan. Even if continued renewal gives in effect a long-term loan, it is probable that the rate of interest charged would vary with fluctuations in interest rates generally and such an arrangement might not be acceptable to a borrower, as it would be impossible to predict the long-term viability of a project with any degree of certainty.

In many cases a purchaser of commercial investment property will borrow money from a finance house, or a body such as one of the larger insurance companies. Insurance companies do lend for such purposes, although in these days the majority of them probably prefer to invest directly in property.

The rate of interest charged on commercial loans will invariably be well in excess of that charged for housing loans. Lenders of money for commercial purposes have the objective of making money themselves, and so the rate of interest charged will not only have to cover their own operating costs but also have to produce a profit. Lenders will normally try to charge as much as the market will bear, and they will in any event require a higher rate of interest than for housing loans because commercial lending is generally riskier. The rate of return will have to be sufficient to make up for any profit that could have been made had the money been used in other ways. This concept is commonly referred to by economists as the "opportunity cost" of money, and unless lenders can make as much profit out of loans as could be made in the next best alternative opportunity open to them, then they will be ill-advised to lend money for the purchase of commercial property. This comparison will dictate the lowest rate of interest at which lenders can offer their money, and over and above that they will also wish to obtain sufficient profit to cover any additional risk which loans may entail. The alternative source of investment open to them may be less risky than lending on commercial property and they would therefore be unwise to lend unless they received an extra payment to cover the extra risk.

In deciding what the opportunity cost of their money is, lenders will consider all the other investment opportunities open to them, and this will include the possibility of investing money in themselves.

Unless more money can be made by lending than could be made by using savings in business ventures of their own they will be unlikely to lend. They are also likely to be influenced by the rate of interest offered by other people wishing to borrow money, and this may entail consideration of the international money market and the rates offered therein, also the rates offered nationally, such as those paid by local authorities and other public bodies who wish to obtain loans from individuals.

There is little doubt that lending for commercial purposes is normally riskier than the type of lending engaged in by building societies. The trouble experienced by fringe banks in the United Kingdom during 1974 illustrates how such risks may be rapidly increased by changing economic trends. The events of the early 1990s re-emphasised these risks, and in that recession in property markets worldwide it was not simply secondary and fringe banks which suffered, but some of the major banks operating internationally also suffered very substantial losses.

The risk in commercial lending for property purchase revolves around a variety of other factors apart from the property. The use to which it is going to be put will be an important factor, and the riskier the use and the smaller the market for such uses, the less secure the loan is likely to be in the long-term. If there is a great demand for property for a particular use and few properties are available, the risk is likely to be reduced. However, even if this is the case, there may be a rapid change if the general economic situation suddenly changes. Should there be a slump in the economy it may soon become apparent that a shortage of property has rapidly become a surplus as a result of potential property users disappearing from the market. Whereas this may happen in the commercial property market during bad times, people will still need houses in which to live, and so, in this respect, commercial property is probably a greater risk than owner-occupied residential property.

As with residential property, the character and ability of the user will also be very important. This may be quite difficult to assess if the purchaser is a company in which a number of individuals are combining in a joint enterprise. It may not even be certain that they have the ability to carry out the enterprise, or even the ability to get on together as people. As the board of directors of a company may change frequently and even rapidly, it may be even more difficult to assess the security to the lender if changes do occur in its constitution.

For these and other reasons the lender of money which is to be used for the purchase of commercial property will "vet" both the scheme and the borrower very carefully and endeavour to ensure that the rate of interest to be received on the loan is adequate to cover all the risks involved. As there is also invariably considerable competition for a limited supply of funds, the commercial money rate is again likely to be relatively high. Often there will be demands for more money than the total available to lend, and the supply and demand situation will in such cases result in lenders being able to demand a high rate of interest, which they will more often than not succeed in obtaining. As long as the borrower considers that the money borrowed will provide greater returns than the cost of borrowing, a loan at a high rate of interest will still be a sound financial proposition.

The term "gearing" refers to the relationship between the money an investor puts into a scheme himself or herself, and the money borrowed for the scheme. The money the investor provides is referred to as their "equity" in the scheme.

Where the equity is a small proportion of the total capital involved, and the greater part is money which has been borrowed, the project is referred to as being highly geared, and where the converse is the case the project will be said to enjoy a low gearing. The degree of gearing has been an important factor in property investment since the Second World War, and the rate of interest charged on borrowed money has also been important. A series of simple examples follows which it is hoped will illustrate the differences which result from varying degrees of gearing and from borrowing at varying rates of interest. It will be seen that there may be terrific, sometimes terrifying, implications to the investor resulting from increases or decreases in the rate of interest, and that this, combined with the degree of gearing, will be critical in determining the viability of an investment. The examples are purposely kept simple, and such items as tax relief on borrowed money are ignored; it is assumed that the cost of the money to the investor is the total cost after allowing for incidental costs and tax relief on the loans.

Example 1

A property which produces a net income of £20,000 pa is purchased for £200,000 the investor providing all the purchase money.

Annual income net of outgoings   £20,000

Yield to investor $= \dfrac{20,000}{200,000} \times 100 = 10\%$

Example 2

As example 1, but the investor borrows £100,000 of the purchase money at a rate of interest of 7%.

Annual income net of outgoings	£20,000
Deduct interest on loan of £100,000 @ 7%	7,000
Net income to investor	£13,000

Yield on investor's equity $= \dfrac{13,000}{100,000} \times 100 = 13\%$

Example 3

As example 1, but the investor borrows £150,000 of the purchase money at a rate of interest of 7%.

Annual income net of outgoings	£20,000
Deduct Interest on loan of £150,000 @ 7%	10,500
Net income to investor	£9,500

Yield on investor's equity $= \dfrac{9,500}{50,000} \times 100 = 19\%$

It can be seen from the above examples that if the cost of borrowing is below the overall yield on the total capital invested, investors can increase the yield on their equity by increasing their gearing. The higher the gearing, the greater will be the yield on equity, and the lower the gearing the smaller will be the yield on equity. Whatever the gearing, as long as the cost of borrowing money remains below the overall yield on the property, the project will remain financially sound.

Example 4

As example 1, but the investor borrows £100,000 of the purchase money at a rate of interest of 15%.

Annual income net of outgoings	£20,000
Deduct Interest on loan of £100,000 @ 15%	15,000
Net income to investor	£5,000

Yield on investor's equity $= \dfrac{5,000}{100,000} \times 100 = 5\%$

Example 5

As example 1, but the investor considers borrowing £150,000 of the purchase money at a rate of interest of 15%.

Annual income net of outgoings	£20,000
Deduct Interest on loan of £150,000 @ 15%	22,500
Loss to investor	£2,500

The last two examples illustrate that although the cost of borrowing money is greater than the overall yield obtained from the property, the project remains financially profitable if the investor's gearing is low enough to provide a return on his or her money, even though that return is below the overall return to the property. In example 5 the investor is so highly geared that, after paying interest charges, there will be no return left to the equity. Indeed on the figures used in example 5 the investor would actually be out of pocket and it would be inadvisable to proceed with the venture.

Example 6 will investigate the break-even point and it will be shown that a point will be reached at which if the investor increases the borrowing an overall loss will be made. It is worth noting that, even in the type of situation illustrated in example 5, an investor may consider going ahead with the project if the view is taken that the income is likely to increase sufficiently at a relatively early date to enable it to become viable and to give an adequate return to the equity.

Example 6

A property producing a net income of £20,000 per annum is to be purchased, the cost of borrowing being 15%. The total interest paid cannot exceed £20,000 if the investor is not to make a loss. Let the maximum capital which can be borrowed = X.

$$\frac{X \times 15}{100} = 20,000 \qquad \frac{15X}{100} = 20,000 \qquad 15X = 2,000,000$$

$$X = £133,333.33$$

These are very simple examples which illustrate several different aspects of gearing and the importance of the rate of interest, but in reality it can be very difficult indeed accurately to predict the constituent parts of such calculations. It may be difficult to estimate the net income which will be obtained. If the property is not yet occupied by a tenant, and a tenant has to be found, it is a matter of opinion as to the rent which may be obtained. In the examples above if the actual rent obtained was £1,000 less or £1,000 more than that which was anticipated, completely different results would be obtained from the calculations. Again, it may be that the investor has to pay various outgoings from the gross income, and fluctuations in the cost of repairing and insuring a property may result in fluctuations in the net income obtained. As we are aware of the investor's liability to pay income tax on the income flow, it should be noted that fluctuations in the rate of income tax or corporation tax will result not only in a variation in the income available for spending after tax has been paid, but in variations in the net cost of borrowing money, as if there is a decrease in the amount of tax payable the amount of tax relief obtained on the loan costs will also decrease, and so the true cost of borrowing will increase.

Occurrences such as those mentioned above will tend to upset the investor's original predictions and calculations. Indeed, possibly the most important of the variables discussed above is the actual rent obtainable, and although for many years rent levels generally increased, this was not the case in many areas in the early 1990s. Falls in rent levels can result from either a change in the supply position of properties, or a change in demand resulting from economic trends generally. Whatever the reason for a change in any of the variables, if one does occur the yield on the investor's equity will be changed at the same time.

During 1974 many British property companies ran into considerable difficulties because of the general increase in the cost of borrowing money. This increase resulted in borrowing costs being considerably higher than they had been a year or two earlier, and in many cases the cost of borrowing even doubled. These increases automatically took some companies from a position such as that illustrated by examples 2 or 3 to the position illustrated in example 5. There were also other factors helping to cause the difficulties experienced by some companies, particularly a freeze on business rents which was introduced as one of a series of moves intended to arrest inflation in the economy generally. The combination of a government imposed freeze on rents which prevented higher rents being obtained even where legally they could otherwise have been collected, combined with an increase in the rate of interest payable on borrowed money, took many companies into a position in which they were losing considerable sums of money on investments. The higher the gearing of a company in such a situation, the greater the problems caused by an increase in interest rates. The reader may care to do a few calculations using progressively higher rates of interest and progressively higher degrees of gearing to illustrate the problems encountered by many companies.

The occurrences of the 1970s were repeated again in the early 1990s in many property markets worldwide. Increased supplies of newly developed commercial properties in particular became available at a time when the demand for such properties was adversely affected by widespread economic recession, the result being that in many markets there were few tenants for vacant accommodation. Those that could be obtained by property owners were generally in strong bargaining positions which enabled them to negotiate very favourable rent agreements, often with several years of rent free occupation before any rent liability would fall on them.

In such circumstances many property owners were in a position where the income produced by properties was far lower than originally anticipated, and if they were also highly geared the income was sometimes inadequate to cover the loan charges on borrowed funds. Their situations were similar to the following example.

Example 7

The development of a property which was expected to produce a net income of £100,000 pa was undertaken at a total cost of £1,000,000. The developer borrowed £750,000 to undertake the development at a cost of 12% pa, a major reason for undertaking the development being that rental levels were expected to rise in the near future. In the event, when the development was completed the developer was only able to let half of the accommodation at half the expected rent level, and was therefore faced with the following situation.

Basis for development:

Anticipated rental	£100,000 pa
Less Interest on loan of £750,000 @ 12% pa	90,000 pa
Return to developer's equity	£10,000 pa

This represented a return to the equity of $\dfrac{10,000 \times 100}{250,000}$ or 4%

Although this is not a high rate of return the developer had been anticipating an increase in rents within three years of the completion of the development which would have created a situation such as below.

Net income	£125,000 pa
Less Interest on loan of £750,000 @ 12%	£90,000 pa
Return to developer's equity	£35,000 pa

This would have represented a return to the equity of

$$\frac{35,000 \times 100}{250,000} \quad \text{or } 14\%$$

Such a situation would have made the development attractive, particularly if still further increases in rent levels could be expected. However, in the event the situation which actually resulted is as below.

Net income	£25,000 pa
(from half the space at half the anticipated rent level)	
Less Interest on loan of £750,000 @ 12% pa	90,000 pa
Net deficit	£65,000 pa

Not only has no return to equity been realised but the developer has been left with the additional problem of having to find £65,000 pa from other sources to cover the costs of loan finance which have not been covered by the income produced by the property.

This is exactly the type of situation many developers and investors found themselves in during the early 1990s which caused the insolvency of many, and considerable problems for many others who managed to survive. The only redeeming feature of the early 1990s was that in most countries interest rates fell, which provided some relief for borrowers who were in a position to take advantage of the lower rates. Where borrowers were "locked into" arrangements at fixed rates of interest this was not always possible. However, the situation for many investors and developers was so bad that even when they were able to benefit from lower interest rates it was in many cases insufficient to save them from insolvency.

If interest rates rise, as they did in the 1970s, the problems encountered by an investor can still be solved if the rate of interest to be paid remains below the yield obtained from a property, but the more the interest rate exceeds the yield from an investment, the greater the problems for the investor. In more fortunate times for property developers, particularly in the 1960s, many made fortunes because they were able to borrow money at low rates of interest to invest in properties which produced a higher yield. It should not be forgotten that the developers who did this had great foresight, and that the majority of them also had considerable development skills which they utilised in their projects. They were also prepared to take what were considered great risks at the time, and even though, with the benefit of hindsight, we may consider their successful projects and say "Well they couldn't have lost", it should not be forgotten that at the time they ventured on them many people considered them to be very foolish and unwise.

The developers who made fortunes by borrowing at low interest rates were in fact borrowing on the basis illustrated in examples 2 and 3, and, because the interest charged was low in comparison with the yield from the property, they obtained high yields on their equity. The higher their gearing, the higher the yield they obtained on their equity. With time, the yield increased even more because incomes obtained from property investments increased almost dramatically following the general increase in rent levels which arose from the seemingly ever-increasing demand for commercial property during the post Second World War years.

Examples to illustrate the positions of two developers/investors following increases in the rent obtainable from their properties are given below.

Example 8

A property which produces a net income of £20,000 pa was purchased for £200,000, the investor borrowing £100,000 of the purchase money at a rate of interest of 7%.

Annual income net of outgoings	£20,000
Deduct Interest on £100,000 @ 7%	7,000
Net income to investor	£13,000

$$\text{Yield on investor's equity} = \frac{13,000 \times 100}{100,000} = 13\%$$

Example 9

As example 8, but the income has increased to £40,000 pa.

Annual income net of outgoings	£40,000
Deduct Interest on £100,000 @ 7%	7,000
Net income to investor	£33,000

Yield on investor's equity $= \dfrac{33,000 \times 100}{100,000} = 33\%$

Examples 8 and 9 illustrate the very considerable improvement in the investor's overall position, as, with a doubling in the income produced by the property, the yield on the equity has almost trebled. The reader may care to do calculations to work out what the yield on the investor's equity would be if the interest rate in each case doubles to 14%.

Example 10

A property which produces a net income of £20,000 pa is purchased for £200,000, the investor borrowing £150,000 of the purchase money at a rate of interest of 7%

Annual income net of outgoings	£20,000
Deduct Interest on £150,000 @ 7%	10,500
Net income to investor	£9,500

Yield on investor's equity $= \dfrac{9,500}{50,000} \times 100 = 19\%$

It will be noted that by borrowing a larger sum of money at 7% the investor has increased the yield on the equity in comparison with the situation in example 8.

Example 11

As example 10, but the income has increased to £40,000 pa.

Annual income net of outgoings	£40,000
Deduct Interest on £150,000 @ 7%	10,500
Net income to investor	£29,500

Yield on investor's equity $= \dfrac{29,500 \times 100}{50,000} = 59\%$

These examples illustrate that high gearing can sometimes produce very profitable results, but it must be remembered that it can equally well be very, very risky. It is easy to be wise after the event, and it is too easy to comment that many developers and investors who encountered difficulties in 1974 and in the early 1990s were foolhardy and deserved all the problems they encountered. While such criticism may have been justly deserved by some, a generous person might comment that in more fortunate times, when the underlying factors of the market moved in their favour rather than against them, they might have been considered skilful and foresighted. However, the vagaries of the economy and the market are problems which an investor or developer has to accept, and one of the great skills needed in property investment is to be able to estimate accurately the necessary financial requirement of a project, and thereafter to borrow on the right terms to obtain an acceptable yield on the equity, so making a scheme viable. If, having made all these predictions and calculations, the investor is unable to borrow on the right terms, it would be inadvisable to proceed with a project.

As shown above, changes in circumstances can improve the investor's lot, but it should not be forgotten by those who might begrudge the high returns that are sometimes made, that such changes can also cause great problems and may result in considerable money losses being made, to say nothing of the mental strain and worry caused when adverse conditions arise.

Chapter 23

The Operation of Property Markets

As has been observed earlier, property markets are comprised of people who bid against each other in the market for a limited range of properties which is available. It is the interaction of supply and demand which ultimately determines the prices at which goods are exchanged in markets. Property prices are determined in the same way as those for other goods, although as has been indicated markets tend to be specialised in nature (for example residential, retail or commercial) and many are relatively local in nature.

Nevertheless, they are affected by the same forces as other markets and it is a basic fact of economic theory that when price rises there should be an increased supply of goods onto the market and a smaller number of bidders for those goods. The new suppliers are encouraged into the market by the prospect of higher prices resulting in greater profits, while some existing buyers are deterred because prices have moved beyond the range of their financial capabilities.

However, in the period since about 1960 in many markets there have been instances in which these patterns have not been followed partly because unsatisfied demand has existed, while price increases have attracted more potential purchasers into the market rather than fewer. This has created boom periods in which the panic action of potential buyers has forced prices up rapidly, the length of time in which it takes to bring increased supplies onto the market also being a cause of some boom periods. This has been an important aspect of many property markets, it being a feature of most that the market alternates between boom and bust, with only short periods of relative market stability.

It is widely acknowledged that a phenomenom known as "the property cycle" exists in which the "boom and bust" periods are clearly identifiable. The demand for property is a derived demand from users who wish to use properties for specific purposes including the production of goods and services, and periods of economic growth invariably give rise to an increase in the demand for property. Developers and investors respond to such demand by planning and developing new property, but because the production period is invariably long, some of the increased supply takes several years to come to the market, by which time the economic boom which gave rise to increased demand for property has collapsed, leaving developers with new properties which no one wants to use.

Such a pattern of events has led to a series of "boom and bust" periods in many countries since the Second World War, only those countries with rapidly growing economies seeming to escape the worst effects of the cycles. In 1993 many property markets throughout the world were still firmly gripped in the downturn of what many regarded as one of the worst depressions property markets have ever suffered.

The period from 1980 through to 1992 was one of considerable activity in property markets worldwide, and a period in which the concept of the "global village" was clearly a reality in terms of property development and investment activity. Almost worldwide there were periods of depressed demand and limited development and investment activity, interspersed with upswings in property markets, with a significant boom period in the late 1980s followed by the almost inevitable "bust" at the end of the decade through into the early 1990s.

The level of demand for property is extremely dependent upon the general state of an economy, be it local or national, while the value of an individual property is very much a function of the overall level of demand and the utility of the particular property to would-be users. The level of economic activity and more particularly the trends in an economy are therefore critical determinants of the level of activity and values (and trends in values) in property markets, facts which events in the 1980s would suggest have not generally been given enough consideration in the past by many valuers. Macro-economic conditions and trends at both international and national levels can have a significant effect on local property market activity and consequently on local property values, a fact which should by now have been brought home to many who previously might not have accepted such a proposition.

If, as a result of events in the property markets of the 1980s, valuers in general have learnt that property markets do not operate in insolation from the general economy but rather are part of that economy and are affected by events in it, then some benefits will have resulted from the traumatic happenings in the property markets of that period.

Property transactions do not take place in a centralised market place, there being a series of local property markets, each local market in turn operating as a number of submarkets dealing with specific classes of property, such as offices, retail properties or residential properties. Despite the existence of these different markets, a consideration of events in the Sydney Central Business District Office market during the period under consideration will provide a picture of events which was mirrored in many other property markets in a number of countries over the same period, with only minor variations in emphasis occurring from market to market.

The similarity in events within the different property markets emphasises the dependency of property market activity and property values on general national and international economic factors. While the Australian property markets of Sydney, Melbourne, Brisbane, Adelaide and Perth are, by virtue of the large distances between each centre and their relative isolation from each other, quite separate markets, they are subject to the same external influences and are dependent upon financial institutions many of which operate on a worldwide scale. It is not surprising, therefore, that there was a marked similarity in the general course of events in property markets throughout Australia during the 1980s. With many corporations and banks now active on an international basis these similarities were also evident in many other countries of the world.

Consideration of the Australian economy in the 1980s reveals inconsistent growth in the Gross Domestic Product, with the annual rate of growth being at or below 1% for much of the time. Similarly, in terms of relative productivity as measured by Gross National Product per capita, Australia slipped from fourth in the world in 1950 to about 13 in the world by 1992. From seasonally adjusted unemployment rates of about 5% or below in the 1970s, in the 1980s the rate was consistently above 5% and as high as 10% at the beginning and end of the decade. Although the high level of unemployment resulted in part from an increasing population, a rate consistently above 5% should perhaps have counselled caution to such as property developers, particularly when considered in conjunction with the national productivity statistics.

During the 1980s there were also uncomfortably high levels of inflation and for most of the time very high annual balance of payments deficits, so high in fact that Australia now has to devote much of its earnings to funding the accumulated overseas debt.

During this period there were therefore indications of problems in the economy with, if anything, a generally adverse trend in a number of important economic indicators. Overall there appeared to be a lack of a sufficiently positive trend in any of the main indicators to support many of the ambitious property development decisions which were made in the 1980s. This is probably an appropriate observation in respect of the United Kingdom and North America also.

There was during the second-half of the decade a very active development market in Sydney, so active that when the recession and eventual depression of the period 1990 to 1992 occurred an extremely large stock of new office developments (and other types of property) existed in various stages of development from the drawing board to completed and ready to let. The advent of the recession resulted in many planned or partially completed schemes being suspended, while many of those which were already completed became either unlettable or lettable only at very much lower rents than were originally anticipated. As a result there were numerous bankruptcies of developers, investors and speculators, while banks and other lending institutions were left holding property loans which often exceeded the values of the properties on which the loans were secured.

In retrospect it appears difficult to understand how such a situation could have occurred. It is probable that in part at least it was attributable to the limited availability in most property markets of reliable market information (this state of affairs being exacerbated by the practice of inserting confidentiality clauses in many contracts). Additionally, the time required to conceive and complete new property developments was probably also a contributory factor.

It would appear that many developments are planned without the developer being fully aware of other planned developments which will compete with his or her own scheme, although it is suspected that well-directed research would in fact reveal a reasonably helpful level of information on such matters. It is possible that, rather than lacking information of this nature, many developers do not pay sufficient regard to its significance, relying rather on an almost blind faith in the quality of their own proposed development and its ability to succeed in the market whatever the competition.

This optimism was probably built into many of the appraisals which were done for prospective developments, the use of discounted cash flow layouts and computer technology enabling complex valuations to be produced which incorporated estimations of such things as rental growth for many years into the future. There is little doubt that many of the growth predictions used by developers were extremely optimistic, and it is suspected that some of the inputs to such valuations were used to support the optimism of the developer rather than to test the soundness of the proposed scheme. It is suspected that many projections included annual rental growth much in excess of that which could be supported by any economic indicators, and that such growth was sometimes projected many years into the future.

A very significant factor in the development boom which occurred in the late 1980s was the lead time needed to plan and develop a modern scheme. A time period of five to six years is not unusual, while the apparently ever-increasing size of modern developments has tended to exacerbate this problem. Similarly, the great size of many developments has created a situation in which the costs incurred in the planning stages are so great and the development period so long, that once construction is started it is often not practicable to suspend work even though market conditions may change dramatically during the development period. It is also extremely difficult, if not impossible, to accurately predict the condition of the economy and property markets at the time a completed development is likely to actually be placed on the market.

The depressed years of the 1970s resulted in there being only limited development activity during that period and through into the first-half of the 1980s, and the eventual result was that by the early 1980s the vacancy rate (that is the percentage of unoccupied accommodation) in the Sydney CBD was down to about 2½%. This is generally considered to be a low vacancy rate and is such that prospective developers are likely to be persuaded that there is in reality an unsatisfied demand for accommodation, to the extent that they are likely to enter the market by commencing developments. In the event, from the relatively low level of building commencements from 1976 through to 1983, there was a marked increase in the annual number of building commencements from 1984 to the end of the decade. With a construction period which would quite often be about three years and sometimes even longer, buildings which were commenced in the mid 1980s did not in fact reach the market until the late 1980s, while those which were commenced in the late 1980s were in some cases not completed by 1993.

The relative inactivity in the development sector in the period 1976 to 1983 had helped to create these low-vacancy rates, which in turn led to increasing rental levels by the mid-1980s and increased confidence among and increased competition between property investors. With a relatively static supply of properties, market values increased and provided the stimulus to developers to increase the supply of developments.

Development activity was in fact further stimulated by an overnight stockmarket collapse on 19 October 1987, the reaction of many investors to that collapse being to transfer their investment activity to property markets, there appearing to be an almost blind faith among many that it is impossible to lose in property investment. Activity in the property investment and development markets was assisted by an apparently similar belief among lending institutions which lent generously for such activities, often to the extent that developers and investors had to put very little of their own funds into their schemes and investments. With only limited amounts of their own funds at stake developers in particular were, it is suspected, tempted to venture into projects which they may well have considered too risky had more of their own funds been at risk, and many lenders were to rue their actions in lending so generously when "the bottom dropped out" of the property market at the end of the decade.

A factor which caused many lending institutions to lend so generously to many developers and over-generously to some was undoubtedly the deregulation of the financial sector by the Australian Government in 1985, similar developments occurring in other countries including the UK. Increased competition resulted between lenders and it is almost certain that the desire of lenders to both retain and increase their market share was a factor in at least some of their lending decisions. Their desire at that time to make their mark in the competitive market must have since been regretted by many.

As observed earlier, it appears that developers and investors did not pay sufficient heed to the lack of any really positive economic indicators in the Australian economy, while few forecasters in Australia or elsewhere foresaw the very significant and almost worldwide recession (which in some countries was of depression proportions) which was to occur from about 1989. A combination of the perhaps over-optimistic attitudes of investors and developers, their failure to adequately consider the significance of general economic factors, the trend towards very large development schemes, the apparent ease with which finance could be obtained from lending institutions, the reliance of many schemes upon highly-geared financial arrangements (that is a high percentage of borrowed funds), and the excusable inability to foresee the impending and significant world economic slump, were significant elements in the collapse of Australian property markets which occurred at the end of the decade. Such observations could also appropriately be made with respect to events in other countries including the UK.

The results of this collapse were: greatly reduced rent levels with high vacancy rates in most types of property; drastically reduced capital values which resulted from a number of factors including lower rents, higher capitalisation rates caused by reduced investor confidence, and reduced competition amongst investors; the bankruptcy of numerous entrepreneurs and companies caused by the collapse of property values; the recall by banks and other lending institutions of loans against property investments and developments where recall was practicable, and their not unnatural reluctance to be further involved in property funding; and the abandonment or suspension of many planned or partially completed developments.

All of these factors created property market scenarios which were a far-cry from the heady days of the boom period of 1988 in particular. Some of the leading lending institutions in fact became major property owners, a role they never envisaged for themselves but one which circumstances forced upon them, it being necessary to take over properties in the hope of retrieving through ownership the finance lent on them.

Faced with situations in which mortgagors were often not only unable to make periodic repayments of capital but also unable to pay interest charges when due, the sale of mortgaged properties was in many cases impractical because of the absence of potential buyers, or at least buyers at what were considered realistic prices. In fact the absence of buyers in part resulted from a change of attitudes on the part of the lending institutions themselves which, having burnt their fingers very badly with much of their previous over-generous property lending, became so opposed to making further loans on property investment that they assisted in restricting the number of potential purchasers in the market.

From having been almost everyone's favourite investment medium, property became a dirty word in many quarters. The poor world economic conditions resulted in poor industrial and commercial prospects which restricted the number of potential tenants, the restricted demand coupled with the glut of property on the market in turn creating a very depressed scene with low capital values. Indeed, so great was the fall in capital values that those for some modern buildings were only about half the equivalent cost of developing a suitable replacement.

The outcome of such a situation is inevitably the virtual cessation of development activity, and in due course as and when world economies revive, it is almost equally inevitable that in some countries and localities there will be shortages of property on the market once existing surpluses have been taken up. Among the difficulties facing valuers and property consultants are predicting exactly when market conditions will improve, to what levels values will in due course rise, and how rapidly they will rise. Many commentators predicted that in Sydney it would take as much as 10 years for the 1992 surpluses of commercial accommodation to be let, while few were predicting less than five years for this to occur. This difference of opinion of about five years is itself of major significance, as with a development period of about five years needed for most modern developments, a five-year error in the commencement of future new developments could in due course help to contribute to a future shortage of accommodation (or another surplus in supply).

Another result of the market collapse was that there were in 1992 many very inefficient property investments. Many developments which were commenced at a time of rising rents and rising capital values, actually came to the market when rental values had collapsed and potential tenants were scarce and only prepared to pay very much lower rents than originally anticipated by the developers and investors. Indeed, not only did many investors have to accept low rents to obtain tenants, in many cases they also had to make other concessions to those tenants. Such "market inducements" included periods of rent-free occupation, generous fitting-out allowances for the new accommodation, and even the purchase of tenants' interests in their old accommodation, the landlord being left with the problem of disposing of the premises to which their new tenants were previously committed. This was a far cry from the market scenario in which most new developments were planned.

Inevitably such happenings left some developers with reduced or even no income from developments, while capital values were regularly far lower than anticipated both because of the low income flow and the absence of potential investment purchasers. Indeed, there were so many unknowns in the market that it was in fact often quite impossible for valuers to determine whether there were any potential purchasers at all for properties, let alone to make realistic estimates of what the market value for a particular property interest might be.

One of the unfortunate outcomes of this situation was that valuers were frequently accused of being incompetent, and in many cases they were often blamed for having caused developments and property investments to occur which later proved to have been unwise and to have produced losses. Many commentators with the benefit of hindsight expected valuers to have predicted the recession which others failed to foresee, while some even held them to be major contributory factors to the recession.

Whatever else may have happened, the property market collapse focussed attention on the role of the valuer and identified a number of major problems for valuers and the valuation profession to consider in the future. Among these factors and problems are:

1. The difficulties caused by the limited availability of reliable market information, particularly at times when there is either limited market activity or when major changes have occurred in the market;
2. The determination of the appropriate method of valuation for specific valuation situations and the correct application of each method;
3. The need for consistency in the application of valuation principles and methods, particularly when valuations are to be made for asset valuation purposes;
4. The implications of modern computing techniques and the application of discounted cash flow approaches to valuation;
5. The deregulation of financial institutions and the growth of international property financing;

6.   The growth of international property investment;
7.   The problems of valuing properties subject to unusual leasing arrangements, eg where inducements have been given to tenants or where there are turnover rents;
8.   The problems of valuing very large property developments;
9.   The problems caused by the development of new "vehicles" for property ownership and property investment;
10. The valuation problems caused by the increased environmental awareness of society and by such specific problems as contaminated (or poisoned) sites;
11. The need for valuers to take macro-economic considerations into account in addition to local considerations;
12. The role of valuers as a profession and the threat from other professional groups.

Each of the above factors needs careful consideration by the valuation profession if useful lessons are to be learned and if valuers are to avoid in the future some of the criticism (often unjustified) which has been levelled at them in the past. The world changes constantly and many changes occur very rapidly, so it is also necessary for valuers to consider the relevance of past practices which may have only a limited role to play in the future, even though their wider use may have been quite acceptable in the past. It may also be the case that a number of fundamental valuation principles may need to be more keenly observed in the future if some of the problems experienced in the 1980s are not to recur.

Having identified the above factors each of them will be briefly considered in the next chapter, although it should be understood by the reader that each of these topics offers scope for very much fuller consideration than is possible in this volume.

Chapter 24

Special Considerations for Valuers

As indicated in the previous chapter, recent developments in property markets have highlighted a number of considerations to which valuers will have to pay special attention in fulfilling their duties as professional advisers, and it is appropriate to briefly discuss each at this stage, even though many of these matters have already been touched upon in this volume.

The limited availability of market information

The relative scarcity of good market evidence makes it important for the property professions to seek to co-ordinate the collection of market evidence and to ensure that enough detail of each transaction is available to make the information truly useful to valuers.

The question of confidentiality has regularly been an obstacle in the past to making market information freely available to valuers other than the one doing a particular valuation, that valuer normally being bound by the contract with the client not to divulge information of the transaction. However, some data has been available from such sources as auction room results and press releases, while a number of commercial organizations are collecting market information and making it available on a regular basis. The announcement that from 1 April 2000 the Land Registry in England will record the price paid for all property upon registration "when it is clear from the transfer that a sum of money has been paid, whether or not there is also other consideration", should make more comparable information available to valuers in England, but such evidence, as in all cases, will have to be treated with caution. As a result there will hopefully in the 2000s be an increasing availability of market information which will be easily accessible through computer networks.

Such data will be of great assistance to the valuation profession, but great care will continue to be needed by valuers in the analysis and use of the available information, and in particular in the steps taken by valuers to add to the data bank information found by personal enquiry and discovery. The continued development and extension of such data banks will be an important task for the valuation profession and will greatly assist the achievement of an improved level of reliability in valuation. However, it will remain true that the real skill in valuation will continue to lie in how available evidence is used, all of it being evidence of past market events rather than evidence of what the future will be.

The choice of valuation methods and their application

Different valuation methods involve valuers making different assumptions and the use of different information, and the choice of one method instead of another can result in a different figure of value being produced.

Where, for instance, there is development potential in a property the valuer can choose either to use comparable evidence of prior transactions in development land to find a value for the subject land, or he or she can choose to value using the Residual (Hypothetical Development) Method of valuation. Both methods can be defended in use and there are different valuation requirements which may at different times make each preferable over the other.

However, it must never be overlooked that development sites can vary enormously in a number of ways including their physical characteristics, the availability and cost of providing services, and the type and density of development which can be placed on each. Time is also a very important factor and underlying economic considerations in particular can vary enormously with the passage of time. Accordingly, the use of comparable evidence for the valuation of development sites carries with it a lot of dangers and such evidence has to be used with great caution. As an example of the problems, some developments actually lose money and the use of the figure paid for the site of such a development as comparable evidence might in reality represent the use of an excessively high land value which resulted from the over-optimism of the developer. To use such evidence could in fact result in the continuation of over-optimistic valuation.

With the large-scale and very expensive developments which have occurred in recent years, with the very large dependence on loan finance, and with the frequently changing economic and property scenarios, the choice of the wrong valuation method or the careless application of a chosen method by the valuer could result in a widely wrong valuation figure being produced. It is therefore stressed that the assumption that where comparable evidence exists one automatically uses the Comparative Method of valuation needs to be carefully examined in the light of actual circumstances. While it may have been true in the past that such an approach was a safe valuation approach, this may not necessarily be true in the rapidly changing world of today.

Just as the skilful batsman chooses his strokes to suit such things as the condition of the pitch, the weather, the quality of the opposition and the state of the game, so in today's complex business world the valuer needs to be a careful and systematic analyst who chooses and applies a valuation method which is completely appropriate for the circumstances in which a valuation has to be made.

The need for consistency in the application of valuation principles and methods

At first glance there may appear to be conflict between this section and the previous section, but in reality there is consistency rather than inconsistency. There are certain situations in which valuations are depended upon by a large number of people who in fact never actually see the properties valued but whose investment decisions and financial fortunes may be influenced by property valuations. The valuation of companies for flotation on the stock exchange and the valuation of the assets of existing listed companies are matters of concern to a large number of potential and existing shareholders, while the methods of valuation adopted and the way they are applied can result in widely differing asset valuation figures being produced. The frequency with which the valuation of assets occurs can also have an important effect on the figures recorded in balance sheets as representative of the value of a company's assets, and accordingly can also affect the recorded value of a company's total assets.

It has been apparent for some considerable time that many past stock market transactions and takeover bids have been stimulated by the inaccuracy of out-of-date assets valuations or assets valuations which are based on the historic costs of acquiring assets rather than on their current market values or values to a company. In many countries there have been moves to legislate to ensure that company reports and accounts do not contain misleading information or information which can easily be misinterpreted. Such moves have contained measures to ensure that valuers involved in the valuation of property interests for accounting purposes value in a way which is likely to produce reliable information and in a way which ensures consistency between the valuations of the assets of different companies. The achievement of these objectives will enable a more reliable comparison to be made of the relative worths of different companies, will produce more realistic and dependable information in respect of each individual company, and will overall enable investors to make judgements based on more reliable information than was available in the past.

These developments have resulted in accountancy bodies and valuation bodies in a number of countries liaising with each other to draw up codes of practice for the valuation of company assets in their individual countries. In many countries there now exist specific guidelines and rules for valuers who are involved in the valuation of assets for accountancy purposes, but the development has progressed even further to the stage where international agreements have been reached and continue to be further developed on such matters.

Guidance Notes and Background Papers on The Valuation of Fixed Assets for Financial Statements have been drawn up by The International Assets Valuation Standards Committee. These contain a wide range of guidance and information on such things as the classification of fixed assets, the valuation of land and buildings, plant and machinery valued with buildings, depreciation of land and buildings and depreciation of plant and machinery. There is a great amount of further information and it is therefore critical that valuers should ascertain whether their valuation figures are intended for use in financial statements, and if so that they should undertake their valuations in accordance with the current guidance notes. Not to do so may result in the production of inappropriate valuations, while it may also leave the valuer open to the accusation of having been professionally negligent.

It is important to understand the objectives of such guidelines, and when The International Assets Valuation Standards Committee issued newly revised standards on 1 June 1994 they were described as representing "the best concensus of 40 participating nations", and the principal objective was described as being:

> . . . to formulate and publish, in the public interest, valuation standards and procedural guidance for the valuation of assets for use in financial statements, and to promote their world wide acceptance and observance.

The second objective was described as being:

> . . . to harmonize Standards among the world's states, and to make disclosure of differences in standards statements and/or applications of Standards as they occur.
>
> It is a particular goal of TIAVSC (*as it was then known*) that international valuation Standards be recognized in statements of international accounting and other reporting standards, and that Valuers recognize what is needed from them under the standards of other professional disciplines.

Familiarity with and observance of the guidelines is therefore essential when valuing assets for financial statements, while it is imperative that the valuer also remains constantly vigilant for changes in and additions to them. The need to do so is important as the guidelines will inevitably change both as a result of the passage of time itself and different underlying circumstances and needs which result, and by the the need to constantly refine rules and guidelines which results from experience in their use.

The implications of modern computing techniques and the application of discounted cash flow techniques to valuation

The incredible development of micro-computing capabilities over the past 10 to 15 years and the ability to purchase very versatile and powerful computers for a few thousand dollars has created a situation in which (as indicated earlier) the average valuer can now do very complex valuation calculations very rapidly. Whereas only a few years ago it might have taken a matter of days to do a development valuation and tests to check its sensitivity to changes in the various inputs, such calculations can now be done in a fraction of the time to a high degree of accuracy.

There is no doubt whatsoever that modern computer hardware and software have greatly increased the capabilities of the valuer, and it is now a relatively simple matter to produce investment valuations using a discounted cash flow format as opposed to the traditional capitalisation format. As a result the valuer can easily incorporate a big range of variables in valuations taking account of factors which before the advent of computers would probably have been ignored simply because of the difficulty of incorporating them in the valuation process.

Clearly the power and speed of computers are great assets to valuers, but it is important for the valuer to remember that the computer should be used to assist the valuer rather than to dictate the way in which work is done. Despite the ability to use the discounted cash flow format it may be that in some circumstances the use of the simple investment valuation (capitalisation) format gives as accurate a valuation as can be obtained. Similarly, it has to be appreciated that with a more complex format and a greater number of variables, the result obtained will only be as accurate as the accuracy with which the method is used and the reliability of the variables incorporated in the valuation.

There may well be occasions on which the valuer is tempted to use the capability to build in a lot of variables in circumstances in which some of those variables cannot be predicted with any great degree of accuracy. If this is the case the answer which will be obtained will be likely to be inaccurate and misleading, and the temptation to use them is probably best resisted by the valuer whose judgement in such cases will be critical. In particular, the future being uncertain, the temptation to incorporate elaborate projections of future rental growth has to be considered with great caution by the valuer.

There can be no substitute for good judgement in situations of this type, and in most valuations the use of sound judgement is very important. Similarly, there is no substitute for accuracy and the more sophisticated discounted cash flow format demands a high degree of accuracy in use and good judgement in determining the inputs.

Having warned of the problems which may be encountered in using powerful computers and more sophisticated valuation methods, it has to be stressed that the modern valuer cannot afford to be without computing skills, neither can he or she afford to ignore the enormous benefits that computers can provide. They offer fantastic data bank facilities, spread-sheet valuation capabilities, report-writing and publishing capacity, all of which are extremely useful to the valuer. It is important that the benefits of computers are embraced by valuers to enable them to offer the type of service and the quality of service demanded in the modern world.

The deregulation of financial institutions and the growth of international property financing

While the deregulation of the financial sector in the 1980s created a volatile market situation which probably had adverse short-term effects, in the long-term the increased market competition should in theory result in more stable markets for finance. Competition between a larger number of suppliers ought to create greater stability in lending rates, although national and international economic considerations could at times counter this.

In retrospect, the short-term effects of deregulation in some countries appear to have included a "price-war" between financial institutions to attract depositors' funds, each institution seeking to offer better deposit rates than its rivals. New financial institutions were encouraged and entered the market, and an extremely competitive but ultimately unstable situation resulted. In the event a number of prominent financial institutions failed in major international markets, their failures contributing to the problems of the property markets to which many of them had lent and in which many of them were actual players.

The valuer should never forget that investment in property is costly in relative terms involving in most cases a lengthy and costly purchasing process, while it is also costly and generally time-consuming to sell a property investment. Most property purchasers intend their purchases to be long-term investments, and the valuer should not be unduly influenced by short-term interest rates, the main concern being the levels of and trends in long-term borrowing rates.

It is a simple but true statement that if interest rates are at very high levels any movement in most circumstances is likely to be downward, while if they are at very low levels they are more likely to rise in the future. All the underlying economic factors must be taken into account when considering such a basic statement and its applicability at any particular time, but it is almost certainly true that during the 1980s many developers and property investors failed to give adequate consideration to what future movements in borrowing rates were likely to be. In considering such matters it is not sufficient to look at what has happened in the past as future trends may be different, while it is also not sufficient to restrict one's study to local or national economic circumstances only.

The ability exists to borrow money on international markets and many take advantage of this facility. Such borrowers are often persuaded to do so by more advantageous borrowing terms particularly lower interest rates, but some who have borrowed abroad have failed to take a sufficiently long-term view of their financial arrangements or to appreciate that their borrowing was subject to economic and financial forces in the country from which they borrowed as well as in their own country. Many have suffered large financial losses, some of which were completely ruinous. Their major problem was that unfavourable changes in foreign exchange rates often resulted in a significant increase in the size of their loan as their currency depreciated against the foreign currency. Many have been left owing far more money than they had originally borrowed with the borrowings also being secured against properties (and other investments) which had depreciated in value because of recession and the collapse in property values.

Borrowers of funds from overseas should be constantly aware of this problem, and if it is possible should ensure that their loan agreement avoids the possibility of this happening, or alternatively they should take other measures to limit their exposure to such risks. People contemplating borrowing overseas should also remember that while short-term overseas interest rates may be preferential at any point in time, it is only if overseas interest rates are likely to be lower than local rates for a long period of time that any real benefit will result from overseas borrowing.

We live in an age of international business and international money markets, and national financial policies are very much influenced by what happens on the international stage. The property valuer needs to be as conversant with the macro-economic scene as with the micro-economic scene, such is the influence of international financial organisations, particularly since the deregulation of the financial sector in a number of important countries.

The growth of international property investment

Just as the post Second World War period has seen the growth of international industrial and commercial conglomerates, so there has also been the development of international property groups and international property investment. Governments have the power to regulate the degree to which they will allow foreign investment in their country, and in some countries it is completely prohibited. However, in many countries foreign investment is possible and the influx of foreign investment money must obviously affect local property markets and all other things being equal will result in increased market values, at least in the short run.

During the 1980s the operations of foreign property investors (especially Japanese investors) were a big influence in international property markets. Some of these foreign investors had large accumulated funds to spend, and coming from countries in which their domestic investment yields were low they were prepared to outbid most others in international markets. Wherever they were active they were an important influence in the increased property values which occurred in some markets.

Foreign investment in property is likely to continue to be important in the future, and there are a number of factors which valuers should always remember when it is under consideration. It is probable that some of these matters were not adequately considered in the 1980s and that they were contributory factors to the problems in a number of property markets.

As with all investors, foreign investors will invest because they see financial benefits in so doing. They will, however, not necessarily have any particular allegiances to the country in which they invest other than those related to financial gain, and they are therefore likely to be what could be termed as fickle investors. If better returns can be obtained elsewhere or if more onerous investment restrictions are introduced in the country in which they have invested, it is quite likely that foreign investors will divert their investment funds elsewhere, with consequences for property values. They will also be very much influenced by the economic situation in their own country being inescapably subject to two sets of economic forces which will be very influential in determining their future investment policy. If there are economic problems in their home economy it is a not unnatural decision for such investors to seek to ensure that all is well on the home front first, which may result in a decision to realize their overseas investments.

For the above reasons it is suggested that it is wise to realize that foreign investors may in fact disappear from the market almost as rapidly as they appear, and that they may do so not because of any problems with their international investments or the foreign economy, but entirely because of problems elsewhere or the existence of better alternative investments.

Properties subject to unusual leasing arrangements

In the depressed market in the early 1990s, in order to let their properties many property owners agreed leases which gave lessees benefits not usually included in leases in normal market times. New lessees were given rent free periods, often several years in length, generous fitting-out allowances and other inducements to persuade them to take leases.

Valuers may as a result have to value properties which have been subject to such arrangements, and if so they will have to determine whether there is a liability attached to the property which reduces its value. Where lessors have taken over leases of other property, generally the liability would be a personal liability on the part of the lessor which would not adversely affect the value of the newly-let property. However, valuers will need to carefully inspect the leases of properties which they are valuing to ascertain whether there are rent free periods to take into account or financial liabilities of any other type which will necessitate a deduction from what would otherwise be the capital value of an interest. Rather than relying upon details supplied by a client or a solicitor with respect to such information, the wise valuer will insist upon actually inspecting lease documents, which should in any case be normal practice.

It will also be necessary to inspect leases to determine the precise implications of any rent review clauses and to check whether there is a "break clause" which gives the lessee the right to withdraw from a lease at any stage. It may well be that the terms of such clauses create uncertainty regarding the level and period of receipt of income flows, and if this is the case the valuer will have to make due allowance in his valuation. Market conditions at the beginning of the 1990s were such that it was no longer safe to assume that income would continue to flow at at least the current rental level, nor was it safe to assume that a property would continue to remain tenanted or that if vacated it would be easy to relet it at or above the current rent level. Although such a situation does not exist in many developed countries in 2000, it is far from impossible that it will not recur in the future.

Indeed, there will be situations in which the valuer will have to value on the assumption that future rent levels will in fact be lower than those currently reserved on a property. The valuer will also have to remain vigilant for possible changes in law which might outlaw rent review agreements which only allow the upward revision of rent under leases, or for changes which might even forcibly permit the downward revision of rents at review dates.

The valuation of large property developments

In the 1980s in particular there were many property developments which were large both in size and value, and there may be difficulties in valuing such properties. Often such developments were what could be termed as "user specific" and "investor specific", that is they were built to suit the specific user requirements of one organisation and were then purchased by one particular investment organisation whose investment needs had been determined before the development was commenced. Developments of this type are therefore very specialised, and the problem often exists that if the original user or the original investment owner no longer wish to retain their interests, there may be very few or even no alternative users or investors interested in purchasing the property interests.

There is no simple solution to a valuation problem of this nature and the valuer will have to research the market thoroughly to try to determine exactly which potential purchasers there are and how much they are able and prepared to pay for a particular interest in a property of this type. It has to be realised that because of the amount of money involved or the characteristics of the property there may at any point in time be no potential purchaser, or alternatively there may be such limited competition among would-be purchasers that only a restricted sale price is likely to be realised.

When valuing very large properties the valuer should remain aware of this possible problem. The long-term value of a property development of this type may in reality be more dependent upon the existence of future potential tenants and investors than upon the existence in the short term of a first-time tenant and first-time investor. The valuer should therefore seriously consider whether there are in the longer term likely to be other potential users and investors over and above those first committed to a new development. If there are not, such a development assumes the nature of a very specialised property which may as a result have very limited value to anyone other than the original user and the original investor.

New "vehicles" for property ownership and property investment

During the 1980s attempts were made to try to make large properties more easily marketable by such devices as "unitising" them; that is creating a number of units in each property with each unit being marketable separately. Although the concept was excellent it did not overcome the fact that units in a property have the same disadvantages that part-shares in any property have, that is the unit owner has rights which are severely restricted in comparison with the rights of a freehold owner. Unfortunately the time when such disadvantages are most apparent is when market conditions are bad and demand is restricted, and at the end of the decade it was quite clear that the value of a unit in a large property was not necessarily even as high as the proportionate part of the current market value in the entire property.

The limitation of the rights of the unit holder took away some of the main attractions of investment in property. The lack of an established market system for trading in units was another disadvantage, while the very specialised nature of units and the fact that in comparison with many other forms of investment they were still relatively expensive, meant that they were very much harder to market than stocks and shares being within the means of far fewer potential purchasers. The problem of how to value units with reasonable accuracy is likely to remain unless many more property interests are unitised and active markets for such units are developed somewhat similar to stock exchanges.

There are other proposals for the division of interests in large properties in the hope of making them more marketable and of spreading ownership, but care will be needed to ensure that any such developments do not in themselves destroy the essential attractions of property as an investment medium. As such developments occur, and they are highly likely to, the valuation profession will have to address itself to the problems of valuation which will undoubtedly accompany them.

Increased environmental awareness and valuation

Two problems which valuers have had to consider in recent years have been the effect on property values of the existence of asbestos in some buildings and the contamination of many sites as a result of previous usage involving the use of such things as chemicals or petroleum. These are considerations which would not even have been contemplated about 15 to 20 years ago when there was not a general awareness of the problems to health and safety caused by asbestos and contaminated sites.

However, an increased awareness in society of such dangers means that the valuer now has to be vigilant to identify the existence of dangerous materials in buildings and sites, and then has to make due allowance in the valuation process for the cost of remedial work. Because the valuation process involves discounting future income and expenditure to the present, the valuer may need to go further than simply allowing for currently known liabilities of this type. Allowances may be needed in respect of probable expenditure which is likely to result from expectations of improved standards of safety and environment in buildings. These may merely be the result of changing consumer expectations, but they could in fact result from the threat of improved standards imposed by legislation. Such a possibility is not unrealistic as the history of the built environment in developed societies reveals ever more demanding planning and building codes in respect of such matters, and the future is likely to be no different from the past in this respect.

Expenditure which proves necessary because of problems of this type can be very considerable indeed, while the blight caused by their existence can make properties affected by them completely unsaleable in their existing conditions. Even when the problem has been addressed by the owner, there may still be a residual stigma which reduces what would otherwise be the value of a similar unaffected property.

The valuer may in fact sometimes be faced with a market situation in which there may be no willing buyer for a property because it has problems of this type, the only course of action being for the current owner to remedy the problems prior to offering the property for sale. If this is done it may be that the cost of remedial works exceeds the cost which is ultimately realised for the property, creating a situation in which the property in its affected state has in reality a negative value.

If this is the case the valuer has to accept that the value in its current state being negative, the concept of willing buyer in the market place has to include the "buyer" of a property interest who will only accept that interest if paid a premium to do so. Put simply, a "buyer" has to be paid by the seller to accept the liability the seller currently owns, there being no buyer who will willingly pay for the property in its current condition.

Valuers will need to be ever conscious in the future with regard to problems of this nature and will need to carry out very careful property and site inspections to identify whether they exist. Failure to identify such problems will be likely to render a valuer subject to a claim for professional negligence, as will failure to make a sufficiently large allowance for rectification works when problems have been identified in a survey. Indeed, when problems of this type are identified by valuers they will be well advised to consult experts in environmental matters or to recommend their clients to do so.

Macro-economic factors and valuation

As indicated earlier property markets are regularly differentiated by geographical differences and by differences in types of property. In this respect property markets tend to be both local in nature and specialised in nature, so there are for example the Sydney Central Business District office market, the Melbourne industrial market, the Paris office market, and the London hotel market. Traditionally the major factors influencing property values have tended to be local factors, but over the past 15 to 20 years there has been a significant increase in the importance of national and international considerations in property markets.

The development and ownership of property require considerable sums of money however small the property may be, and most developers and owners are dependent upon borrowed funds for at least part of their property commitments. Often the majority of money invested in property development or purchase is borrowed, in recent years borrowed funds often contributing 90% or sometimes even 100% of the required capital amount. The cost of borrowing money and the terms on which it can be borrowed become particularly important in circumstances such as these, but both are very much dependent upon national political and economic considerations rather than upon local considerations.

In turn national economies have become very much interdependent with the growth of international trade, the development of international business conglomerates, and the increase in the number of financial organisations which operate internationally rather than just on a national basis or with only limited international activity.

The result is that investment in property is very much affected by international and national economic and financial factors. Changes in interest rates in Europe which result from changed economic scenarios there, may in fact have an impact on property market activity in other parts of the world by affecting the availability of money and the cost of borrowing, and also the level of foreign investment activity in those markets. Similarly, measures introduced by a government in an effort to control such things as inflation and the unemployment rate are likely also to affect the property market, especially as they are likely to affect willingness and ability to invest in property.

Other macro factors may also affect the property market and if, for instance, it becomes easier or cheaper to invest in the stock market or if the Government or any other organisation introduces attractive alternative investments, people may be less willing to invest in property and property prices may be adversely affected.

The levels of commercial and industrial activity are also important as they can affect the ability and willingness of people to buy property, while such things as the make-up of the population and its purchasing-power and trends in population will be critical factors in determining the viability of proposed new property developments.

As already indicated it is therefore important for valuers to be conversant with economic and financial matters at local, national and international level if they are to identify those factors which are critical to current and future values in property markets. Regular reading of the financial pages of newspapers and of financial and business publications will provide invaluable background information, while there are now numerous sources of statistical data in government departments and commercially organized research companies.

While valuers may have been able to survive in years gone by on good local knowledge alone, this will not be possible in the future. As mentioned earlier there were in the 1980s a number of economic indicators which forewarned of likely problems and yet many in the property professions appear not to have responded to those warning signs. In mitigation, it can be be said that those in other professions also appear to have ignored them, but the lesson has to be learnt from these past mistakes if the valuation profession is to operate effectively in the future.

The effect of modern technology on property usage and property values

Within a matter of a relatively few years, modern technology has developed in a way that now enables many business activities to be carried out completely differently to the practices of ten or twenty years ago. We now have incredibly compact portable computers which are as versatile as many main frame computers were about twenty five years ago; compact and versatile photocopiers; facsimile machines which permit very rapid transmission of documents; mobile telephones which are incredibly compact and which are also capable of performing other functions over and above telephoning, and the internet which gives access to a vast amount of information and which provides a versatile and rapid means of communication and information transfer. The result of all this technical development is that business operations which once needed a lot of people and a lot of space in which to carry them out, can now be done much more rapidly than previously using a fraction of the space.

Many business activities can now be carried out using very little accommodation – indeed much work can be done from the seat of a motor car – and many quite complex business activities can be carried out from one room of a residence. The effect that such trends are likely to have on long-term demand for office space is difficult to predict, but it is essential that valuers should carefully monitor the development of modern technology and the resultant effects on the property market, particularly with regard to any effect on the total demand for space, the type of space demanded, and the effect on rental and capital values of new methods of doing business.

The development of new technologies is also likely to have an impact on retailing activity, and there are indeed already signs in some countries of the effect of internet marketing on traditional retailing activities. The ability to use the computer screen as the "shop window" may reduce demand for some existing retail units in some types of retailing, the retailer being able to market via the computer direct from warehouse type accommodation located almost anywhere. Indeed, in some types of retailing it may be possible to even dispense with the warehouse, marketing via the computer screen direct from production unit to consumer. There has been recent evidence of this being done with respect to high value Italian motorcycles, in particular the limited production run of Ducati MH 900 Evoluzione motorcycles.

Quite what the impact of internet retailing will be ultimately, no-one really knows at present, but, as with many modern developments, its impact is likely to be significant and it may even be substantial. What is certain is that it is likely to have some impact on the demand for and the value of many existing retail properties, particularly those which are currently in marginal positions for retailing.

New technology is likely to continue to have an impact on industrial and warehouse property values also. Just as older pre-Second World War properties have been superseded by newer style buildings, the development of new methods of manufacture and new methods of storing completed goods may well lead to demand for new types of accommodation or the updating of existing accommodation, with resultant implications for property values.

The competent valuer will have to be ever alert to the effect on values of new technologies and new methods of doing business, which may well bring about changes even more rapidly than over the past twenty to twenty five years.

Property values and the long-term

In the relatively recent past it was not uncommon to regard the long-term in property ownership as being a period of fifty years or more, and most valuers would have regarded the long-term as being a period at least longer than a minimum of about twenty or twenty five years.

Indeed, it was common for lessees in the United Kingdom to take leases of twenty one years or even longer in order to secure their short- and medium-term property needs. When such leases existed, much of the uncertainty relating to property valuation was removed, as with many properties a valuer could study a lease to determine, for example, that there was a lease for twenty years or so with a substantial lessee and an almost guaranteed rent, plus likely regular increases in rent. In addition outgoings would often be easily assessable for the freeholder because responsibility for most outgoings was accepted by the lessee. There was therefore in many cases, real risk attached to such a property investment only in the relatively long-term, that is from about twenty years or so in the future. Because of the effect of discounting and the relatively low amount of total value which would be attributable to a period so far in the future, the effect of any increase in risk on that period was also limited, sometimes even negligible.

From a situation in which the basic structure of many High Street shops and much office accommodation, and indeed even factory and warehouse accommodation, had not changed much in a period of sixty or seventy years, or even longer, changes in technology and the development of new practices have created a situation in which many properties become obsolescent, or even obsolete, in as short a period as twenty years. Under the old ninety-nine year ground leases which were common in the United Kingdom in the nineteenth century, long leaseholders would build properties which lasted the length of the long lease and beyond. Such properties were subsequently regularly used by the freeholder following the reversion of the lease in basically the same form as when originally built, apart from improvements such as the addition of electricity. However, with retail, industrial, and office properties in particular, this is not likely to be the pattern in the future. It is now normal to see provision for major updating and refurbishment of properties at twenty year intervals or less, and it is not unknown for modern shopping centres to be built, knocked down, and redeveloped within a twenty year period.

In such a rapidly changing environment lessees are also likely to be very reluctant to enter into long-term leases, not wishing to commit themselves in an uncertain and changing environment. Valuers are therefore likely to have to do much valuation in an environment in which:

(i)   the long-term is in fact a much shorter period than used to be the case, possibly not much more than fifteen or even ten years distant;

(ii)   changing technology and changes in business environments result in typical leases being very much shorter than in the past, perhaps only about four or five years in duration, or even shorter;

(iii)   the changes referred to in (ii) above mean that property owners must allow regular expenditure for updating their properties or even for complete redevelopment after a period of only about twenty years or so;

(iv)   the overall risk attached to the ownership of property is therefore very much harder to assess than in the relatively recent past when many properties were subject to low levels of risk for twenty years or longer.

Today's "state of the art" property might in fact become a "dinosaur" in as little as twenty years! If the above seems to be suggesting too much change may occur too quickly, how many would have thought that a computer which required a relatively large amount of office space only twenty years ago would be replaced today by one with probably greater capabilities which measures about the same size as a large box of chocolates?

The biggest challenge of all for valuers in the future may be in comprehending and keeping up with the pace of change.

The role of the valuation profession and the threat from other professions

During recent years the valuation profession has come under threat from other professions seeking to provide similar services to those traditionally performed by valuers, especially the assessment of the value of investment properties. With the collapse of property values at the end of the decade, valuers were also subjected to a considerable amount of criticism from many quarters, and many tried to blame them for the collapse in values and for not having foreseen it. It is not altogether surprising that those hurt by the collapse should seek scapegoats, and in the light of these events it is apposite for valuers to seek to ensure that individually and as a profession they offer so high a quality of professional service that they are able to resist the competition from other professional groups while being free from blame when things go wrong.

Valuers have a big advantage over such professions as accountants and investment analysts as their education provides them with much knowledge of the specialist aspects of property with which other professions are not familiar. Valuers should have knowledge of planning control, building technology, aspects of law relevant to property, and property management considerations together with several other relevant areas of knowledge. They should therefore be able to give advice on many aspects of property ownership and use as well as on value, and where the latter is involved they should be the specialists who are best equipped to detect matters which will create or affect value. It is therefore important that valuers should seek to obtain a high quality of initial education and training, and that they should also seek to maintain this high level by regular, relevant updating.

When acting for a client the valuer should ensure that right at the outset precise instructions are obtained. Receipt of a clear, unambiguous brief is important in helping the valuer to provide the client with exactly the service required, while it is also critical in the unhappy event of a claim that the valuer has been professionally negligent. Having clear instructions should help to avoid such a situation and may also provide a good defence in the event of a claim being made. When valuing properties as security for the provision of loan finance, the valuer should only accept a brief from the lender as this will ensure that the valuation is done on the basis required by the lender while at the same time it should avoid the possible suspicion that collusion could have occurred between valuer and borrower.

We live in a world in which there seems with the passage of time to be an ever greater tendency to sue for negligence for the least thing. The best way to avoid such a claim is for valuers to provide so high a quality of professional advice that such a claim is impossible, but even the most skilled people sometimes make mistakes. It is therefore important that all valuers should carry adequate professional indemnity insurance to cover the possibility of claims and to give potential clients the added confidence of knowing that in dealing with a valuer they have the added security of adequate insurance indemnity. The provision of an overall indemnity scheme by the valuation profession is also desirable to avoid the possibility of loss where a miscreant member has not provided insurance cover or has taken out inadequate cover. The credibility of a profession is threatened by its least competent and least scrupulous members, and it is therefore wise for a profession to take action to protect both its clients in general and itself.

The quality of service provided by the valuation profession is what it will be judged by, and it is important that valuers should always act objectively and with complete integrity. They should seek to adopt realistic assumptions where assumptions are necessary, and should be wary of the temptation to incorporate speculative or unrealistic elements of value in their valuations. Meticulous research of background information and careful attention to legal detail is called for at all times, and with regard to these tasks valuers have a great advantage over other professions in that they are conversant with the factors which affect value and with the physical aspects of properties.

The high level of knowledge of property and the property markets possessed by valuers should place them in a better position than other professions to select the best valuation method for any particular valuation task, and also to apply the chosen method. It is likely that on occasions in the 1980s fundamental principles of valuation and investment were sometimes ignored particularly when the discounted cash flow format of valuation was used. The valuer should be best able to adapt this technique to the valuation of property interests in a realistic way to ensure that any dangers in the use of such approaches are avoided while capitalising on the benefits they offer. Above all the valuer should be best able to detect threats to the security of future income, and should also be best equipped to predict future income flows which, it should never be forgotten, are attached to a future which is inevitably uncertain.

While the basic concept that today's value is the sum of the anticipated future income flows discounted to the present remains the fundamental underpinning of all valuation, predicting those future income flows becomes ever more difficult in a rapidly changing world. Records of past transactions will remain the major source of evidence for valuers, but because of the speed of change in the modern world it will be essential for valuers to analyse such records with precision and to use the evidence with circumspection. Comparable evidence is evidence of the past and the future may be very different, and valuers will in the future need to avoid the almost slavish dependence upon the comparable which some valuers seem to have demonstrated in the past. As is to be expected with professional advisers as opposed to technicians, the role of analysis and judgement will be a crucial part of the process of utilising known facts and researched evidence in the valuation process.

The major problems with modern valuation are not really that available valuation methods are inadequate; they lie in the difficulty of determining the inputs to the particular method adopted in any given situation. The assumptions adopted by the valuer and the opinions influencing his or her valuation are in many cases as important to the client as the actual valuation figure determined by the valuer.

It is therefore important that if the profession is to resist threats from rival professions, a high quality of reporting should be provided by all valuers. The report is the only evidence many clients have of the quality of the valuer, while it is important that a client should have a full appreciation of the merits and disadvantages of any property interest which has been valued or inspected for any other purpose. It is especially important that they should be informed of the circumstances in which a valuation figure is appropriate and of any limitations attached to a valuation or a report. Reports should have a high-quality presentation and should be informative, concise, unambiguous and lucid.

While threats to the valuation profession through competition from other professions will undoubtedly continue, such is the level of competition in the modern world, the particular areas of expertise covered by valuers in their education and training should give them a very substantial competitive edge. It is up to valuers in general to learn from the problems of the past and of the 1980s and 1990s in particular, to modify and improve their techniques in the light of the lessons learned, to adopt modern technology and methodology where appropriate, to increase the level and quality of research undertaken, to operate in a professional way with objectivity and a high level of integrity, to provide high quality reports for clients, and above all to respond to the changing demands of society while supplying a high quality of professional service.

If valuers strive to do these things the valuation profession will have nothing to fear from other professions and it will firmly establish itself as an invaluable contributor to modern economic and social development.

Chapter 25

Conclusion

The valuation menu has been studied and it is hoped the reader considers he or she has also tasted an appetiser. If enough has been gleaned from this book to encourage further pursuit of the study of valuation, and if it has helped him or her to acquire a reasonable understanding of the topics covered in it, the author will consider it to have been well worth writing.

Appendix

Valuation Methods
Reconciliation of Various International Valuation Methods

Great Britain	Australia	U.S.A.
Comparative Method	Direct Comparison	Market Data or Sales Comparison

The most widely used method of valuation entailing a valuation of property by direct comparison with similar properties that have been sold. This method is used where sale property and property to be valued are sufficiently similar to enable a value to be applied to the subject property on the strength of sales evidence.

Great Britain	Australia	U.S.A.
Contractors' Method	Summation	Cost or Summation Approach

This method is employed in the absence of comparable sales or earning rates. Usually a last resort method which assumes a relationship between cost and value. *Equation*: Cost of site + cost of building – obsolescence – depreciation = value of property. (Used in valuation of hospitals, schools, police stations, etc where no comparable sales exist.)

Great Britain	Australia	U.S.A.
Investment Method	Capitalisation	Income or Residual Earnings Approach

This method is used where there is a direct relationship between anticipated annual income and sale price for comparable properties sold and the property to be valued. Sales analysis will provide a factor or multiplier by which anticipated annual income is multiplied to give the capital value. Discounted cash flow techniques can be used.

In Great Britain and Australia the factor or multiplier is known as the "Years' Purchase".

Income can be capitalised net or gross.

Capitalisation rate = Income expressed as a percentage of Capital Value.
Equation: Value = Income × year's purchase

"Direct Capitalisation":
Value = Income × Factor.
"Yield Capitalisation":
Future benefits converted to present value with a required profit rate.

Great Britain	Australia	U.S.A.
Investment Method	**Capitalisation**	**Income or Residual Earnings Approach** "Residual" techniques. Here a component or part is known or can be estimated. Income for this part is deducted from total income leaving a residual income which is capitalised to give the value of the unknown portion. "Building Residual" technique is applied when land value is known. "Land Residual" technique is applied when building value is known.
Residual Method	**Hypothetical Development**	**Land Development or Cost of Land Production Approach**

This valuation approach is employed for property with development or redevelopment potential. Equation: Value of completed development — total expenditure (including developer's profit) = Present value of land. DCF techniques can be employed.

Great Britain	**Australia**	**U.S.A.**

**Profits (or "Accounts")
Method**
Used where there are no
comparable sales and there
is some degree of monopoly
either legal (licenced, etc)
or factual.
Equation:

Gross Earnings
Less Purchases
= Gross Profit
Less Working Expenses
(except rent)
= Net Profit

Rental value determined
from net profit. This method
used for hotels, cinemas, etc.

Productive Unit
The market may provide
an indication that buyers
will pay a price for certain
units, ie the sale of orchards
may indicate a certain price
per tree plus structural
improvements; with rural land
"Sheep Area" values are
produced by the market.

Compiled by David Lloyd, Avle (Val).

Index

"Okay, that's fair enough. I just need to hear your input. Then you can tell me how disappointed you think my very religious parents will be."

"I don't think your parents will be disappointed at all, especially since they like Jaylin."

"You mean as much as they used to like Jaylin. I told Mama about what happened and she told me to move on with my life. Daddy came by to see me the other day, and when I cried on his shoulder about our ups and downs, he wasn't too happy. Actually, he said he was going by Jaylin's place to have a few words with him about how he's been treating me lately. Of course, I stopped him."

"Your parents love you. They'll understand. You're a grown woman, and I don't think they're going to be disappointed in their thirty-year-old daughter for having a baby out of wedlock. You have a good job, and you'll definitely be able to provide for this baby. As for you and Jaylin, don't tell him."

"Why not?"

"I mean don't tell him right now. Wait a while. And if he starts showing you some love without knowing you're pregnant, then work things out with him. If he doesn't call or come around, then raise this baby by yourself and do the best you can. The worst thing you can do is let him think you had this baby just to trap him. If he thinks that, you're going to hear about it for the rest of your life."

"But, Pat, you know I didn't get pregnant on purpose. When he finds out, Jaylin is going to be excited."

"I'm not saying he wouldn't be. But you know how some men are. Always thinking somebody's trying to trap their ass when they're right there making that baby with you."

called my doctor to make an appointment, and then called my boss and told him I would be late. The thought of food poisoning crossed my mind because for some reason, the chicken didn't taste right to me.

I arrived at Dr. Beckwith's office in the Central West End about nine-thirty in the morning. Immediately, the nurse called my name, so I didn't have to wait long. When Dr. Beckwith came in, he asked me all kinds of questions. When I told him what my symptoms were, he said it didn't sound like I had food poisoning and told me he wanted to give me a pregnancy test. Since I hadn't missed my period, I knew it wasn't possible. Besides, Jaylin and I only had sex one and a half times. Still, I knew one time is all it takes, so I anxiously waited for the results.

Dr. Beckwith came back into the room and pulled a chair next to me. He had a smile on his face as he slid his pen along the side of his ear.

"Nokea, I have good news and more good news. Which one would you like to hear first?" I'd been with Dr. Beckwith since I was a little girl, and he always joked around with me when something was wrong.

"Well, Dr. Beckwith, if it's double good news, then let's hear it."

"First, you don't have food poisoning, and second, you're going to have a baby."

The grin on my face vanished.

"Wha . . . what did you say?"

"Yes, Nokea, you're pregnant. And we're going to do everything possible to make sure you have a healthy baby."

I was speechless. When Dr. Beckwith left the room, I dropped my head and burst into tears. I never thought I would have a baby out of wedlock. Mama and Daddy wouldn't be happy about the news. I knew they'd be disappointed in me and Jaylin.

For the last few months, I had really been a disappointment to myself. Why did I have to make so many messed-up decisions? Decisions that cost me big-time.

Dr. Beckwith's nurse came in and congratulated me. She gave me a hug and immediately noticed that I'd been crying.

"Are those tears of joy?" she asked, helping me off the examination table.

"No . . . I don't know. I'm confused right now. Really, I don't know how I feel."

"I know it comes as a shock today, but once you get home and think about how much happiness this baby is going to bring to your life, you'll feel a whole lot better. It's normal for you to feel the way you are. Just don't go making any decisions until you've had time to think about it."

"Thank you," I said, giving her another hug.

I called my boss and asked for some personal time off. Hearing how anxious I sounded, he didn't seem to have a problem with it.

I drove down Euclid Avenue and thought about how Jaylin would feel about this. I knew how much he loved his daughter, who disappeared with her mother years ago, so I was positive he wouldn't have a problem loving the baby I carried. Would this baby finally change our lives? Was this a sign from God we needed to be together as a family? The big question was, when would I break the news to him—or would I do it at all?

I needed advice, so I went to Barnes Jewish Hospital on Kingshighway, where Pat worked, to see if she would take an early lunch with me. She told her boss

it was urgent, grabbed her purse, and we heade
Pasta House.

"So, why are you dragging my butt out of the
like this couldn't wait until I got home?" Pat a
The waiter poured our water and handed us men
wanted to wait until he was gone to answer Pat's qu
tion.

"If you don't mind," I said, "give us about ten min
utes and we'll be ready to order." The waiter nodded
and walked away.

"Okay, Nokea, out with it. He's gone, so what's on
your mind?"

I reached my hands across the table and held hers.
"Pat, I'm having a baby. The doctor confirmed it this
morning, and I'm confused about what I need to do."

She squeezed my hands tighter and yelled. "Girl,
I'm so happy for you!" Her voice lowered. "But please
don't tell me it's Jaylin's baby. I know he's the only one
you've been with, but just make up somebody, please."

I laughed with her.

"Girl, you know I can't lie like that. You know its
Jaylin's. The question is, what am I going to do? I
haven't called him in weeks, and I've been work
hard trying to get him out of my system. And
when I thought things were going well, bam—
pregnant."

"I know I'm your best friend, but when it come
Jaylin, I'm not one to give you advice."

"Yes, you are, Pat. You've always given me good
vice. I just never do what you tell me."

"Well, I'm going to tell you how I see it. If yo
cide to listen to me, then fine. If you don't, I wor
mad."

"I really don't care what Jaylin or anybody else thinks. I didn't get pregnant on purpose to trap him."

"Okay, Nokea, do what you want to. If you want to tell him, by all means, do. If it's meant to be, then things will work out." That was the best thing Pat said to me all day. Her advice wasn't what I wanted to hear, but I always appreciated her input.

After lunch, I decided to stop by the barbershop to see Stephon. Since he knew Jaylin better than anybody, I thought he might be able to offer me better advice than Pat did.

When I walked in, he was on the phone while working on somebody's hair. He looked at me and smiled, then hurried to end his call.

"What's up, Shorty? I know you didn't come in here to get your hair cut."

"No, I didn't. I wondered if you had a minute to talk."

"Yeah, let me finish this young man's hair and I'll step outside with you to chat. In the meantime, get a soda out of the machine," he said, handing me a dollar bill. "While you're at it, get me one too."

I went to the machine, got us some sodas, and then put his drink on the counter of his workstation. I looked at the pictures lined on his mirror; it was all about him and Jaylin. He had a few pictures of some females, but you'd have thought Jaylin was his girlfriend.

I looked at one picture where Jaylin had on a black gangster hat with a toothpick in his mouth. He looked ghetto, kneeling down with a peace sign held up. It was a good picture, but trying to put on the ghetto look wasn't working for him.

As the fellows in the shop rambled on about women, I found myself a chair and took a seat. They didn't care if I was around; they dissed women so badly that I was almost forced to say something. Just when I was about to intervene, Stephon finished his customer's hair.

"Come on, Shorty. Let's go to my car," he said, opening the door to his BMW so I could get in.

"So, what's so important that you came to see little ole me on the job?"

"First, I want to know if you've talked to Jaylin."

"Yes, that was him I was talking to when you walked through the door. He's in the Ba—" He shut his mouth before finishing.

"Don't stop now. Where is he?"

"Nokea, why you always making me tell you shit about Jaylin? You know how tight we are. I don't want to be caught in the middle of this chaos between the both of you."

"I don't want you caught in the middle either, but I have a serious problem I'm trying to work through right now. So, the more I know what's going on with him, the easier my decision is going to be."

"What kind of serious problem do you have?"

"Where is he? Once I know, I'll be happy to tell you about my problem."

Stephon hesitated to tell me, but I begged and pleaded with him. I expressed how it was in my best interests to find out as much as I could about Jaylin and his new woman. I knew that he and Jaylin were close, but I also knew he thought of me as a friend. Stephon had always shown me that he had a good heart. That made it much easier for me to trust him with my secret. I believed that no matter how close he was to Jaylin, he wouldn't betray me by telling Jaylin the news that was mine to tell or not tell.

He scratched his head then put his hands in his pockets. "He . . . he's in the Bahamas. He'll be back on Monday night. So, what's your problem?"

"Who is he with? I know it isn't Felicia because she came to see me the other day and said she was out of the picture. And I also know he isn't there alone."

"Then I guess you answered your own question."

"Tell me, what is it with him and Scorpio? Is he in love with her?"

"Nope, don't think so. I just think she got a hold on him right now. If you know Jaylin, this phase will be over soon."

"I don't know, Stephon. I see more to it than just that. He's different. He's had this I-don't-give-a-shit attitude about everything lately, and that's a side of him I've never seen before."

"Yeah, he has changed a little bit, but men always get excited about something new in their lives."

"Well, I hope he gets excited about this baby I'm carrying."

Stephon's eyes bucked and his mouth opened wide. "What? Nokea, are you pregnant?"

"Yes, and . . . and I don't quite know how I'm feeling about it. I just found out this morning, and since then I've laughed and I've cried—don't know if I'm happy or sad."

"I . . . I'm very happy for you and Jaylin," he said, reaching over and giving me a hug. "And if you came to ask me if Jaylin is going to be excited about the news, hell yes! He's going to be ecstatic. He misses the hell out of his daughter, and your news will be like music to his ears."

"Yeah, but since things haven't worked out between us, do you think he'll feel differently about having a baby with me?"

"No. And things are going to work out for you two. There's no way you can let another woman stand in your way. When he gets back on Monday, you go right over there and tell him."

"I'll make a nice dinner for him and then tell him. Thanks so much, Stephon. You don't know how much you've put me at ease."

Stephon kissed my forehead and said he had to get back to work. His advice was definitely the kind I needed to hear. He lifted my spirits up so high that I stopped by the mall and bought two outfits for my son. I had a deep feeling it was going to be a boy. I also threw in a bib that said I LOVE MY DADDY and some pacifiers. I couldn't wait to tell Jaylin about the baby. Monday couldn't get here fast enough for me.

18

JAYLIN

Scorpio and I had the time of our lives in the Bahamas. She was good, relaxing company for me; exactly what I needed to get my head on straight. The moment we arrived on the ship, it was on. Men checked her out like she was some kind of beauty queen or something. I had the women all checking me out, too, but not like the men rode Scorpio. And it seemed like the harder they looked, the closer she clung to me. She didn't leave my side. She didn't complain about anything, and she definitely had enough sense not to bring up Nokea or Felicia on our vacation.

Before we left St. Louis, I'd gone to Saks Fifth Avenue and bought her two beautiful evening gowns to wear for dinner. One of them was black with pearls that gathered around the neckline. The back was open and the bottom had a tail-like flare. The other was short and red with a sheer scarf that draped on her side as she walked. The edges were trimmed with rhinestones and perfectly matched the sexy red shoes I bought. The dresses and shoes cost a fortune, but when I saw her in them, she was definitely fit to be in Jaylin Rogers' world.

At our first dinner, we got so many compliments as a couple that she lied to people and told them we were on our honeymoon. For me, that took shit a bit far. When Scorpio said it, I didn't correct her right then and there, but while we were in our cabin, I did.

Other than that incident, she really knew how to keep a brotha happy. Rubbed my feet at night, massaged my body with oil, and washed me up in the shower. Of course, I returned the favor and rubbed her body too, but she did it to the extreme. And the sex— whew—it was on. Every time we stepped foot in our room, we were at it. Could barely get the door open before we ripped each other's clothes off. I had even gotten some at four o'clock in the morning on the upper deck of the ship while mostly everybody else was 'sleep. Scorpio didn't mind being creative, and that was a positive thing.

I enjoyed being with this woman, and frankly, I hadn't thought about kicking her to the curb anytime soon. She'd been taking damn good care of me, and I had no problem splurging my money on her.

Our second to last night in the Bahamas, I went to a jewelry store and bought her a diamond Rolex she seemed to be infatuated with when we browsed earlier. She didn't know I'd purchased it until we got back to our cabin. After I showed it to her, her eyes filled with tears. She made love to me that night like sex was going out of style. Fucked me so good, I damn near cried myself.

But after tonight, our last night of vacation, it would be time to get back to reality. I had to decide where I was headed from here. I was tired of my situation with Felicia and Nokea, and was ready to try something different. Maybe settling down with one woman wasn't a bad idea. For now, Scorpio was giving me all the things I needed, with the exception of cooking for me. But with sex as good as it was, I could hire a chef to cook for me, or very well do it myself. I had some serious thinking to do, and this vacation allowed me time to think about my future.

Scorpio and I put on our swimming gear and went for a late night swim on the upper deck with some of the other couples. She laid her pretty self between my legs as we looked up at the sky and tried to count the stars.

"I counted six hundred twenty, Jaylin. How many did you count?" she asked.

"I only counted ten. Ten over here, ten over there. I don't know. . . . Why don't you help me count?" I picked up her hand and we reached for the sky. We counted together and when we got to twenty, I took her hand and kissed it.

"Jaylin, what's on your mind? You've been awfully quiet today."

"Nothing much. Just thinking about how much I've enjoyed these past several days with you. Thinking about how good it's been to get away from all the bull-shit at home. That's all."

"Well, I'm glad you asked me to come along. I never imagined the Bahamas being so beautiful. And Paradise Island, it's to die for. Just amazing." She leaned back and I put my arms around her waist.

"I'm glad you had a good time too. Next time, though, I'm going to take you somewhere even better."

"Jaylin, please, it doesn't get any better than this. I don't care where I am, as long as I'm with you."

"Aw, trust me, it gets a whole lot better than this. There are places we can go that are more beautiful than you've ever imagined."

"And I'll still say anywhere is great as long as I'm with you. It's just so funny how well I've taken to you. I thought I was in love with you before, but now I definitely know I am."

I looked away, wanting to change the subject. I was feeling Scorpio too, but it was too early for us to be talking about falling in love. I had no control over her feelings, but they were making me a bit uncomfortable.

Scorpio noticed that I hadn't responded so she continued. "I know you didn't want to hear that, but what else am I supposed to say? If I feel a certain way in my heart, why should I have to hide it because those words frighten you?"

"Scorpio, I didn't say that I didn't want to hear it. All I'm saying, as I said before, is love complicates things. I'm just not ready for that kind of relationship, baby."

"You also said you haven't had a woman in your life that has made you love her. If I'm not making you love me, then you tell me what else I need to do. I'm trying, but eventually, my energy is going to run out."

She had touched a nerve. This was one subject I didn't want to delve into on our vacation. "Baby, please. We're having such a good time. Don't spoil it talking this nonsense that don't really matter right now. Can I please just enjoy my last day here with you?"

"Sure. But since you don't want to talk about that, can we talk about what happens when we get back to St. Louis? I really would like to know where things stand between us."

"What do you mean, where things stand? I thought things were cool just the way they are."

"They are, but . . . but I want to be in your life at all times. In fact, I'd like to move in with you. Be with you around the clock to take care of all your needs."

I fell silent again. I hadn't given much thought to allowing any woman to move in with me. Scorpio was really pushing it, and I was concerned about her urgency. I tried not to ruin our vacation, so I remained calm and spoke truthfully. "I don't know if I'm ready for that. When we get back, we'll talk about it then, okay?"

Scorpio laid her head back on me and closed her eyes.

When we got back to the room, I guess she was a bit upset because she went right to sleep without upping no booty. I couldn't sleep, so I slid on my sandals and some shorts and went for a walk on the deck to clear my head. The wind blew a strong breeze, and the high waves splashed against the boat, making it rock.

I rested my arms on the rail and thought about what I planned to do about my situation. I loved being with Scorpio better than anybody, and I knew that giving up my relationships with Nokea and Felicia would give me a fresh start. Still, I had to be careful about my approach because they had always been there for a brotha—up until lately.

Nokea talked about all this time she needed to get herself together, and as far as I was concerned, time wasn't on her side. I was the one who called the shots, so by now, if she hadn't figured out what she wanted to do, then fuck her.

And Felicia? I just wanted to make sure I had some back-up booty if things weren't going cool with Scorpio and me. They were, and even though she was currently upset with me, I knew our problem could be resolved.

I had to seriously think about Scorpio moving in, and not just her, but her child as well. How would I be able to deal with a woman and her child in my home? The downsides would be not having my privacy, not having female company when I wanted to, having a junky house, more mouths to feed . . . my list went on and on. On a positive note, she wasn't a nag, she cared about my needs, and she hadn't forced me to be with only her. Her daughter could possibly help ease some of the pain I still had about not being with my own daughter, but I wasn't sure how well she'd take to me.

Was a man like me ready for this kind of change? This . . . this was too premature, I thought. But then I asked myself, *If not now, when?*

If I didn't like how things were going, I could always just ask her to leave. We could part ways and I could at least say I gave it a try. She couldn't be mad about that.

So . . . I'd let her move in with me. I hoped like hell that I wouldn't regret my decision.

As I stood and watched the water rock the ship,

Scorpio came up from behind, wrapped her arms around me, and rubbed my chest in a circular motion.

"I guess I know what you're thinking about," she said.

I turned around and held her in my arms.

"I'm sure you do know. First, I want to apologize for snapping at you earlier. You didn't deserve that, and I know exactly where you're coming from." I kissed her nose. "Second, when we get back to St. Louis, I want you to move in with me. I know I still have some unfinished business to take care of, but I'll work that out when I get back."

"Jaylin, are you sure? I mean, I do have a daughter, too, you know. She'd have to move in with us as well."

"Yeah, I know. I thought about that too, but it's okay. I can hire a nanny to come in and take care of her during the week. She can have the room I fixed up for my little girl before she moved away with her mother."

"Again, are you sure you want to do this? It's going to be a big change for you, and I don't want you to do this unless you're ready."

"Scorpio, it's time for me to make some changes. I can't make you any promises about being faithful, but at least I'll try."

"I'm not worried about you being faithful to me. I'm going to make you so happy that you're not going to have enough time to think about another woman."

I picked her up and laid her on one of the recliners behind us on the deck. I covered my ass with a towel and gave her insides a tickle, not even caring if other people were around. There were only a few other couples on the deck, but they seemed to have the same idea as we did at four o'clock in the morning.

I hoped I hadn't made a bad decision by telling Scorpio she could move in with me. The worst thing that could happen was I'd fall in love with her—but I had my guards up, and there was no way for a brotha like me to slip.

19

FELICIA

After not hearing from Jaylin, I tried to move on and put my plans on hold to get him back. I left him a few messages just in case he decided to return my phone call, but then decided to stop. I'd become banging buddies with Damion and was making the best of it. After all, he wasn't that bad. I pretended he was Jaylin, and everything was cool.

Because Damion didn't completely fill the void, I'd gotten so lonely that I also made a connection with this white man name Paul from work. He was fine and had entered my life at the right time. I invited him over to my place on Saturday night and he was good company. Our conversation was interesting and flowed like we'd known each other for years. He was such a gentleman and treated me with a lot of respect. Unlike the way Jaylin criticized me for every little thing, Paul paid me many compliments. He talked about how he'd had his eyes on me for a long time, but he wasn't sure if I was interested in dating white men. Told me many of the white men who I worked with wanted to get to know me, and I laughed because I never thought none of them would be interested in me.

After dinner, I thought Paul would try to lay me, but he didn't. He walked himself to the door, gave me a kiss, and said he'd call me the next day.

And that he did. I admired a man who kept his promises. He even had five dozen yellow, pink, and red roses delivered to my house the next day and thanked me for such a wonderful time. I was flabbergasted. The only thing I worried about was rushing things with him. I'd heard the rumors about white men not being as competent as black men in the bedroom, and I was a little afraid of taking our relationship to the next level.

One thing about me: a man must be able to display talent in the bedroom. If not, he'd eventually be history. That's what I liked so much about Jaylin. There was no way a woman would leave his bedroom unsatisfied. He went above and beyond the call of duty. Did whatever he had to do to make sure a woman's needs were met. And if you did go home feeling unfulfilled and he knew it, he'd be sure to make it up to you the next day.

I just couldn't understand why things had to change between us. Our sex life was off the chain until that bitch Scorpio came into the picture. If he hadn't met her, we'd probably be sexing each other up right now. Just the thought of him being with her upset me, but I knew for the time being, I had to keep myself occupied elsewhere.

For lunch, Paul and I went to Café Calimino not too far from work. We tried to keep things on the down-low so no one would find out about us. People always seemed to make a big fucking deal about mixed relationships, and I wasn't prepared to answer any questions. Seemed like everybody and their mama from work came in the café and spoke to either Paul or me. They didn't question us, but we could see and hear all the whispers going on.

I asked Paul if we could leave and invited him over to my place for dinner that night. He didn't seem the least bit bothered by all the attention we got and asked me to stay.

I stood up and left my tray of food on the table. "No, I'm ready to go. I don't like to be watched when I'm eating."

"Felicia, would you please take a seat. Who cares what other people think? I'm enjoying our lunch together, and wish you were too."

"I am too, but would you mind if we take it with us? Maybe be can go to a park or something, but I don't feel comfortable being here."

Paul didn't seem pleased by my actions, but instead of putting up a fuss, he got our to-go boxes and we left.

I left work early so I could go home to prepare a scrumptious dinner for us that night. I stopped by Saveway on Broadway and picked up some coleslaw and catfish nuggets. Everything was perfect. I set the table and put on some reggae music to set the mood. I changed into a red silk thigh-high dress and left off my panties just in case Paul decided he wanted some action.

As I lit the peach-scented candles on the table, the phone rang. I knew it was probably Paul calling to tell me he was on his way, but when I answered, I heard Jaylin's voice.

"Surprise, surprise," I said, filled with excitement just to hear his voice.

"Hey, Felicia, I need to see you tonight. Do you mind if I come over?" He sounded like it was important.

"Uh . . . sure, why not? What time should I expect you?"

"Give me about an hour and I'll be there."

"Okay, I'll see you in an hour."

I hung up and rushed to call Paul to cancel our dinner plans. Before I could, he rang the doorbell. I had to get rid of him. I missed Jaylin too much to turn him away. I'd make it up to Paul some other time, but tonight, my ass belonged to Jaylin.

"Pauuul," I said, smiling as I opened the door. He had a red rose in his hand and gave it to me as he entered.

"Hello, Felicia." He kissed me on the cheek. "You look wonderful."

"Thanks. You look nice too," I said, trying to think up a lie to tell him so he could leave.

"Dinner smells delicious. What are we having?" He took off his jacket and hung it up.

"Some fish and slaw, but I . . . I just got a call from one of my girlfriends. She had an argument with her husband and asked if I could come by and talk to her.

She's one of those emotional-type women, and I'm afraid if I don't go, she might try to do something to herself."

"By all means, Felicia, go. Would you like for me to wait until you come back?"

"No. I'm not sure how long I'm going to be, and I'd hate to have you here waiting for me all night. Can we make plans another time?"

"Sure. No problem. I hope everything works out for your friend. She's very lucky to have a friend as caring as you are."

"Yes, she is. I tell her that every day," I said, handing him his jacket so he could hurry up and leave.

"I'll see you tomorrow at work."

"Okay, Paul. Thanks for being so understanding."

I waved good-bye, shut the door, and ran upstairs to change into something more sleazy for Jaylin. What I had on hid all my good body parts, and if I wanted to compete with Scorpio, I had to reveal something. I put on my purple see-through nightie with a purple silk bra and thong underneath. I sprayed myself with a dash of the Chanel No. 5 that Jaylin bought me a while back, and put the food on the table so we could talk over dinner.

When the doorbell rang, I grabbed the phone and pretended as if I were in deep conversation with somebody important. He walked in looking out of sight. Had a deep tan that made his gray eyes glitter even more. I could tell he was coming from work because he still wore his navy blue tailored suit and multi-colored silk tie.

As he waited in the hallway for me to end my call, I saw him check out the five dozen roses Paul bought me that were all over the living room. I turned and walked toward the kitchen so he could get a glimpse of my butt that he could see so well through my nightie. When I turned to face him, I could see the come-fuck-me look in his eyes. I gave a few more laughs on the phone and told no one on the other end that I'd have to call them back because I had company.

"Jaylin, don't just stand there. Come have a seat. I fixed you a little something because I figured you'd probably be hungry by the time you got here." He stepped into the dining room and took a seat.

"Felicia, how did you cook dinner that fast? It only took me an hour to get here."

"Please. It don't take me long to cook. When you called, my fish was almost finished. All I had to do was prepare the slaw."

He shrugged his shoulders. "If you insist. But look, I really didn't come over here to eat dinner—"

"I know you didn't. It was supposed to be a surprise. I've missed hearing from you. Not only that, I miss being with you. Do you ever think things will be the way they were before between us?"

"Felicia, that's what I came over here to talk to you about. I just got back from the Bahamas, and when I got home, I got all twenty-nine of your messages. Baby, this gotta stop. If a brotha don't return your phone calls, that means he doesn't want to be bothered."

"Ya see, I had no other choice. You kept telling me you were going to call but you didn't. I guess it was because you were in the Bahamas. And I guess I don't have to ask you with who, do I?"

"And I guess I don't have to feel any shame when I tell you I was with Scorpio."

"So, what is it with you and her? Don't I mean anything to you anymore?"

Jaylin looked sternly into my eyes. "It's over, Felicia. I can't continue to see you because it just ain't enough of me to go around. I'm tired of being pressured by you, and I need to start making some sense out of my life. More than that, I need a woman who understands me. You used to be that woman, but lately, you're starting to create too much drama. Drama I can't and won't deal with."

"You mean drama that you've brought on yourself? I can't believe that just like that it's over for you. Do you really think you can go without being with me, Jaylin? Every time you meet somebody else, I get set aside like a week-old piece of bread. Do you think I'm going to let you continue to do this to me?"

"This time is different. I'm with a woman who I have a good feeling about. I didn't feel that way about the other women, and I've never come to you before and asked you to end this either. So, if we end this tonight, then no, I won't continue to do this to you."

I shook my head from side to side. "I can't believe your sorry ass. This bitch got your mind all fucked up and you have the nerve to come over here and tell me it's over? Just tell me one thing: are you in love with her?"

"I ain't in love with anyone. I'm just trying to live decent for one time in my life, that's all. Scorpio gives me something you or Nokea have never given me, and that's a peace of mind. She don't nag, she don't bitch, and anything I ask her to do for me, she does it. She's been patient with my situation and has never forced me to choose. You, on the other hand, are the opposite. Your mouth . . . it's been good for some things, but it's foul. I don't like for my women to use the type of language that you do, and it irritates the hell out of me.

"For now, Scorpio's the kind of woman I need in my life. Will that change? I don't know. But if it does, I won't be coming back your way any time soon. We're done."

I was lost for words. Here I had spent four years of my life putting up with this son of a bitch and his bullshit, and he had the nerve to step up in here and brag about another bitch? And as for my mouth, how dare he complain when my mouth was responsible for making his eyes roll to the back of his head! I had been nothing but patient with Jaylin, and who in the hell was he to toss me aside like I wasn't nothing? I felt like getting a gun and killing his ass right then.

As I felt my emotions about to take over, I took a deep breath and tightly clinched my hands together. "So, I guess you'll be having this same conversation with Nokea? That's provided you haven't already."

"Yes, I will. I haven't seen or heard from her in a while, but I'll be sure to let her know as soon as possible where things stand between us."

Jaylin didn't seem to be bullshitting this time. I was desperate to somehow make things right between us. I knew that pussy was his weakness, so I stood up and removed my lingerie. I walked over to his side of the table.

"We can end this, but only after you make love to me. Just this last time, please. If you do, I promise you I won't interfere with your relationship with Scorpio.

And whenever she fucks up—because she will—I'll be here for you." I placed my ass on his lap and leaned in to kiss him. He rubbed my ass, but avoided my kiss.

"Felicia, I said it's over. Don't make a fool of yourself, all right?"

"Please don't do this to us," I begged. "Don't leave me like this. We've been through too much."

I leaned in to kiss him again, but got the same response. I couldn't hold back any longer; tears started to trickle down my face. All the years I'd been with him, I'd never let him see me cry, but this time, I couldn't help myself because no matter how bad things had gotten between us, he'd never turned down making love to me.

He ignored my tears and forced me away. "Save the tears, Felicia. It ain't like our relationship was all that anyway. You know you played second best to Nokea for a long time, so cut the act and let me get out of here."

He stood, but I grabbed at his jacket so he wouldn't go. By the time he made it to the door, I managed to pull it off and ripped his shirt.

"Damn, Felicia! What did I tell you?" He ripped the rest of his shirt from his chest and threw it on the floor. "Come on! Let me fuck you! Even though things aren't going to be different tomorrow, let me just give you what you want so you can get the fuck off my back!" he yelled, and pushed me against the wall.

He pulled my braids back and gave me a wet kiss on the cheek. I knew he didn't really want to be here, but his angry aggression excited me. He pulled the string on my thong so tight that it tore and hit the floor.

Then he unzipped his pants, and as they fell to his ankles, he lifted me and forced himself inside of me. I held his neck tight as he pounded my back against the wall.

"Are you happy now, Felicia? Is this what you wanted? Is this all the fuck you wanted?" he said, pounding me harder.

"I want you, baby! That's all I want is you." I continued to cry as the feel of him sliding against my slippery walls excited me.

He took charge, and after I came, he slowed down his pace. He rested his sweaty forehead on my shoulder while he held my legs apart in his arms. He then gave me a quick peck on the cheek.

"I know how badly you want me, but you can't have me anymore." He dropped my legs and I eased to the floor and watched as he slid back into his pants. He stood proudly with a smirk on his face. He had no sympathy for me.

"Get out, Jaylin!" I yelled as I helped myself off the floor. "Get the fuck out!"

He nonchalantly gazed at me and then continued to straighten his clothes while looking in the mirror. I rushed to the dining room, picked up the plates on the table, and threw them at him. I wanted to cut his motherfucking face up.

"I hate you!" I yelled as I continued to throw damn near every piece of china that was on my table. My aim wasn't worth a damn, but he ducked a few times as he tried to unlock the front door.

By the time he slammed it, food was everywhere. I'd made a complete mess, but it was certainly a good way to let go of my frustrations. I dropped to a chair behind me and pressed my knees closely to my chest. I cried like I'd just lost my best friend, and deep down, I really thought I had. I knew Jaylin wasn't coming back my way anytime soon. All I was left with were memories.

20

NOKEA

I probably jumped the gun, but I'd been to the mall about five more times looking at clothes and furniture for the baby. According to Stephon, Jaylin was back from the Bahamas, but I got nervous and didn't go talk to him right away. I decided to wait until the weekend to surprise him with the news. He was probably exhausted after leaving the Bahamas then going straight to work on Monday. I knew that the more rest he had, the better my news would be for him. I still contemplated when I would break the news to Mama and Daddy, but Jaylin had to be the first to know. If I didn't get it out of my system soon, I'd probably listen to Pat and never tell him.

Either way, when Saturday morning rolled around, I didn't hesitate. I got up, showered, and cooked a fulfilling breakfast so I wouldn't feel lightheaded when I talked to him. Then I hugged the teddy bear he'd given me and put it in the room that would soon have our new baby in it. I drove slowly down Wild Horse Creek Road to his house, thinking about what he would say. We'd probably sit around all day thinking about what to name the baby or discuss who it would look like.

There was no doubt in my mind that Jaylin would be happy—until I turned the corner and saw Scorpio's car in his driveway. My heart raced. It was nine o'clock in the morning, so she must have spent the night with him.

I started to call him on my cell phone, but I was there to tell him about his baby, and so that's what I intended to do. I rang the doorbell because I didn't want to use my key and walk into what I did the last time.

It took a minute for someone to come to the door, and it was Scorpio. She had on Jaylin's burgundy silk robe and looked at me with a blank stare on her face.

"Hi, Scorpio, is Jaylin here?" Since I was uninvited, I tried to show her a little respect.

She didn't say anything; she just opened the door and let me walk in. I went into the living room and sat down on the couch. I immediately noticed toys all over the floor. Moments later, Jaylin came from the swinging kitchen door with what looked to be a four or five-year-old beautiful little girl on his shoulders. I was shocked.

"She's here to see you," Scorpio said as she walked back into the kitchen. Jaylin put the little girl down, and after she kissed his cheek, she ran back into the kitchen after Scorpio.

"So, what's up, Miss Lady?" he said, picking up the toys on the floor.

"No. You tell me. Seems like you got yourself a new family over here."

"Something like that. So, what brings you by?"

I wasn't about to tell him about the baby until I found out what was up with him, this little girl, and Scorpio. Maybe he wouldn't be as happy about my news as I'd thought he'd be.

"Is that her daughter?" I asked.

"Yes. Her name is Mackenzie."

"So, what is she doing over here? And why are all these toys here?"

"Because, Nokea, she lives with me. After we came back from the Bahamas, both of them moved in with me. Is there anything else you'd like to know?"

"What's going on with the two of you? I mean, I see you've made a commitment to her, but you were so unwilling to commit to me."

"Put it like this: it's a commitment I'm making to myself. See, when you were busy trying to decide if you were going to give yourself to me or not, I decided for you. I'm moving on, Nokea, moving on to bigger and better things. Leaving all bullshit behind me, baby."

"In other words, you're ending our nine-year relationship to be with this woman and her daughter?"

"If that's how you want to look at it, feel free. You didn't make my decision any harder. You were the one playing games like you didn't want this, and holding back on the sex didn't help us much either. So like I said, it's time to move on. Now, if you don't mind, I was in the middle of eating breakfast. Do I need to show you the way out or can you find it yourself?"

I couldn't believe Jaylin's tone. He'd never talked to

me like that. He had an attitude like he just didn't give a damn. There was no way I would tell him about the baby now. He'd definitely think it was a trap right about now, and I didn't want my baby being raised with a father who had such a horrible attitude. I got off the couch, unable to even look at him.

"Hey, Nokea?" he said as I walked to the door. "Do you still have the key to my house?"

"Yes." I turned and swallowed the huge lump in my throat.

"On your way out, leave it on the table." He walked back into the kitchen.

My eyes filled with tears as I dug in my purse and removed his key from my key ring. I laid it on the table and shut the door behind me.

I barely made it to my car before I gagged and threw up all over myself. I started my car, jetted down the street, and thought about my only other option: abortion. There was no way I would raise this baby alone. How could I have been so foolish to think he'd be happy about me having his baby? He already had his homemade family and seemed to be just fine with it.

I was so miserable and needed someone to talk to. I didn't feel like talking to Pat; I wasn't in the mood to hear "I told you so." I decided to stop by the barbershop and give Stephon a piece of my mind, since he was the one who assured me Jaylin would be happy.

I went home first and cleaned myself up, and then headed to the shop to see him. By the time I got there, my eyes were so puffy from crying that he stopped cutting his customer's hair to come outside and calm me.

As soon as he walked out, I grabbed him by his shirt.

"Why did you tell me to go see him, Stephon? You

knew how he would react, didn't you!" I said hysterically.

"Hold on, Shorty." He grabbed my arms. "What are you talking about? Calm down and tell me what happened!"

"He doesn't want me anymore! He doesn't want this baby! He asked me to leave and told me to give him his key back!"

"Did you tell him about the baby? I know he wouldn't have told you to leave if you told him."

"After he told me he was moving on, I couldn't tell him about the baby." I wiped the salty tears from my face. "He basically said he was with who he wanted to be with, so there was no sense in me hanging around."

"You should've told him about the baby, Nokea. I knew Scorpio had moved in with him, but I thought once you told him the news, things would be different. If you don't tell him, I'm going inside right now and call him."

"No, Stephon! Please don't call him. Please! I wish like hell you would've told me about her moving in with him. I never would've gone to his house and embarrassed myself like that. Now," I begged, "I don't want him to know anything. You can't tell him, Stephon, please!"

"Then when are you going to tell him?" he asked. "He needs to know that he's got a baby on the way. And I can't make you any promises that I won't tell him."

I looked at Stephon with serious hurt in my eyes. "Stephon, please. Let me decide how to handle this. I don't even know if I'm going to keep this baby."

"Now, that ain't even an option. You're definitely going to keep it. If I'm going to keep my mouth shut, you gotta promise me you won't have an abortion. If Jaylin finds out you did, he's really going to be upset. Not only at you for doing it, but at me for not telling him. So, go home and think about being a good mother to the baby you're carrying. I'm kind of looking forward to having a little Jaylin around."

"Jaylin could care less if I had an abortion, Stephon. He doesn't even care about me, and I don't care what you say, his actions show it."

"He cares, and deep in your heart, you know he cares. I care too, and I want you to relax and think about this before you make any drastic decisions. If you need me, Shorty, I'm here."

Stephon wrapped his arms around me. Whenever I had problems with Jaylin, Stephon always seemed to be there for me. He was the only one who could turn a bad situation into a good one.

I thanked him for lifting my spirits and headed for home with a better attitude. An abortion was out of the question. If Jaylin didn't want anything to do with me, then maybe it was time to move on. I had a bigger thing in my life to worry about now than trying to get him to be with me. This baby was going to need all the love and support a mother could give. If I continued to stress myself with his mess, I'd probably have a miscarriage or something. From what I heard, those ain't no picnic.

I felt so confident about continuing my pregnancy that I stopped by my parents' house to tell them about the baby. I stood on the porch, as they'd just come home from a prayer meeting at church. Seemed like they lived in church. I didn't mind going, but my Saturdays always kept me occupied.

Mama got out of the car and put her arms around

me; she'd said no matter what, she could always tell when something was wrong with me.

"Nokea, are you all right?" she asked as Daddy opened the door.

"Yeah, Mama, I'm fine. I just came over to see how you and Daddy were doing."

"Well, you could've called to see how we were doing," Daddy said.

Daddy and I sat at the kitchen table while Mama poured us each a glass of orange juice. I took a few sips.

"So, how was church? Did Reverend James preach today?"

"Church was good, Nokea. And Reverend James always preaches a good sermon. Even on Saturday. Now, you know that.

"Everybody's been asking about you. Asking when you're going to come back to church. You haven't been there in a while. Why don't you make plans to go with us tomorrow morning?" Mama said.

"I don't know, Mama. I've been going to Pat's church with her. Kind of like hers a little better, that's all."

"Okay, well, when you're ready to visit, let us know."

"I will. I promise you I will." I finished up the orange juice and got up to pour another glass. Mama and Daddy looked at each other.

"Nokea, what's on your mind, girl? You've always been able to talk to us about things, so come on, out with it," Mama said, seeming tired of all my fidgeting.

I stood in front of the sink and took another sip of orange juice. I put the glass on the counter and looked directly at Mama first, then Daddy.

"Mama, Daddy," I said, swallowing. "You're going to be grandparents."

Immediately, Mama smiled and I was relieved. Her reaction was a lot better than I thought it would be. She stood up and ran over to embrace me. Daddy, however, didn't say a word.

"Oh, Nokea, I'm so happy for you," Mama said. "I thought you'd never say those words to me. How far along are you? Do you know what you're having yet?" She leaned back and looked at my stomach. She didn't notice Daddy's demeanor like I did.

"Daddy, is everything okay?" I asked.

"Where's Jaylin?" he asked in a deep, strong voice.

"He's at home, I guess."

"Are the two of you planning on getting married?"

"No. As a matter of fact, we're not. Sorry to tell you this, Daddy, but Jaylin and me never got back together. Because I'm pregnant, I'm not going to force him to be with me."

"No, Nokea, that's nonsense. You're not going to raise this baby alone. If Jaylin helped you make it, then it's his responsibility to take care of it too," Daddy said, raising his voice.

"I'm gonna have to agree with your father on this one, Nokea. You and Jaylin need to try and work things out so you two can raise this child together," Mama said.

"Mama, you and Daddy both know women raise children by themselves all the time. I've already talked to Jaylin, and he doesn't want to have anything to do with me. He's with somebody else, and I can't make him be a father if he doesn't want to."

"Well, whether he likes it or not, I'm going over there to talk to him tomorrow—"

"Daddy, don't. Let me handle my own business. I'm a grown woman and I'm quite capable of making the right decisions."

"Nokea, I thought you and Jaylin always talked about waiting until you were married to have sex. Not too long ago, he promised me that he would wait until you all were married. What happened? Was it all just a bunch of lies?"

"Daddy, believe me when I say we tried to wait. We waited a little over nine years, and one day, things just happened. I don't regret giving myself to him because I'm going to have a beautiful baby. And the two of you are going to be wonderful grandparents. Just please be happy for me, okay?" I gave him a hug.

Daddy smiled and when Mama saw he wasn't tripping, she smiled at me too. By the time I left, they were planning for everything, even for the baby's education. I just threw my hands up in the air, thanked them, and told them good-bye. I thanked God for giving me understanding parents. I wished that my attempt to tell Jaylin about the baby had been this simple.

21

JAYLIN

The nanny didn't show up, and since Scorpio was still asleep from coming in late last night, I worked from my home office. Last night, Scorpio had left to go spend some time with her sister, saying that I could probably use some quiet time. They'd stayed up late playing cards and watching movies with the kids, and I did enjoy a little time by myself. I liked having them around, but I was still adjusting to having a woman and a child in my house.

I couldn't get much done in my office because Mackenzie drove her Barbie around in a toy car on my desk. My papers dropped on the floor, and as I talked to one of my clients on the phone, she disconnected it.

"Mackenzie! What are you doing, sweetie?" I tried

to calm down as I looked at her cute little face that looked exactly like her mother's. I sat her on my lap and she reached out her arms for me.

"Nothing, Uncle Jaylin. I just want somebody to help me play with my dolls."

"Well, as soon I get off the phone, I'll play with you, okay?"

"You said that the last time, but you didn't. Do you promise this time you'll play with me?"

I felt so bad about lying to her that I decided to play with her dolls in the middle of my office floor. I felt like such a damn fool as I put on different outfits for the dolls to look like they had some real money.

When the doorbell rang, it saved me. Mackenzie asked me to drive her Barbies to the mall to buy some more clothes.

"I'll be right back, Mackenzie. When I get back I'll take them."

"Okay," she said. "I'll get their purses so we can go."

I smiled at her and went to the door. When I looked out, I saw that it was Stephon. I hadn't seen him since I returned from the Bahamas.

"What's up, my nigga?" he said as I opened the door.

"You got the best go. What brings you by?" I said.

"I had a few things on my mind that I want to holla at you about. You got a minute?"

"For you, always, my brotha." I walked back into my office. Mackenzie was still on the floor playing with her dolls. I sat in the chair behind my desk and Stephon sat on the sofa. Mackenzie looked at Stephon and climbed on my lap. She put her arm around my neck.

"Uncle Jaylin, who is that? Is that your brother?"

"Naw, sweetie. That's my cousin. We like brothers,

but he's my favorite cousin." She looked at Stephon and looked at me again.

"He looks like your brother. But you're a lot cuter than he is."

Stephon and I laughed.

"Fool, what you laughing at? The girl got good sense. She knows a fine brotha when she sees one," I said.

Stephon chuckled. "Only in the eye of the beholder. You know damn well you ain't got nothing on me." I threw one of Mackenzie's Barbie dolls at Stephon and he ducked. Mackenzie even tried to help, but she aimed at my expensive lamps and damn near broke one.

"Thanks, Mackenzie. Why don't you go upstairs and try to wake up Mommy," I said, taking her off my lap.

"Okay, but will you still play with my dolls when your cousin leaves?" I looked at Stephon and I could tell he was cracking up inside.

"Sure, Mackenzie. I'll play with your dolls." She ran out of the office, excited about my answer.

"Man, I thought I would never see the day when you played with dolls. I knew you had a thing for them when you were little, but ain't you a tad too old for that shit?"

"Yeah, right. Don't be over here talking that bullshit. You know damn well I don't like playing with dolls. I'm just trying to make a little girl happy."

"By the looks of things, you seem to be doing a pretty good job at that. She's crazy about you already. But what's up with the proper talking and Uncle Jaylin stuff?"

"For a five-year-old, she speaks very well. And the Uncle Jaylin thing, I asked her not to call me that. I told her to call me Jaylin. I don't want her thinking we're family. She's the one who insists on adding the uncle. It bothers me a bit, but eventually she'll learn to call me by my name only."

"You know, you've really changed. I haven't figured out if it's good or bad. I dig what you're doing for that little girl, but I think you're rushing things a bit."

"I know you do. And sometimes I think I am too, but so far, I have no regrets. My house ain't been nothing but peaceful. The things Mackenzie says and does remind me of my little girl. Remind me how much I could kick Simone's ass for taking her away from me. So, if she can fill that void for right now, I'm okay with it."

"What if things don't work out with you and Scorpio? She's going to take Mackenzie and you'll be right back where you started. Personally, I think it's a bad idea that you're getting so attached to her."

"Who says Scorpio and I aren't going to work out? Man, believe it or not, that's a good woman up there. I mean, she might not have all the glamorous material shit like the other women I've messed with, but her personality counts for everything."

"So, honestly—and I mean honestly, Jay—are you falling in love with this woman?"

I hesitated to respond and got up and shut the door. Then I sat on the edge of my desk and looked Stephon directly in his eyes.

"Honestly, my brotha, I can't say that I am. I mean, I dig the shit out of her, but I . . . I don't understand why I can't love a woman like I should. You of all people know I had some type of love for Nokea, but lately, I'm not sure what it was."

"So, what about Nokea? Are you ever going to try and work things out with her?"

"Nope, not right now. I'm going to play this out for a while and see where it leads me. If things don't work out, I won't sweat it. Maybe I'll see if she's available then. I think we needed a break anyway. I need this time away to figure out who or what I really need in my life. And in the meantime, I'm going wherever my heart and my dick lead me. Right now, that's with Scorpio."

"If you say so. I really wish you'd reconsider your position with Nokea, but whatever you decide to do, you know I support you all the way."

"Thanks. You know I appreciate it. It's good to know I at least got some kind of family who supports me in my decisions."

Scorpio woke up and cooked Stephon and me some hamburgers and fries. She had my kitchen in a mess, but she was learning slowly but surely how to be a better housekeeper. She and Stephon got along well. She laughed when he told her about a few good times that we had growing up. Everything from the ass-whippings he gave me to the girls we tried to sneak in the basement. I was embarrassed. And since Stephon exaggerated some shit, he made the stories sound even more dramatic than what they really were.

When Stephon was ready to go, he pulled me aside in the bonus room where we had just finished up a game of pool. Scorpio was in the kitchen washing dishes, while Mackenzie helped her.

"Man, you know what I said to you earlier about how I didn't know if this was a good thing or a bad thing?" he said.

"Yeah, I remember. Why?"

"Let me just say that I like this change in you. Scorpio's a lot better than I thought she was, and if that little girl can bring joy to you like she has, then it's got to be all good. There is one thing I want you to do for me, though."

"What's that?"

Stephon had a serious look on his face. "Go see Nokea. Just tell her how happy you are and why. I think she deserves to know a little more than what you told her the other day—and let her know, just maybe, there's still a chance for you and her."

I wasn't sure why he seemed so concerned about my relationship with Nokea, though I knew he always had both of our best interests at heart. This time, though, I couldn't live up to his request.

"I'm sorry, my brotha, I can't do that. I already told Nokea what was up, and I'm afraid if I go see her, I'm going to wind up sleeping with her like I did Felicia when I tried to end it. Right now, I'm leaving well enough alone."

"Not even for me?"

"Not this time, not even for you. You know better than anybody when I stand my ground, it's hard to make me change my mind."

"All right," he said, giving me a hard handshake. "I'm going to let you get back to your beautiful woman and her daughter. I've taken up enough of y'all time today already."

I walked Stephon to the door, and then went into the kitchen to check on my sweet ladies. Mackenzie had water all over the floor and was trying to mop it up. Scorpio was wiping down the counter with a wet rag and didn't hear me come into the kitchen. I picked up a jug of cold water that was on the table and whispered for Mackenzie to be quiet. She smiled because she knew I was getting ready to pour the water on Scorpio.

"Mommy!" she yelled. "Watch out!"

Scorpio turned around so fast that she knocked the jug out of my hand and the water splashed on me.

"See, that's what you get for playing so much," she said as she and Mackenzie laughed. "Now, go upstairs and take off those wet clothes."

I held her waist and kissed her. "Only if you come and help me out of them. After all, you're the one who wet me up."

Scorpio looked down at Mackenzie still trying to mop up the water on the floor.

"Now, you know she's not going to let us be alone," she whispered. "Since I was so tired last night, I'll make it up to you later."

"You, tired? When did you start getting tired?"

"Ever since you've been making love to me two and three times a day. You know a sista gotta have some down time, Jaylin."

"I guess I'll let it slide this time, but only this time, beautiful. As a matter of fact, why don't you slide some clean clothes on you and Mackenzie so we can go shopping? I saw this cute little pink toy car for her to drive. It was in the paper today and I want to go get it."

"Really? Her mother needs a cute little car too, you know," she said, running her fingers through my hair.

"I know. So, like I said, why don't you go change

clothes and maybe I can help you out in that department too." I hit Scorpio on the ass and then she and Mackenzie left the kitchen.

No woman of mine should be driving around in an old beat-up car, and I wanted my lady to have the best. Besides, her jacked-up car didn't look good in my driveway, and the appearance was a no-no in my neighborhood.

I cleaned up the kitchen to my satisfaction and went upstairs to get out of my wet clothes.

By ten o'clock that night, Mackenzie had her pink 4X4 Jeep wagon and Scorpio's old car was hauled away by J's Towing Service to make room for her new convertible red Corvette. She was in tears most of the night. And Mackenzie was right in bed with us as we watched TV and talked.

"Jaylin, I don't know what I would do without you. Why are you so good to me?"

"Scorpio, I ain't no stingy brotha. I take care of those who take care of me. You and Mackenzie do a damn good job of that, so like I said, when you take care of Jaylin, Jaylin takes care of you. It's as simple as that."

"But this is too much. First the cruise, then the dresses, and then the watch. The watch must have cost you a fortune. Now, a car? And not just any old car— an expensive car. I don't know how I'm ever going to be able to repay you."

"Woman, please. You don't have to repay me anything. All you have to do is keep making me happy, that's all. Now, it doesn't get any easier than that."

"I'd like to buy you some nice things too, but how do I compete? The only time I make decent money is when I get a call from Jackson. He looks over my scripts, and if he likes them, he pays me. If he doesn't, then I don't get a dime. I love to write, but it's not getting me the money I need to buy nice things for us. I don't want this relationship to be all on you.

"Maybe I need to give up writing and go back to school to study business. I've been thinking about it for a long time, and since you've hired a nanny to take care of Mackenzie, this might be the perfect opportunity for me."

"Sounds like a plan to me, Scorpio. An education never hurt anybody. If you're sure that's what you want to do, I'll even front you the money."

"I'm sure, but you don't have to pay for it. I will. Not that I don't appreciate the offer, but I want to do this on my own."

I truly felt as if Scorpio was sincere. She wasn't coming off as a gold digger. I did question how she intended to get the money, since she didn't have a full-time job.

"Jackson pays me pretty good money for my scripts. I'll use that money to pay for school."

"All right, Miss Lady. I'm not going to force you to take me up on my offer, but if there's anything I can do, let me know."

Mackenzie had fallen asleep in my bed. I picked her up and carried her into her bedroom. Just when I got ready to close the door, she stopped me.

"Uncle Jaylin, are you and Mommy going to get married?" I smiled and walked back into the room.

"It's Jaylin, Mackenzie, not Uncle Jaylin. And no, right now we're not. If we do, you'll be the first one to know. Okay?" I tucked her into bed.

"Would you read me a bedtime story?" Her eyes searched the room for a book.

I opened the closet and pulled out Cinderella. I sat on the bed next to her and started to read. I tried to hurry so I could go make love to Scorpio, but Mackenzie held me up, asking questions as I tried to finish the book. After I read it four times, she was finally sound asleep.

When I got back into the bedroom, Scorpio was already in the tub waiting patiently for me. I slid in behind her, but the water was slightly cold.

"I'll warm you up, so don't worry about how cold it is," she said, pecking my lips.

"Then stop talking and start warming."

"Jaylin, I . . . I love you," she whispered in my ear as I rubbed her silky smooth body.

"And you know how I feel."

22

FELICIA

I thought Paul was all I needed to get my mind off Jaylin, but when he tried to make love to me the other night, I didn't feel a thing inside of me. I was so disappointed, but after how kind he'd been to me, I couldn't find it in my heart to dismiss him.

He was constantly all over a sista at work and at home. Once I told him we had to cut the chatting at work, he backed off a little. I didn't mind him coming over to my house to see me, but when he showed up without calling, I had to bring it to his attention.

Damion was still coming over from time to time and I didn't want the two of them meeting up at once. It was okay for a brotha to get caught in his game, but a sista—we had to play it cool, pretend like we were only with one brotha at a time, knowing damn well some of us be knocking two or three behind closed doors.

I had learned a lot from Jaylin. Like him, I had specific days I asked Paul to come over and specific days I asked Damion to come over. Everything was right on schedule. I had Damion over when I needed some good loving, and Paul over when I needed a good friend to chill with.

The only problem was that Paul was falling in love with me. During dinner last night at the Macaroni Grill, it was his second time telling me since we'd met. I know I put it on him the other night, but this love shit was too soon for me.

Hell, I still hadn't gotten over Jaylin. I knew it would be a matter of time before he came back to me, but the question was when? I called his house a few times, but when his bitch answered, I hung up on her. She had to be living with him, because every time I called, she answered.

When I called him at work, Angela made up excuses for him, Saying he had just gone to lunch or he was in a meeting. I wanted to pay him another visit on his job, but I decided to just sit back and let this mess play itself out. Anyway, it wasn't like I didn't have two other men occupying my time.

I even thought about Nokea. The other day, I saw her at the Quik Trip pumping gas on New Halls Ferry Road. She looked a mess. Looked like she had put on a few pounds and her clothes didn't even match. I really felt bad for her. If she'd had another man on the side, she might not be in the situation she's in now. I know she didn't expect Jaylin to settle down and marry her. Then again, shit . . . I thought some day he would marry me.

Either way, I wasn't writing him off just yet. Whenever he was able to be alone with me and not make love to me, then I'd write him off completely. Until then, the door was always open.

Paul picked me up at 7:00 P.M. so we could go to the movies. His dark brown hair was slicked back and he had on some loose-fitting Levi's with a black button-down shirt. His dark tan almost had him looking like a brotha. His black shades covered his pretty green eyes.

He opened the door to his SLK 230 Mercedes Benz and I felt like royalty riding with him. At every stop, people checked us out. I didn't know if they noticed us because we looked good together or because he was white and I was black. And when we got to the movie theater, the stares continued.

Paul always liked to hold hands and kiss in public, but I was uncomfortable with everybody checking us out. I loosened my fingers from his hand and pretended I had to sneeze. When nothing came out, I dropped my hand by my side.

As we stood in line waiting to get some popcorn, Paul looked at me.

"Felicia, why are you so uncomfortable with me?"

"I'm not uncomfortable with you; I'm just uncomfortable with all these people looking at us like they ain't never seen a mixed couple before."

"All you have to do is pretend they're not there. Just focus on me. If you do, you won't even know they exist."

"So, really, why don't all the stares bother you? One thing about black folks, we can't stand to be stared down. It's harder for me to ignore them than it is for you."

"No, it's not. I refuse to give them the attention they want. Besides, I'm here with a beautiful woman, and if I'm happy, who cares what other people think?"

I smiled because I knew Paul was right. I even reached over and gave him a kiss. Some girls behind us in line looked at each other and rolled their eyes. Paul saw it, and then he embarrassed the hell out of me.

"Excuse me, everyone," he said, with an English accent. "As you can see, I'm white and my stunning woman here is black. We've been noticing all the frightful stares we've been getting and would like to say thank you. You've made us feel like celebrities. So, my name is Paul and this is Felicia. We just got married yesterday, so would you all be so kind and give the gorgeous bride a big round of applause?"

Paul clapped and so did everybody else. When he kissed me, the applause got louder. I couldn't believe how he embarrassed us, but it was quite funny. As we walked through the theater to our seats, some people smiled and told us congrats. Even though it was a stupid thing to do, it worked. I felt more at ease with him, and the stares had turned into smiles.

During the movie, I gripped Paul's muscles every time a scary moment came on the screen. He laughed but held me tightly in his arms.

"Felicia, are you really that scared? If so, we can watch something else."

"No, Paul, this is cool," I said, chewing on some gummy bears. "If you don't mind, I just like snuggling up with you."

"Of course I don't mind. Whatever you want, my dear." He held me tighter.

After the movie, we headed back to my place. The first thing I did was check my messages to see if Jaylin had called. He hadn't, so I went into the living room and entertained Paul. When it came to sex, if he was a bit more aggressive, maybe I would like him more. I always had to be the one to initiate it. It was like he was scared to touch me.

But when I changed into my black teddy and straddled his lap, he couldn't keep his hands off me. He laid me back on the couch and laid my coochie out with his tongue. Now, in my opinion, this was one thing white men sure knew how to do better than black men. For the first time, Jaylin had nothing on him. I could barely keep still. Damn near broke his neck as I squeezed it tightly with my thighs.

Once he finished, he slid himself inside me and the excitement faded. Damn, I thought, why couldn't he just keep licking? I moaned and groaned like it was the best thing ever.

When he left, I called Damion over to finish the job. In the meantime, I'd have to work on Paul until he got better. Eventually, I hoped, he would, but only time would tell.

23

NOKEA

I ate everything in the house, from the rooter to the tooter. I had been pigging out even when I wasn't hungry. The stress from not talking to Jaylin added to my bad eating habits. I was only a few months pregnant and already had picked up fourteen pounds. Dr. Beckwith said if I didn't slow down, I was headed for a difficult pregnancy.

When I left his office, I stopped by Burger King on West Florissant Avenue to get my last taste of a double Whopper with extra cheese and some French fries. Then I had the nerve to stop at Krispy Kreme and get three glazed donuts. By the time I got to the office, I felt like a pig. I had to struggle just to make it up the stairs. I went to the bathroom and looked at myself in the mirror. I could still see my curves, but if I kept at it,

I knew they would soon disappear. I decided to take my doctor's advice and cut back on the fattening foods. From now on, it would be just salads and Jell-O. If I splurged, it would only be on the weekend, or if I felt like going out to dinner with Pat and her husband.

She was a charm. Since she was the closest friend I had, I confided in her a lot. She even started to pick up weight with me, forever bringing me ice cream and a bunch of other fattening foods. Every Saturday she took time away from Chad to come by my place and keep me company. She knew I missed Jaylin and tried to do everything in her power to make me forget him. But no matter how hard she and I both tried, he could never leave my memory. We had too much history together, and since we had a baby on the way, it would be even more difficult to get him out of my system.

On Saturday, Pat was right on time with some salads she made and movies she brought from Blockbuster. She rented *The Best Man* and *Training Day*, movies we had seen time and time again. She knew that Morris Chestnut and Denzel Washington were definitely a way to snap me out of my misery. We sat in my den mesmerized by the fineness in both of them and munched on our salads.

"Girl, I'm so hungry that I'm going to imagine this is the hamburger and fries I ate earlier this week," I said, picking all the meat out of the salad first.

"Nokea! I thought you said you weren't going to eat any more fast food."

"That was after I stopped at Burger King. Since then, I've been doing pretty good. And since it's Saturday, and I can splurge on the weekend, can we please order a pizza with everything on it?"

"No, Nokea. You know you need to eat a little healthier. If not for you, then for the baby."

"I didn't think it would be this hard for me to watch my weight, but this salad stuff is driving me crazy."

"Look, if you want to order a pizza, you go right ahead," Pat said, chewing slowly and picking at the salad like she really wanted me to order a pizza.

I reached over to the coffee table, picked up the phone, and ordered a supreme pizza with extra mushrooms and olives from Pizza Hut.

"Okay, don't blame me when your ass gets all fat. Then I'm going to have to listen to you gripe about that. In the meantime, what I don't understand is . . . why didn't you tell them to put extra cheese on the damn thing? Girl, call them back and tell them to add more cheese," she said.

We laughed as I picked up the phone and called Pizza Hut back. The man gave me a new total and I hung up.

"I'm sorry, but that salad was just not cutting it. And I made it. Shame on me for bringing that bullshit over here," Pat said.

"Well, at least you tried. And as hard as I tried to make it taste like a hamburger, it wouldn't."

"At least we got a good laugh out of it. I haven't seen you laugh like that in a long time. So, seriously, how have you been? And don't tell me what I want to hear, tell me the truth," Pat said, putting *Training Day* on pause.

"It's been tough, Pat. Really tough. Sometimes I want to pick up the phone and curse Jaylin out for doing this to me. And other times, I thank God he's out of my life. Then there's a part of me that thinks this is some day going to work itself out. How? I don't know, but I really wish that it would. I cry myself to sleep almost every night, torturing myself over seeing him with Scorpio. Wondering why it couldn't be me living there with him. With our son.

"He seemed so excited about her little girl, and I don't even know if he's going to be excited about his baby when he finds out—"

"What do you mean when he finds out? Are you planning on telling him?"

"No, but what if he does find out? I can't keep this from him forever. It would be impossible."

"Look, Nokea, nothing's impossible. You tried to tell him and he didn't want to hear it. So, fuck him. Raise this child by yourself. You never know; somebody decent might come along and be a good father to him. He doesn't need a father like Jaylin setting bad examples for him, especially when it comes to how to treat women."

"You're talking like you know it's a boy too. I hope it is. And I hope he looks just like Jaylin. If I can't love the big one like I want to, then the little one will just have to do," I said, rubbing my belly.

"You are out of your mind. All I can say is go with your heart. It'll take you places no one else can."

"But my heart is with Jaylin, Pat."

"For now it might be, but you'll find somewhere else to place it. Just give it time." She took the movie off pause. "Who knows? Maybe somebody like Denzel Washington will come your way. And if he does, will you be so kind to a friend and share him with me?"

We laughed. "Now, I'll share Jaylin, but Denzel, or any man like him, I will not. Denzel is the kind of man you want to keep all to yourself."

"Okay, fine, keep Denzel. But when I show up at

your door next week with Morris Chestnut, don't be mad at me."

"Only in your dreams, Pat. Only in your dreams."

The Pizza Hut driver came and we ate the pizza so fast that we ordered another one. When he came back, he laughed because I had pizza sauce all over my white T-shirt from the first one. Pat only had two slices of the second pizza, and I nearly ate five slices all by myself. I hated the thick edges, so when I pulled them off, it actually only accounted for four slices. No matter how many it actually was, I paid for it, and so did Pat. We lay on the floor in the den with cold rags on our bellies to cool them.

"You are a mess. Look at you, Nokea. I told you this was a bad idea." Pat rolled over on her side.

"No, you didn't, Pat. You told me to call them back and add more cheese. And in case you forgot, you were the one who suggested a second pizza."

"Damn, I was, wasn't I? And I have to drive home and face Chad looking like a big fat pig. You know, he's noticed this sudden weight gain I've had since you've been pregnant. He told me if I gained one more pound, he would leave me."

"Girl, please. That man loves you. I don't care how fat you get, he isn't going anywhere."

"I know, but girl, we've got to slow down. We got six more months to go and if we're eating like this now, we gon' be some fat chicks by the time you deliver."

"Okay, starting Monday. No more of this pigging out after Monday. I promise."

After *Training Day* was over, Pat helped me clean up the den and headed home. She called Chad to tell him she was on her way and gave me a sista-hug before she left. I didn't know what I would do without her in my life. Mama and Daddy tried to be there for me, but I couldn't talk to them about everything like I could Pat—even though she would never be accepting of Jaylin the way I'd wanted her to be.

I went to the kitchen and poured a soothing cup of the hot Chinese tea Mama had given me to relax. It didn't have any caffeine, so I had no problem going to sleep.

The phone woke me at one o'clock in the morning. When I heard Stephon's voice on the other end, it scared me.

"Stephon, is everything okay?" I asked, holding my chest.

"No. Not really. I just wanted to call and let you know that my mother passed away last night. You know she's been battling this drug addiction for years, and last night . . . she, uh . . . she decided to take her life."

I could hear the pain in his voice.

"Stephon, I'm so sorry. If there's anything I can do, please let me know. I know how you felt about your mother doing drugs, but at least you had the courage to make peace with her years ago. God will bless you for that, and you'll be able to go on knowing that you did all you could for her."

Stephon was quiet. "Yeah, but I guess it wasn't good enough. She called me last week and asked for some money, but I wouldn't give it to her. I knew what she wanted it for, but . . . but now I feel bad because I didn't give it to her."

"How could you feel bad about not contributing to

her habit? Look at all the good things you did for her. Out of all her children, you were the one who stood by her. Please don't go dumping on yourself because you did the best you could."

He let out a deep sigh. "I'll call you in a couple of days and let you know about the arrangements. Sorry to call you so late. Go back to sleep, Shorty, and get some rest."

"Stephon, before you go . . . how's Jaylin taking the news? I mean, I know he's probably thinking about his mother at a time like this."

"He's doing okay. I think he's more hurt because I am. He never really cared too much for Mama anyway, but I think it bothers him knowing he wasn't there for her either."

"All right, Stephon. Thanks for calling, and call me as soon as you find out the arrangements."

I couldn't sleep a lick as I thought about Stephon's mother. The memory of her leaving them at home alone, night after night when they were kids, kept coming to mind. Several times, my mother even stepped in and fed them when Stephon's mother was out on one of her drug binges. And Jaylin, she treated him like crap. One time, one of her boyfriends badly beat him with an extension cord and he came to school with whip marks all over him. Some kids made fun of him, but I was always by his side. I couldn't blame him for not being hurt about her death, but I knew he would always be there when it came to Stephon. And so would I.

24

JAYLIN

The funeral was torture. I sat in the front pew with Stephon and his brothers, who I didn't get along with. We wore black suits and dark black shades. My glasses helped to hide the tears as I found myself thinking about Mama when she left me years ago to take her place in heaven.

Stephon took it the hardest. When he fell to his knees in front of his mother's casket, I damn near lost it myself. I walked up and put my arms around him. He grabbed my leg and asked the Lord, why? Why did He have to take her?

My other cousins sat there with a few tears here and there, but that was it. I wanted to kick all of them straight in the ass. One of the assholes was a crackhead his damn self and had the nerve to sit there like his shit didn't stink. And the other two were broke as hell, didn't have nothing going on but thug-ass women and a street corner. Basically, didn't have a pot to piss in or a window to throw it out of. Stephon was the only decent one. They couldn't stand either one of us because we didn't turn out like they did.

Stephon's girlfriend and I helped him back to his seat. When I got back to mine, Scorpio was right there, waiting to comfort me. She held my hand and rubbed my back. I held back the tears because I definitely didn't want my woman to see me cry. I didn't care how bad things got.

When the funeral was over, there was a dinner in the lower level of the church, in remembrance of Aunt Betty. Stephon sat in a chair in the corner all by himself. It was obvious that he didn't want to be bothered. When I looked up, I saw Nokea talking to him. She looked beautiful in a knee-length, sky blue dress and some sexy high-heeled shoes. Her eyebrows were perfectly arched and her short hair had been freshly cut. She even looked as if she'd thickened up a bit. She wore the extra weight well, especially in the breast area.

Before my mind went into the gutter, I had to think about where I was. I couldn't be in church thinking about sexing up women, could I? But with all the fine women running around, it was hard not to. Since Scorpio watched my every move, I backed off. Especially on this nice li'l tender who mugged me from far across the room. She stared me down like she wanted to break a brotha down right then and there; however, now wasn't the time or the place.

After we chowed down, I walked to the water fountain and got a drink. When I lifted my head, I saw Nokea standing close by, waiting for me. Since I saw Scorpio looking at me, I kept on walking like I didn't even see Nokea.

She grabbed my arm. "Listen, I'm not trying to come between you and your woman, but I wanted to tell you how sorry I am about your aunt. I know this has probably been a difficult time for you, and I wanted to tell you if you need anything, call me." She handed me a piece of paper.

"Thanks." I leaned forward to give her a hug just so I could feel her in my arms again. "I appreciate it. And . . . and take care."

I went back over to the table and sat next to Scorpio. She smiled and encouraged me to go talk to Stephon because he seemed like he was out of it. I tried to give him some space because men don't like all that attention when they're feeling down. We like to get our thoughts together and deal with it whenever. But since Scorpio pressed the issue, I went over and pulled up a chair next to him.

"Man, are you going to be okay? I know it hurts; I've been there before. But it gets easier, my brotha. In due time, it gets easier," I said.

"I know, man, but why did she have to kill herself? If she was going to do that, she should've done it years ago. Just don't make sense. And then to make us suffer because she didn't want to anymore."

"Those suffering days were over a long time ago. For us anyway. Now, Aunt Betty had a choice. She could've cleaned up her act, and you gave her the opportunity to do so. I know when you asked me for that fifteen grand you gave it to her. I know every time that you borrowed from me you put it right into her hand. All I'm saying is you can't feel responsible for something she done to herself. And if you do, then that's too bad. But she knew out of everybody, you were the one who took care of her."

There was silence. Stephon closed his eyes and swallowed.

"Jay, I don't know what I'd do without you. After the way Mama abused you when you were growing up, you knew the money was for her all along, didn't you?"

"Yeah, but I also knew if I didn't give it to you, I'd lose the only family I had. I wasn't willing to lose you."

"Naw, you knew if you didn't give it to me, I would kick your ass like I did when we were little," he said, laughing.

"Whatever, nigga. I held back because I didn't want your brothers jumping in it, trying to help you kick my ass. But now I wish I would've fucked all y'all up." We both laughed again. Just to get one smile out of him made me feel good.

I looked over at Scorpio and blew her a kiss—only to see Nokea reach out her hand and catch it. I smiled and went back to conversing with Stephon.

"Say, man. Did you see Nokea? She's looking good, my brotha. Looking damn good," Stephon said, sounding like he wanted to hit that.

"She looks a'ight. I mean, if you ask me, she look like she's picked up some weight," I said, though there was no doubt in my mind that Nokea looked spectacular.

"Weight is good, bro, especially if you're wearing it like she is."

"So, what are you trying to say? You trying to get a piece of that action?"

"Naw, I'm just telling you like I see it. That woman got it going on. She's pretty, she's smart, got a good job, a good heart—she's the full package. Right about now, she'd give Scorpio a serious run for her money."

"Negro, please. She might be the full package, but when it comes to looks, Nokea don't even compare to Scorpio. They both fine, but Scorpio gets a ten-plus in my book. Nokea only gets a nine. With the exception of today; today I'll give her a ten."

"Well, I have that reversed. You're only thinking about the pu-tain but I'm strictly talking about appearance. And appearance-wise, Nokea looks much better than Scorpio."

"You are out of your mind! Look at them. Take a look at both of them now and tell me Nokea looks better than Scorpio."

We checked out both women. Stephon hesitated for a moment.

"All right. You got me convinced, but Nokea is still the bomb."

"Like I said, she's workable, but she ain't got my baby beat. Besides, why you riding Nokea so tough? Are you finding yourself a bit attracted to her?"

Stephon licked his lips and gave me a stern look.

"If I was, would you be mad?"

It didn't take me long to think about it. I gave him the most serious look he'd probably ever seen on my face.

"Man, we like brothers, but that's one woman we'll never share."

"So, are you saying she's only hands-off when it comes to me? Or to anybody?"

"All I'm saying is that's one woman we'll never share." My palms started to sweat. I could tell where this conversation was headed.

"But she deserves to have a good man in her life, don't she?"

"I'm saying that man won't be you, so let's drop it."

Stephon hopped up and grabbed my hand. We gave each other a pat on the back.

For the rest of the evening, I felt a tension between us—tension I had never felt before. When Nokea said good-bye to everyone, Stephon got up and offered to walk her to her car. Deep down, my insides burned, but I managed not to let it show.

Scorpio said she was ready to go, so I told her to wait in front of the church while I went to get the car. Outside, I noticed Stephon in the car, talking to Nokea. Looked like they were in a deep conversation, but once again, I kept my cool. When they saw me, Stephon got out and waved good-bye to her, then walked toward me.

"Man, what you looking all uptight for? I was just thanking her for coming," he said.

"Hey, that's cool. I ain't tripping. If you want my leftovers, that's all on you," I said, getting into Scorpio's Corvette.

I drove off, and then saw that sweet li'l tender that had her eyes on me all day. She flagged me down and I stopped the car.

"Hey, sexy. Can a sista get your phone number to call you sometime?" she asked. I looked at her shiny, thick thighs that begged me to open them.

"Let me get yours and I'll call you when I get time."

She wrote down her number as I watched for Scorpio. I quickly took her number, put it in my pocket, and then drove around the corner to pick up Scorpio. I didn't know if I would ever call the girl, but it was always good to keep my options open in case things changed between me and Scorpio.

25

JAYLIN

The stock market was a serious blood bath. Everybody was calling like crazy trying to sell out. I'd even sold a few of my own stocks since they'd done so badly. If anything, losing money wasn't the name of my game.

A little after noon, Mr. Schmidt stepped in my office to talk. I thought it was about how badly the market was doing, but when he talked about my performance, he quickly got my attention.

"Jaylin, you're still my number-one producer, but lately you haven't been as dedicated to this company as you were before. And since you haven't, I've been losing money. If we could bring in some more business, it would make up for some of the losses when the market drops like it has today. So, I'm bringing in Roy to help you. I want you to train him to be like you, and then we can go from there."

"What!" I said angrily. "Train him to be like me? Mr. Schmidt, there's nobody on this planet like me. And there never will be. I don't care how well I train him, he's not going to be as productive as I am and you know it." It seriously sounded like this sucker was trying to replace me.

"You're right, Jaylin. That's why I need a second man. When you don't feel like giving one hundred percent of yourself, he can step in and fill the gap. Trust me; it'll all work out for the best. You'll see."

"Look, Mr. Schmidt, you're running the show around here. And whatever you say goes, so let's roll with it. But I hope you understand, if the situation gets sticky, I'm packing up and going elsewhere. Remember, I really don't have to be here," I said as matter-of-factly.

"I understand that, Jaylin, but at least give it a try. Will you do that for me?"

"It's whatever, Schmidt. Again, let's just roll with it."

For the rest of the day, I was in a shitty mood. I asked Angela to hold all my personal calls so I could finally get some things done. Besides, Roy watched my every move. He listened in on my conversations with my clients and followed me around, trying to learn "how to be like me."

I took a break and went downstairs to Barb's Coffee Shop to get a cherry Danish. Something about eating the filling out of the middle excited me. Roy came with me and tried to make conversation, but when I found myself thinking about how good Nokea looked at the funeral, I ignored him. I thought about calling to take her up on the offer she made: *if you need anything, call me*. But I didn't want to start complicating shit between Scorpio and me.

I was proud of myself for being with only one woman, and as much as I'd been thinking about Nokea, I knew I'd made the right decision. The only problem I had lately was that things between me and Scorpio had been a little less perfect than they were in the beginning.

Many nights, by the time she got home, Mackenzie and I would already be asleep. Sometimes I was so tired, I didn't even hear her come in. If I rolled over and tried to get some sex, sometimes she would and sometimes she wouldn't.

Mackenzie kept me busy, though. She had me running around the house like a slave: "Jaylin, cook me this, read me this, write me this, comb my doll's hair, and play hide-and-go-seek with me." I did everything for her and I wasn't even her daddy! I had no clue where her biological father was, but he was a fool for not being a part of her life. Not only was she a beautiful little girl, but she was smart and funny as hell. The best thing about her was that she filled the void in me from when my own daughter was taken away from me.

As I sat in the coffee shop thinking about Mackenzie, I took my cell phone out and called to check on her. When Mackenzie answered, I pretended to be somebody else.

"Jaylin, I know it's you. When are you coming home?"

"Mackenzie, I'm going to be late tonight. It's kind of busy today, so I'll see you when I get home."

"Well, what if I fall asleep? When you come home, will you wake me up and read a story to me?"

"Yes, I will. And I have a surprise for you too."

"What! What! Tell me what it is."

"Now, it wouldn't be no surprise if I told you, would it?"

"No, so I'll see you later. Nanny B wants to talk to you."

Mackenzie put the nanny on the phone, and I told her I would be late, but she could leave if Scorpio came home. She was cool because we paid her a fortune to watch Mackenzie, and I kicked her out extra because she kept the place spotless, since it had gotten to be such a mess when Scorpio moved in.

I finished my Danish and my half-ass conversation with Roy. When we got back to my office, there were a dozen yellow roses on my desk.

"These came in for you while you were out," Angela said, standing with her hands on her hips. "So, who are they from?"

"None of your business, Miss Secretary. Roy you have to excuse my secretary; she tries to be my mother sometimes." I had to clear things up because Angela was tripping like we still had something going on. And since Roy was tight with her husband, I'd have to really watch it.

I opened the card. It read: *I know you thought these were from one of your li'l breezies, but I just wanted to say I'm sorry for reacting the way I did at the funeral. Love always, your only true brotha, Stephon.*

I chuckled. I figured Stephon had sent the flowers as a joke. Angela rolled her eyes and walked out the door. I wanted to call Stephon to thank him for trying to clear up our differences, but I didn't want Boy Roy all up in my business, so I decided to wait until later.

Roy and I didn't shut down until eleven o'clock. My eyelids were heavy and my bed was calling my name. Roy looked tired too. From what I could see, he seemed like a pretty cool person. Seemed to really know the business, and could possibly be what I needed to get things flowing again.

After he left, I called Stephon to thank him for the flowers. I had never gotten roses from anyone, and if I ever did, I was sure they'd come from one of my ladies. When Stephon answered the phone, he sounded like he was asleep. But when I heard a moan, I knew he was fucking.

"Damn, dog, if you were in it, why did you pick up the phone?" I said, wishing I was doing the same.

"Because I saw your number on the caller ID."

"I'll let you get back to business, but I wanted to say thanks for the bitch-ass flowers you sent today. I always knew you had a feminine side to you."

"Yeah, that feminine side of me working it right now. So, you're welcome, and I'll holla at you tomorrow." He rushed me off the phone.

"Hey, man?" I whispered.

"What?"

"That ain't Nokea over there, is it?" I said jokingly.

"I wish. Damn, I straight up wish."

He hung up on me.

I smiled, though I knew Stephon wasn't playing. We pretty much had the same taste in women, so I know if I was thinking about tagging that ass, he was too. As a matter of fact, when we were growing up, he was crazy about her. Since she always had her eyes on me, she never gave him a chance. But who was I to tell her who she could or couldn't date? I was doing my thing with Scorpio and doing it well.

I stopped at a twenty-four hour superstore and bought Mackenzie a Barbie that was bigger than her.

My purpose was so she could have a friend to play and sleep with when I had to work late nights. I'd even thrown in a small CD player so they could listen to music together and Mackenzie could show her how to dance. And Scorpio, she liked books. I went through the book section and tried to find her one she didn't already have. I picked up Carl Weber's latest book and tossed it in the cart, planning to read it as well.

By the time I put the bags in the car and drove home, it was one o'clock in the morning. I didn't see Scorpio's car in the driveway, so I was a bit worried. She'd been spending a lot of time working at Jackson's place on her scripts, but her late nights were starting to bug me. According to her, either she was at her sister's house, or she was at Jackson's place. Still, one and two o'clock in the morning required me to start paying more attention.

Nanny B had fallen asleep on the couch in the bonus room while watching TV. I didn't even wake her. I went to the closet, got a blanket, and covered her. I took Mackenzie's doll and put it in my room. Since the house was freezing, I turned up the heat, and then I went to Mackenzie's room to check on her. When I opened the door, I found her sitting up in bed, crying.

I rushed in. "Mackenzie, are you okay?" I sat on the bed next to her. She continued to cry. I held her in my arms and moved her long, wavy hair away from her eyes. "Tell me, what's wrong?" I asked again.

"I didn't think you were going to be this late. I got up three times and went in your room to look for you." She hugged me back as the tears rolled over her cute little cheeks. I wiped them away.

"I'm sorry. I told you I had to work late. I promise

you if I work this late again, I'll call and talk to you until I get home, okay?"

"Okay," she said, wiping her face. "Now, where's my surprise?"

"It's in my room. Come on." I took her hand.

We walked into my room. When I opened the door, she had a fit. She saw the big Barbie box, but when she looked at the doll, her smile vanished.

"What's wrong, Mackenzie? Don't you like her?"

She scratched her head and looked at me curiously. "She's too big, Jaylin. I wanted smaller ones. Lots of them."

"Oh, I see. But . . . but do you think you can give her a chance to be your friend? And if you don't like her, I'll take her back to the store in a couple of days."

"Okay." She frowned.

She didn't even play with the damn thing. For the rest of the night, it sat on the floor while she lay in bed next to me, sound asleep. I put my reading glasses on and started reading Carl Weber's book. I'd gotten so into it that I didn't even notice the time. When I heard the front door shut, I looked at my alarm clock on the nightstand. It was four-fifteen in the morning.

When Scorpio walked in, she looked surprised to see me up in bed. I tilted my reading glasses down and gave her a hard stare.

"So, you're reading now," she said, taking off her coat and laying it across the chaise.

"Hang it up!" I yelled and then calmed myself. "I mean, would you please hang it up instead of laying it there?"

"Excuse the heck out of me." She walked over to the closet to get a hanger.

"What's up with you strolling your ass in here at this

time of the morning, Scorpio? You haven't called or anything. How's a brotha supposed to know where his woman at if you don't call?"

"Jaylin, you don't call me when you stay out late. Besides, I thought this was supposed to be an open relationship. You do your thing and I do mine. Right?"

"I can't recall saying all that, but if you're doing your thing with somebody else, why don't you pack your shit and go live with him?"

Her eyes shot daggers at me. "Are you putting me out?"

"No, I'm not putting you out. All I'm saying is show a brotha a little respect. Don't be strolling up in here at four in the morning like you don't owe me an explanation."

"Listen, I'm sorry. It's been a long and trying day. I had tests today in all of my classes, and from there I went straight to Jackson's place."

"Tell me more about Jackson?"

"You know, the one who reviews my scripts for me."

"Aw, that's right. So, what did he say?"

"Honestly, he said he didn't think I had what it took to be a full-time playwright. He went over my script several times and made a lot of changes to it. Finally, he gave up. Said he'd call me when he had some new ideas." She sounded disappointed.

"Sorry to hear that. Don't give up, though. Why don't you take your work to somebody else to look at? Sort of like get a second opinion?"

"I don't know, Jaylin. I think it's time to give up on writing and focus on my education. I'm spending too much time away from Mackenzie and you. The last thing I want is to come in here every night arguing with you."

"That's the last thing I want too. And since it's our first real disagreement, let's make it our last," I said, getting out of bed. I wrapped my arms around her. "So, uh, are you tired?"

"Oh, I'm tired. But never too tired to make love to my man, especially since I've been thinking about his sexy ass all day. Let me take Mackenzie to her room, hop in the shower, and then give you this loving you've been waiting on."

"Now, you've been waiting just as much as I have."

"I concur. So, I'll hurry."

She kissed Mackenzie and carried her into her room. Then she took off her clothes and stepped into the shower.

My dick was so hard that I couldn't wait to feel her. I slid off my silk pajama pants and opened the shower door. We stood face-to-face, and I expressed once again how upset I was with her about her late nights.

"I'm sorry," she said. "It won't happen again. I never intended to upset you."

I pulled her hair back tightly and spoke with authority. "Don't make me be concerned about you again. If you're going to be late, you need to call and state your exact reason why. Better yet, coming in here after midnight might bring about confusion. Try your best to avoid it."

Scorpio nodded.

I grabbed her wet body from the shower and we made our way to the floor. I massaged her soaking wet breasts and held them together as I licked her nipples one by one. Then I positioned my curled tongue inside her pussy and rubbed her clitoris with the tip of my finger. When she seemed ready, I took my goodness and rubbed it up against her walls. She begged me to give it to her.

"Jaylin!" she said, grabbing my hand. "Stop teasing me, baby. Don't make me wait when I've waited all day to feel you."

I ignored her and continued my foreplay. And when I did give it to her, I waited until her body responded then I pulled out. She squeezed her fingernails in my ass and tried to force me back in, but I wouldn't let her.

"Why are you teasing me like this tonight?" she asked

With my hands holding me up on the floor, I lay over her. My eyes stared deeply into hers. "Because I don't want you to forget how good I am to you. And if you ever think about fucking this up, you'd better correct yourself."

She moved her head from side to side. "Never. Not in a million years."

I carried her to the edge of my bed, and when she bent over on her stomach, I separated her round, juicy ass and straddled it from behind. I stroked her insides so good, I could hear major juices flowing. The sound of her pussy excited me, but she was the one who hollered my name.

"Jaylin what, baby?" I said, continuing with my strokes. "What do you want from Jaylin? Whatever you want, Jaylin got it right here for you."

"I . . . I want you to love me," she strained as I kept the fast-paced rhythm going. "I want you to fuck me all day and all night, but I want you to love me too."

It wasn't that simple, I thought. I wanted to love her, but there was something about Scorpio I just didn't trust. I couldn't love a woman I didn't trust, and I couldn't stop thinking that our relationship was based purely on sex. No doubt, she had won my dick over, but my heart still wasn't in it like I expected it to be.

As we continued our sex session, I felt her body getting tired. I didn't care. I turned her on her back, rested one of her legs on my shoulder, and pounded her insides with nine-plus hard inches of my loving. She couldn't hang as I rubbed, licked, and teased many of her hot spots all at once. She screamed as if she'd lost her mind, and professed that she'd never been to that level before.

I didn't finish my business with Scorpio until damn near eight o'clock in the morning. She lightly kissed the ridges of my six-pack and then rolled her pretty self over and went to sleep. I knew she was probably upset with me for sexing her up all night, but when I have to wait two days for some sex, it's a fucking crime and she knows it. I was glad it was Saturday because if I had to go to work, there was no way I would've made it.

At 11:00, I woke and paid Nanny B extra for staying overnight. She offered to cook us some breakfast before she left.

After Nanny B was gone, I sat on the kitchen stool with a piece of toast and read the *St. Louis American* newspaper. There was a picture of Felicia shaking the CEO's hand at her architectural firm as he handed her an award. It was probably for fucking him, knowing her, but the article said it was for the best creative design.

I wanted to call and congratulate her, but I didn't want to start up the bullshit with her again, especially since she'd chilled out. She still called every once in a while, but I hadn't returned any of her phone calls. When I say I'm done, I'm done—until I get ready to come back.

Scorpio slept most of the day. Mackenzie and I went back to the store and exchanged her big Barbie doll for fifteen small ones. She had the nerve to throw in outfits and shoes for each one of them. Damn dolls were dressed better than I was. I couldn't believe I spent my money on this bullshit, but seeing the smile on her face made my day.

After we left the store, we went to Wehrenberg Theaters on Manchester and watched a kiddie movie she was dying to see. The theatre was packed, and women were all over me, telling me how adorable me and my child were. Like me, Mackenzie enjoyed the compliments, but I think she enjoyed being at the movies with me more.

On the way home, she played with two of her Barbies in the car. When I turned up the music, she yelled, "Jaylin, that's too loud. My dolls are trying to sleep."

"They don't look like they trying to sleep to me. They actually look like they getting ready to go clubbing or something."

"No, they don't go out. They go to work like you and Mommy do."

"Well, that's good. Then I'll turn off the radio so they can get some sleep."

Mackenzie laid her dolls on her lap like they were asleep.

"Jaylin?"

"Yes, Mackenzie."

"Are you really my daddy? I heard you tell those women you were, but you told me you wasn't."

"Mackenzie, I'm not your biological father. I just told them that because they were being too nosy."

"If I be nosy, would you be my daddy then?"

I laughed. "It's not nice to be nosy, Mackenzie. Being nosy can sometimes get you in trouble."

"Okay, then I won't be nosy. But will you still be my daddy?"

I didn't want to confuse her, but I knew if I told her I wouldn't, it would probably hurt her feelings. I also knew if I tried to avoid the question, she'd find another time to ask me. For a little girl, she was smart, and had smoothed me over better than anybody had done before.

"Jaylin, you didn't answer my question," she said, looking up at me with her big, light brown eyes. "Are you going to be my daddy or not? If not, I'll find me another one," she said, pouting.

"Stop pouting, Mackenzie. I'll be your daddy only if you stop pouting when you don't get your way."

She cheesed and showed her pearly white teeth. "So, I can call you Daddy now instead of Jaylin?"

"No, Mackenzie. Continue to call me Jaylin. Daddies have names too, okay?"

I was so glad when we got home. Mackenzie helped me get the bags from the car and we carried them in the house. There was a note on the kitchen table from Scorpio that said she went to Jackson's house to pick up her script and she would call later. She even left his number for me to call, if I needed to reach her. I threw the number in the trash because I wasn't the type of brotha to check up on his woman.

Mackenzie ran to her room and played with her Barbies. When I heard the loud music from her CD player, I went to her room to tell her to turn it down. I stood in the doorway and watched her dance like she was damn near twenty years old, twisting and turning her body like she was putting on a show.

"Mackenzie!" I yelled. "What are you doing? Little girls aren't supposed to dance like that. Who taught you how to dance like that?"

"My mommy. She dances like this all the time. She makes a lot of money when she dances like this," she said, doing another one of her mother's moves.

I picked up Mackenzie and took her into my room. I sat her on the bed and put one of her Barbie dolls in her lap to keep her busy. Then I ran downstairs and searched the trashcan for Jackson's number. When I called, no one answered; the phone just rang and rang. I tried three more times, only to get the same thing. *No, this bitch ain't a stripper*, I kept thinking. She couldn't be.

I jogged back up the stairs and went straight into my closet. Most of her clothes were neatly lined up on one side. I went through her clothes piece by piece, until I came to a gray garment bag that looked thick and full. It was in the far back of the closet, hidden away from all the rest of her clothes. I laid it on the bed next to Mackenzie and unzipped it. The evidence hit me right in the face: all kinds of slutty outfits, leather and lace two-piece sets, belts and whips. There were even a few pictures of her with next to nothing on. I guess she gave them to her fans, because her signature was on the bottom.

I crumbled the photo in my hand. The only thing I could think about was packing up her shit and kicking her ass out. But when I looked at Mackenzie, I didn't know what to do. I put Scorpio's belongings back in the garment bag and back into the closet.

As I searched for more evidence in the closet, the phone rang. Mackenzie answered and talked, so I knew it was Scorpio. When Mackenzie gave me the phone, I took a deep breath.

"Hey, baby, you miss me?" she asked, like shit was all good.

"Miss you so much. I need you, right here and right now," I said, trying to show Mackenzie a little respect by not dissing her mother.

"You know, I've been thinking about that good loving you gave me last night. I hope it's on again for tonight."

"Oh, you'd better believe it's on." I wanted to snatch her ass through the phone and beat the shit out of her. "How soon can you get here so I can give you some more?"

"I was calling to tell you I'm on my way. Jackson made some changes to my script and it looks pretty good. I'll tell you all about it when I get home."

"Sure, can't wait."

I hung up.

Mackenzie stood in front of me and dropped her head. "Daddy, are you mad at me for dancing like that?"

"No, Mackenzie, I'm not. But do you think you can do me a huge favor?" I said, lifting her chin. She nodded. "Do you think you can take a bath and go in your room and play with your dolls?"

"Yes. I need a nap anyway. Can I borrow your pillow? Yours is thicker than mine is," she said, taking the pillow off my bed and heading toward the door. "Good night, Daddy. Thanks for the dolls and the movie today. I almost forgot to thank you."

"You're welcome, Mackenzie. Anytime." She ran

back over and gave me a squeezing hug. I picked her up so she could kiss my cheek. She wiped the spit off my cheek with her hand and smiled.

"Daddy?"

"Yes, Mackenzie," I said, putting her down, trying to hurry her off.

"I love you, Daddy."

I was silent. She caught me completely off guard. She looked at me and waited for a response. And for the first time in my life, I felt good about saying, "I love you too."

She strolled out of my room with two of my pillows, as if her dolls needed one too. My heart ached; if I put Scorpio out, Mackenzie would leave too. I was in a no-win situation. I had no idea what I would do. Stephon warned me about getting too attached, but in no way did I see this shit coming.

No sooner had I tucked Mackenzie into bed than Scorpio was on her way up the steps. I met her at the top of the stairs right after I closed Mackenzie's bedroom door.

"Did I hurry enough for you?" She dropped her purse on the floor, put her arms around me, and puckered up to give me a kiss. I moved my head back to avoid her kiss and took her arms from around my waist.

"We need to talk," I said, walking into my bedroom.

"Jaylin, I hope you're not upset with me. I left Jackson's number so you could call me. And this time, I did call to check in."

She followed me into the bedroom. I sat on the chaise and slowly rubbed my hands together. Scorpio stood on the other side of the bed. I knew she could see the fire burning in my eyes.

"You got one chance to tell me the truth. If you fuck

up, I'm gonna knock your ass straight to Egypt," I said sternly.

"Jaylin, you're scaring me. Are you that upset with me—"

"What in the fuck is your occupation?" I yelled.

"What? What do you mean by what's my occupation?"

I hopped up and stood right in front of her. "Please, don't make me do this," I gritted. "What in the hell do you do for a living?"

She looked down at the floor and removed her jacket. She laid it on the bed, and that's when I lost it.

"Hang the motherfucker up!" I grabbed her hair. "Stop leaving shit around the damn house and hang the motherfucker up!"

"Jaylin, what is wrong with you? Let my hair go!"

She tried to move my hand from her hair, but I grabbed it tighter and slung her ass on the bed. When she fell down, I hopped on top and pinned her hands down above her head so she couldn't move.

Soon, she started to cry. "Stop, Jaylin! Please stop! I'll tell you, if you just please get off of me!"

I continued to hold her hands down tightly, and the bitch tried to kick me in my motherfucking balls. I grabbed her ankles and pulled her off the bed as hard as I could. She went flying and her head hit the floor.

"Oh my God! Would you please stop and just listen for a minute?"

I sat on top of her again and held her hands because she tried to scratch me. We tussled for a while, and I couldn't believe it when my hand went up and smacked her face. She covered it and cried, but she had finally calmed down.

I saw my red handprint swelling on her face, and I

got up and went into the bathroom. I could see her in the mirror as she sat up beside my bed and continued to cry. I felt awful for what I had done. I'd never laid my hands on a woman. I took a towel from the closet and wet it.

"Here," I said, giving her the towel. She looked up and took it from my hand.

"Are you ready to listen to me now?" she said, looking at me with tears flowing down her face.

"No, I don't want to hear anything you have to say. All I want is for you to get your things and go." I calmly poured myself a drink.

"Jaylin, don't do this. Things have been going so well for us. The reason I didn't tell you about being a stripper is because I was ashamed, and I didn't think a man like you would be interested in someone like me. Since we've been together, I've slacked up on putting myself out there like that. I've been trying to better myself by going back to school, and I'm working hard at getting my script together."

I took a few sips of the Remy Martin and looked at her with disgust. "So, are you saying, all these late nights you've been coming in, you haven't been out doing your thing?"

"Twice, Jaylin, that's it. I did two parties just so I could pay for my tuition. I didn't want to ask you for the money, so I did what I had to do to get it myself. They were just parties, Jaylin. And after I made my money, I left."

I had firsthand experience with strip parties, so I couldn't do nothing but stand my ground. "Scorpio, just go. Listen, you don't owe me an explanation. Just get your shit and leave."

I left the room because I didn't want to hear any

more of the bullshit. I went downstairs to my office and sat in my chair. She followed.

"Baby, I'm sorry. I never wanted to lie to you, but if I told you what I did for a living, there would be no way to earn your respect. You would've thought that sex was all I was good for."

"Am I missing something, Scorpio? Sex is the only thing you've shown me you're good for."

"You know that's not true. I . . . I hoped that maybe one day I would be able to look back at this and laugh. Laugh with you when I told you what I had to do to get my education paid for and make a better life for Mackenzie and me."

"Cut the fucking act, all right? You flat-out lied to me and that's all there is to it! One thing I can't stand is a lying, conniving-ass woman. You are not the type of woman I want to be with. When you get that education you're talking about, or when you become a real playwright, holla at me. Until then, let me go help you pack your shit."

Scorpio stood at the end of the stairs with tears streaming from her red eyes. I moved her ass aside and went back upstairs to my room. I pulled her things from the closet and laid them on my bed. She walked up and stood with her arms folded, as if she wasn't going anywhere.

When I put her last piece of clothing on my bed, she broke down and pleaded with me again. "Why do you have to be so stubborn? You know people make mistakes sometimes. All I'm asking for is another chance. I promise I will never take off my clothes for men again. All I care about is being here with you, and if I would've known you'd react this way, I never would've done it."

"And all I care about is you getting the fuck out of here. Now, if I have to tell you one more time, Scorpio, I'm gonna put you out of here my damn self."

I guess she got the picture because she started to make progress and got her clothes off the bed. "Jaylin, what about Mackenzie? If you want me to leave, you go wake her and tell her she's got to leave."

I stared at the wall. I had briefly forgotten about Mackenzie during our heated argument. She wasn't going anywhere, and that's all there was to it.

"Come pick her up tomorrow. Better yet, send somebody over here to get her. I don't want to see your face, so your cousin or sister will be just fine. Tell them to come late because I need time to tell her what happened."

"I need time too, but you won't allow me that. I'll wake her and take her with me tonight."

"No, you won't. I told you to send somebody over here tomorrow to pick her up, didn't I?"

"And I'm telling you she's my daughter and she's coming with me tonight."

Since Mackenzie liked me so much, I had to say something that would deeply hurt Scorpio's feelings. "If I wake Mackenzie and ask her if she wants to go with you or stay here with me, you know damn well what she'll say. She's well aware of your trifling-ass ways and doesn't take too well to you as it is."

"What did you say?" She walked over to me. "Trifling? Did you have the nerve to call me trifling?"

"I call it as I see it. Now get the fuck out of my face." I pushed her away.

She smacked my face. I felt a scratch and my face burned. I put my hand on the scratch, only to see a dab of blood on the tip of my finger. Before I knew it, I grabbed her hair again and slung her ass damn near across the room. I sat on top of her and smacked her around a few times, then tried to push her down the steps to get her the fuck out of my house. I threw bunches of her clothes and shoes over the handrail and they landed in the middle of the foyer. Then, I ran downstairs and forced her and her clothes out of the front door. I wanted the bitch to take a cab, but since I wanted her out so badly, I didn't even care that she took the damn Corvette.

When I threw the last piece of clothing on the front lawn, I went back into the house and slammed the door. I watched as she walked back and forth, loading up her car—no, my car—with her belongings. I wanted to go outside and kick her ass again. I had never stooped to this level with anybody, and if a bitch had made me go out like this, I didn't need her.

When I went upstairs to my room, I heard the Corvette speed off. I picked up my drink and stood at the bar cart, thinking about how this bitch had lied to me and then made me kick her ass. Flat-out tried to provoke me, when all she had to do was tell the truth. I felt like a fool for ending my relationships with Nokea and Felicia for a woman who was nothing but a liar. I picked up a wine bottle and threw it at my glass bedroom doors. They shattered, and the wine splashed on the walls and all over the carpet.

Seconds later, Mackenzie came down the hallway. "Mackenzie, watch out for the glass!" I yelled as she tried to step over the shards on the carpet. I picked her up to make sure she didn't cut her feet.

"Daddy, what happened?" She rubbed her eyes. "Is Mommy home yet?"

"Mackenzie, Mommy won't be home until tomorrow. Get in my bed and go back to sleep."

I was glad that she didn't ask any more questions because I wasn't prepared to answer her. I went into the bathroom and ran some bath water to relax me. I closed my eyes and thought about the drama with Mackenze that would surely arise soon. As far as I was concerned, she wasn't leaving my house with no one. If Scorpio sent her sister over to get Mackenzie, she'd be wasting her time.

26

NOKEA

My belly looked as if I'd swallowed a watermelon. I was busting out of everything in my closet. When Pat and I went the Galleria to buy some maternity clothes, I saw Stephon there with one of his boys in the food court. He told me Ray-Ray was getting married and said they were at the mall to get fitted for their tuxedos. When I asked about Jaylin, he didn't comment. He pretended as if he didn't hear me and quickly changed the subject. I knew something was up, so after talking for a while, I invited him to my place for a friendly dinner.

What Jaylin and I shared was probably over, but I still thought about him a lot. I wondered how his new relationship was going, and I wondered if he was connecting with Mackenzie. The biggest thing I wondered about was if he was happy. Was there anything he'd told Stephon about missing me, and was Felicia still in the picture? I was sure that during our dinner, Stephon would entertain me by giving me the scoop on Jaylin.

Later that day, Stephon came in with what looked to be a bottle of wine. When I took the bottle from his hand, I saw that it was actually a bottle of Welch's sparkling white grape juice. I laughed as I carried it to the kitchen, where I was in the midst of finishing dinner.

Stephon looked remarkably well. He always kept his head clean-cut and shaven, and his goatee was trimmed and shaped perfectly. He had on a Sean Jean blue jean outfit and some clean white tennis shoes. When I asked for his jacket, I could see his thick muscles busting out of his oversized white T-shirt.

"So, Shorty, what you cooking?" he asked, standing over me and looking into the pot on the stove.

"I'm cooking some spaghetti with cheese and some garlic bread. I was going to fry some chicken, but I forgot to thaw it."

"Naw, this is cool. Actually, it smells pretty good."

"Here, would you like to taste it?" I put a dab of sauce on the tip of the spoon. "It's hot, though, so be careful."

Stephon blew on it and I put the spoon into his mouth. He moved the sauce around in his mouth and then nodded.

"Pretty good, Shorty. Jaylin told me you couldn't cook." He laughed.

"Oh, no, he didn't. If I put my mind to it, I can cook. Besides, we always used to go out. But whenever I cooked, he loved it."

"Before you go getting all upset, I'm just playing. Actually, he said you were a very good cook."

"You better had cleaned that up. Besides, don't go talking about a woman's cooking when she's in the midst of cooking for you."

Stephon set the table, and then he took the garlic bread out of the hot oven so I wouldn't burn myself. By the time the spaghetti was good and ready, I was tired from being on my feet. He pulled up a chair in front of the one I sat in, so I could elevate my feet. Then he served our dinners.

I had questions about Jaylin, but I didn't want to rush into it. Stephon went on and on about his job, how he was ready to do something different, and the women in his life. He and Jaylin had so much in common; only he seemed to treat his women with a little more respect than Jaylin did.

As I sat listening to him, I observed his muscles tighten every time he lifted his fork. I couldn't resist; I reached over and gave his arm a squeeze.

"You've been really working out, haven't you?" I said, sliding up his short sleeve so I could see the Q-Dog symbol on his arm.

"Yeah, Jaylin and me both. We've been meeting at the gym every morning trying to relieve some of this stress we've been under."

"Really? What kind of stress? I thought you all searched for women to relieve that."

Stephon chuckled. "Now, you know that ain't so, Nokea. A piece of ass doesn't relieve stress. If you ask me, it only adds to it. If you took me in your room right now and fucked my brains out, I'd still be stressed."

"See, if I wasn't pregnant with Jaylin's baby, I'd take

you up on that offer," I flirted. "Stress is all in the mind. When sex takes your mind to a different level, you forget about it. Wouldn't you agree?"

"Depends on who you're having sex with. Maybe if it's somebody like you. But for the most part, I'm stressing no matter what."

"So, what are you trying to say? You sound like you got something on your mind."

Stephon was quiet. He stared at me while holding in his bottom lip. He got up and put his plate in the sink. I checked out his bulge as he walked back toward the table. Looked like he was hard to me, so I turned my head to look away. And if he wasn't hard, some sistas know they be in trouble.

He took my hand and led me into the living room. "Come here, Shorty. Sit. I have something I've wanted to tell you for a long time."

His tone made me a bit nervous. I took a seat on the couch and he sat next to me. He still hadn't cracked a smile. By the serious look on his face, I knew he wasn't joking.

He took my hands and closed them tightly with his. "Shorty, I think you're a beautiful woman and I'm finding myself attracted to you. I've always been crazy about you, but recently, my feelings have been beyond my control. Since you and Jaylin have been apart for a while, I think it's time you move on. You need to find happiness in your life; and not only that, you need a good father figure for this baby." He reached over and rubbed his hand on my stomach.

"I can give you both of those things and so much more. I've seen the way you've been checking me out too, and I say if you want to, then let's do this. Time is of the essence. I don't want to sit around wasting any more time not being with the woman I've always cared deeply for."

I blinked my long eyelashes and pulled my hands away from him. When I reflected on it, I realized that the signs had always been there. Stephon would always console me during my troubled times with Jaylin, and now I knew what he meant by, "If you ever need me, just call." My mouth wouldn't even open, and my body was still.

He was right about me checking him out, but what was so wrong with a woman checking out a good looking brotha when she saw one? Maybe my interest in Stephon was due to the changes in my hormones during my pregnancy, I thought. I wasn't sure, but one thing I knew: Stephon made me feel wanted.

Before I could respond, he leaned forward and placed his lips on mine. I put my hand on the back of his head and pulled him closer to me. Our tongues explored each other's mouths, and then I backed up and softly touched my lips.

"Stephon, this isn't right. I can't do this with you."

Ignoring me, he stood up and removed his shirt. The sight of his dark brown skin, broad, thick shoulders, and tight six pack weakened me. His jeans hung low on his thick waist and fit him very, very well.

He sat back on the couch, and before I knew it, I eased my way back so he could lie on top of me.

"We can do this, Shorty," he whispered as he leaned in close to my ear. "Just between you and me, we can do this."

I rubbed his back and closed my eyes as he kissed down my neck. He lifted my shirt and pulled it over my head. I needed this so badly, and at this point, there was no turning back. He put his mouth between my breasts and unhooked my bra with his teeth. I smiled as he dangled it in his mouth like a tiger and growled. He stared at my breasts, and then slowly manipulated them with the tip of his tongue.

My insides steamed. I couldn't wait to see what else he had to offer, so I unbuttoned his jeans, just enough to reach my hands down his pants to massage his butt. It was solid as a rock, but smooth as a baby's bottom.

When he stood up and took off his pants, I damn near fainted. Body was cut in all the right places and dick hung longer than Jaylin's. I had no idea what I was about to get myself into, but I had some needs that had to be met.

Stephon pulled off my jeans but left my panties on. He sat up on the couch and put me on top of him. He massaged my breasts with his strong hands and held them like they were the most precious things he'd ever held before. I closed my eyes as he sucked them and rubbed one of his hands on my back. Then he reached his hands inside my panties and rubbed them across my hairs. I moaned to let him know how excited I was.

He took both sides of my panties and tore them. Then, with all of his strength, he lifted me and placed my goodness right on his face. I straddled his face and held on for dear life.

"Relax, Shorty. I got you." He soaked my walls. As I got more excited, it was hard for him to keep me balanced. He kissed my thighs and laid me back down on the couch.

"You taste good, Shorty. I wish you would keep still, though, until I'm finished with you."

"Stephon, I . . . I can't keep still. I'm just not used to all of this," I said, looking into his hazel eyes.

"Well, get used to it. I don't think this is going to be my last time here."

As he pressed himself up against me, I reached down and put his thickness inside me. I tightened my eyes and felt every bit of him. He worked from side to side and continued to hold my legs close to his chest. I figured he was trying not to lean on the baby, but when he opened my legs and leaned in forward, I knew the baby felt him because he was in my guts. He rubbed my hips with his hands and lowered his eyelids, as my body responded powerfully to his. He sucked in his bottom lip and quickly changed to another rhythm—one I kept up with very well.

"You feel good, Shorty. I knew it would feel like this. That's why I couldn't get you off my mind," he whispered.

"Stephon, what if Jay—"

"Shhh . . . I don't want to talk about Jaylin right now. We can talk about him later. Just for the record, I ain't stressing at all right now. My stress is a thing of the past."

I held Stephon's hips and stopped my motion.

"What's wrong?" he asked, and then gave me a kiss.

"Stephon, we don't ever have to talk about Jaylin again. Feeling as good as I do with you, you'll never hear his name come out of my mouth again."

Stephon smiled and the lovemaking went on. He was a gentle lover and definitely knew how to excite the heck out of me. By the end of the night, he'd taken me to my room and loved me all over again. I felt like a new woman. And by the time he left, which wasn't until Monday morning, I never wanted to see Jaylin again.

27

FELICIA

It had been months since I'd last slept with Jaylin, and I couldn't believe how well everything flowed without him. Paul and I made the best of our relationship. I had even kicked Damion to the curb. His baby's mama called here with some of that, "When's the last time you seen him?" bullshit. After I told the bitch what he had on, he had the nerve to call and curse me out. I told him, "Don't ever pick up the phone and call me again if you're going to put this out-of-shape, baby-having skeeza before me!" Since then, I hadn't talked to him.

Paul, though, had been a true gentleman. He had done anything I asked him to do and had gotten a little better in the bedroom.

There was no way to keep our relationship a secret

at work. Since we spent so much time together, of course, people started asking questions. I wasn't ashamed of Paul's fine ass, so I didn't deny nothing. And since I was like some kind of prize for the white men, he was definitely treated like the winner.

For the weekend, we planned a fast trip to Chicago. Deep in my heart, I thought he would ask me to marry him because he hinted about having beautiful children together and about us living together. I didn't know what my answer would be. I still had concerns about being with a white man, let alone marrying one. Plus, as much as I hated to admit it, Jaylin was still locked up in my system. Until I freed him, I wasn't marrying anyone.

When I was at home packing my things to go, I got a call. It was Jaylin, and it was definitely a setback for me.

"I was just sitting here thinking about you," he said softly. "I hadn't heard from you in a while and I was wondering how you were doing."

I was so shocked to hear his voice I could barely get any words to come out. "I . . . I'm doing fine. I'd be better if you'd invite me over."

"Nah, not today. Maybe some other time. It's good to know that you haven't given up on me, and I apologize for how things went down."

On that note, I figured things didn't work out for him and the Playboy bunny. This was the moment I'd been waiting for, so I jumped right on it. "Apology accepted. It's obvious that you want to talk, so I'm coming right over."

"No," he snapped. "Not today. I'll call you soon, but just be there for me, all right?"

"No problem."

He hung up, and when I tried to call him back, he didn't answer. When I called him at work, Angela said he hadn't come in that day.

I was confused, so I put on some clothes and drove to his house. When I got there, I rang the doorbell but got no answer. His cars were in the driveway, so I knew he had to be there. After banging for a few more minutes, I left.

I didn't know where else to go for answers but to Nokea's house. If anybody knew what the hell was going on, she would; she didn't have the guts to stay away from Jaylin as long as I did. But if he really needed me, there was no way I would turn my back on him.

I rang Nokea's doorbell. She opened the door in her work clothes.

"Felicia, why do you keep coming over to my house like we're the best of friends?" She looked as if her face had swelled. When I looked down at her stomach, I could tell immediately what was going on.

"You're pregnant?" I asked, shocked.

"Why? What's it to you?"

"I just want to know. If you are, then maybe that's what's troubling your man."

She walked away from the door and I walked in and shut it behind me.

"Girl, look, I ain't here to start no trouble. I came over to find out if you've talked to Jaylin. And evidently you have. Looks like you've done more than talk to him."

"So, now what, Felicia? Yes, I'm pregnant, but it's not Jaylin's baby."

"Girl, shut your mouth! Since when did you become a little hoochie mama?"

"Since I learned from your trifling butt, that's when. So, if you don't have any more questions, would you please go?"

"Ouch. Now, that hurt coming from a self-righteous woman who pretended to love Jaylin so much. If you did, you wouldn't be knocked up by another man. I thought you had a little more class about yourself, but I should've known better. Once you lost your virginity, you just couldn't get enough, could you?"

"Felicia, why don't you take your butt home? If you came over here to find out anything about Jaylin, I don't have any answers for you. I haven't seen or heard from Jaylin in months. And frankly, I don't think I ever will. So, if information is what you're looking for, I suggest you find Scorpio. I'm sure she's got all the answers you want."

I shook my head as I walked toward the door and got ready to leave. "So, who's the lucky man, Nokea? Is it anybody I know?"

"No, Felicia, it's not anybody who you've had the pleasure of giving yourself to."

"I don't know, now, you never know. I do get around, bitch, so watch your back." I slammed the door behind me.

I jumped in the car and backed out of Nokea's driveway. As I got ready to make a left onto New Halls Ferry Road, I saw Stephon in his white BMW. He looked directly at my face, but when he saw me, he turned his head like he didn't see me. I looked in my rearview mirror and watched as he drove by Nokea's house. I turned the corner like I was on my way out of her subdivision and drove around to another street.

I could see the front of Nokea's house, and I watched as Stephon pulled in her driveway. He got out with a box of Kentucky Fried Chicken in his hand. She opened the door, smiled, and let him in. Instantly, he leaned forward and gave her a kiss.

My mouth opened wide and my eyes went buck. Where in the hell had I been? This bitch was pregnant by Stephon! No wonder Jaylin sounded upset; he probably just found out about the betrayal. Damn. Now would be the perfect time to get back with him. He would definitely need me now. My relationship with Paul would have to be put on hold.

I drove back to Jaylin's house. This time, I wasn't leaving until I got some answers from him. I knocked and banged until he finally cracked the door to talk to me.

"Jaylin, would you open the door so we can talk? I know something is wrong; I could tell by the sound of your voice. Please open the door. All I want to do is talk."

He cleared his throat. "Not right now, Felicia. I'm not in the mood for any company," he said softly.

"Then why did you call me? All you wanted to know is if I'd be there for you when you needed me, and yes, I'm here. So open the door."

I pleaded for a few more minutes before he finally let me in. I damn near melted when he gave me a hug. He had his shirt off and wore only some black Jockey boxer shorts that gripped his package. I could tell he'd just gotten out of the shower; he smelled delicious, his hair was wet, and so was his body.

When he walked up the steps, I followed, wishing like hell he'd put it on me tonight. But the sadness in his eyes told me wasn't nothing going on with us— tonight, anyway.

Outside his room, I noticed that the glass doors

weren't tinted anymore. Looked like he had some new ones put in because they were clear, and I could see straight into his room. He held the door for me and I walked in. There was a cute little girl sitting on the floor with some dolls. She looked up at me and smiled.

"Daddy, who is she? Is this your new girlfriend?"

"No, Mackenzie. This is Felicia. Felicia, this is Mackenzie."

I said hello to Mackenzie, but I was busy thinking about how she'd called him Daddy. I'd definitely been away too long if this was his daughter.

"Mackenzie, would you go in your room and watch TV? I'll come in and read you a story in a minute," Jaylin said. She picked up her dolls and went into the other room. I looked at Jaylin with wrinkled lines on my forehead.

"What the fu—hell is going on?" I was careful to watch my foul mouth.

"Look, Felicia, I know you got questions, but right now, I'm not in the mood for them. I just called you because I hadn't heard from you in a while."

"Jaylin, don't bull—mess around with me. Tell me what's going on. And is that really your daughter? I mean, she's cute, but she doesn't look like you."

"She's not my daughter. She's Scorpio's daughter, and since we fell out the other night, she's been here with me."

"So, what is she doing here with you? Did Scorpio just run off and leave her with you? I knew she was a trifling bitch." That time, I couldn't help myself.

"No. Her sister came to get Mackenzie, but I didn't open the door. I've really gotten attached to her, and I'm not ready to give her up yet."

"But you have to. That's kidnapping. You can be in a lot of trouble for that."

"Yeah, I know. But she's brought so much joy to my life these past few months. She reminds me of myself when I was little; how curious I used to be about shit. I can't let her go. Besides, she doesn't want to leave me anyway. She told me she wanted me to be her daddy and she wanted to live with me forever."

"But she's just a kid. Before this stuff gets out of hand, you need to take her back to her mother. Where's Scorpio at anyway? I thought she was here with you."

"She's gone, but I don't want to talk about her right now. I'm waiting to hear from my lawyer so he can tell me what I need to do. I know he's going to probably tell me nothing, but there's got to be something I can do to keep Mackenzie."

Jaylin sat on his bed looking pitiful. I couldn't believe how much love he had for this little girl. I was kind of jealous. Four years with him and the brotha still didn't have no love for me. I couldn't offer him any advice because what I said to him, he wasn't trying to hear. He was going to do things his way whether I liked it or not.

It seemed like he didn't know about Stephon and Nokea because if he did, he probably would've mentioned it. But maybe he was just trying not to talk about it. I had to dig for some answers.

"Hey, have you seen Nokea lately?" I asked.

"Ye—yes, why?"

I could tell he was lying.

"So, you know she's pregnant?"

Jaylin's thick brows went up. He looked as if he'd seen a ghost. "She's what?"

"Yes. I saw her today and she's pregnant. I asked her who the baby's father was, and she wouldn't tell me; however, she did tell me it wasn't yours."

"Really?" He paused and rubbed his goatee. "How many months is she? Or did she tell you?"

"Naw, she didn't say, but the moment I saw her, I could tell she was pregnant. You might want to ask Stephon," I said, spicing things up a bit. "Since he's been kicking it with her, I'm sure he knows something."

"What do you mean, kicking it with her?" This time, Jaylin's brows scrunched in and his forehead wrinkled. He was mad.

"I mean, I just left her place not too long ago and he was over there. Looked like they were a couple or something."

Jaylin jumped up and went into his closet. He put on his jeans and a white wife-beater. He grabbed his keys off his dresser and asked me to stay there until he got back. Mackenzie came running out after him, but he kissed her and begged her to stay with me. When she cried, he kissed her again and wiped her tears. Motherfucker didn't even give me a kiss, and I was the one who gave him the scoop.

I picked up Mackenzie and we went back in his room to watch cartoons.

28

JAYLIN

I was on a rampage. I was already upset about the shit that went down with Scorpio, and now this. In my Mercedes, I drove about ninety miles per hour down Interstate 270. When I hit a bump, I almost lost control of the wheel.

Felicia had to be bullshitting. I knew damn well Nokea and Stephon weren't kicking it like that. I had just seen him Friday at the gym and he didn't say nothing about Nokea. And pregnant? That didn't sound right to me. If she was pregnant by Stephon, that would mean he'd been screwing her at the same time I was. But she was a virgin, so that couldn't be true.

I racked my brain, trying to figure out what the fuck was going on. I thought about Stephon's and my conversations about Nokea over the last few months. I made it perfectly clear that out of all my ladies, I didn't want him to ever involve himself with Nokea. He could've fucked with anybody he wanted to, but she was off limits.

I continued to speed, and when I flew past a police officer, it was too late to slow down. He pulled me over. He took nearly twenty minutes checking my license and registration, then hit me with a ticket for speeding. The motherfucker had the nerve to ask if I had any drugs in the car. As fired up as I was about Stephon and Nokea, I almost went to jail.

"Do *you* have any drugs in *your* car?" I asked disrespectfully.

"Sir, I just asked you a simple question. Do you or don't you?"

"And I just asked you a simple question too. If you don't, then I don't." I looked at my watch. "Look, this has been really fun, but I need to get going. If there's nothing else, have a good day and move on to the next black man. There are plenty of us to harass, and you seem to be awfully good at what you do."

"Very good," he said and tapped the hood of my car. "You have a swell day too."

"I will. And in the meantime, I'm gon' say a little prayer for you. Today wasn't your day, but you just never know . . . tomorrow may be. Be careful out there, sir. These streets can be a dangerous place for you."

"Don't I know it." He winked and walked away.

I shook my head and calmly drove the rest of the way to Nokea's house.

As soon as I turned into Nokea's subdivision, I saw Stephon's BMW in her driveway. I swung my car into her driveway and almost hit Stephon's car in the rear.

I slammed my car door. I was about to get my clown on.

I didn't even knock on the door; I tried to damn near kick the motherfucker down.

Nokea pulled open the door.

"Jaylin, what are you doing?" she asked in a panic.

I pushed the door open and let myself in. I looked down at her stomach and was shocked at how round it was. Felicia wasn't lying. Nokea was for damn sure pregnant.

My fists were already gripped tight to do some damage. "So, what's up, Nokea? Who's the fucking father of your baby?"

Her eyes watered as she looked at me in fear. Just then, Stephon walked out of the kitchen.

"It's me, man. The father of her baby is me."

I looked back at Nokea. "Is this true? Is Stephon the father of your baby?"

She started crying, and Stephon walked over and put his arms around her.

"Jaylin, man, just go. Can't you see she's upset right now? Once I leave, I'll come over to your place and explain everything to you."

My eyes shifted back and forth from Nokea to Stephon; somebody was about to get fucked up. Wasn't sure which one it would be, but I wanted some answers.

"Nokea, how long have you been fucking my cousin?"

She ignored me and buried her face in Stephon's shoulder.

"Man, look, it ain't even like that. Why don't you just go home and chill out? I said I'll be over there to explain things to you in a minute."

"Shut the fuck up talking to me, you sorry mother-fucker! You always had to have every damn thing I had, didn't you? Couldn't stand to see me with any-thing I called my own. Your ass had to have a piece of it, didn't you?" I said, laughing. "Even when we were kids, you had to take everything, just so I could be mis-erable. You and your fucking selfish-ass brothers. And your mother—aw, now, we ain't gonna talk about how she raised such a lowlife asshole when she was one her damn self, are we?"

"All right, man, that's enough!" Stephon said, grab-bing my shirt and pushing me back. "Get the fuck out of here! And if I—"

I didn't give Stephon a chance to say another word. I slammed my fist into his jaw.

"Son of a bitch!" he yelled. "You wanna fight? Is that what you want? You want my lady to see you get your ass kicked?"

I punched his ass again. Without any hesitation, he punched back, and it was on. We scrapped so hard that we broke two of Nokea's vases and her glass table. I wasn't letting up and neither was he. She tried to break us apart, but I pushed her down. I didn't give a fuck about Stephon's baby she carried.

When he saw her on the floor, the asshole really tried to get tough. He grabbed my waist and pushed me into the wall with every ounce of strength he had left. My body put a dent in the wall, and the picture above it came crashing down on top of Stephon. I picked it up and tried to bust his damn head open with it. He dodged it a few times and then tried to grab my legs and knock me on the floor. When he realized he couldn't do that, I pushed him back into Nokea's baby grand piano and flipped his ass over the top.

Nokea screamed and ran to the phone. As Stephon lay on the floor, holding his back, I ran over and snatched the phone out of her hand. I pulled the cord out because I knew she would call the police.

As I stood in front of her, she cried hysterically and shook like a leaf. I balled up my fist, just about ready to punch her in her damn stomach. She put her hands over her face and screamed.

"It's your baby, Jaylin! Please don't do this. It's your baby!" She fell to the floor and cried out loudly. I slammed the phone down, and it broke into pieces.

"You lying bitch! You'll just say anything, won't you?" I wiped the sweat from my face. "I hope you and my cousin have a happy fucked-up life together.

"You had me fooled, though, baby. I thought you had a bit more respect for yourself. But I guess you ain't no better than all the other fucked-up women in my life. . . . Excuse me, did I say women? Naw, you're not women. You all are undependable, unpredictable, and unreliable bitches. Every last one of y'all, including my damn mother, who left my ass years ago.

"Shit actually started with my mother. And then his sorry-ass mother stepped in and tried to pick up the pieces, when the only thing she could pick up was a crack pipe."

In a rage, I pounded my fist into the wall. Nokea screamed for me to stop, but I couldn't. I couldn't let the situation go. "Let's see who else there is. . . . There's Simone, who left town with the only fucking thing I had to love in my life. And let's not count out Scorpio. Boy, she was a charm. Had me paying for her shit—big dollars—only to find out she had other motherfuckers paying her too. And now you, little innocent-ass, virgin-bodied Nokea. Pretending all along that she loves me so much she can hardly see straight. And now, 'Guess what, Jaylin? I'm pregnant and it's yours, but I'm fucking your cousin too.' Bitch, get real! You were the worst one of them all because you stabbed me in my back purposely. You were the worst and I expected so much more from you, Nokea. So, you know what?" I walked over to her and lifted her face with my hand. "If it is my baby, you raise it your damn self. I'd be too ashamed to tell it that it's got a whore for a mother."

Nokea rocked back and forth on her knees and covered her ears so she couldn't hear me. I wanted to kick her in her face, but I caught myself. I didn't want to hurt her. Instead, I knocked one of her expensive statues on the ground and jetted.

As I was leaving, I backed up my car then went full force into the back of Stephon's BMW. I saw the front end of my shit fold like a piece of aluminum foil. His car was severely damaged.

By the time I pulled in my driveway, my car was smoking and the front dangled off. Felicia came rushing out of the house.

"Where's Mackenzie?" I asked because I didn't see her come after Felicia.

"Jaylin, her aunt came over here and got her. I wasn't about to go to jail for your ass. You were wrong for keeping that little girl, and I did you a favor by letting her go."

"You opened my damn door and let her take Mackenzie? Is that what you're telling me, Felicia?" I grabbed her neck. "Where in the fuck did she take her?" I yelled, though I knew Felicia couldn't tell me because I had a strong grip on her throat and she couldn't talk.

She gagged and tried to pull my hands away. When

I let go, she fell to the ground like a piece of paper. She lay there for a while and coughed. I didn't give a fuck. I shut my front door and turned off the light so she couldn't see.

In hopes that Felicia had lied, I went into Mackenzie's room, but there was no sign of her. I sat on the edge of her bed and dropped my head. Everything seemed to be hurting on my body, but nothing as much as my heart.

I didn't know how I got myself into such a fucked-up situation. I had depended on too many people for my happiness, and every last one of them failed me. I was so mad that I could have killed somebody. But wasn't no telling who I would go for first.

I lay back on Mackenzie's bed and felt very sleepy. It wasn't long before I was asleep and dreaming.

During my dream, I looked up at Mama and apologized for disrespecting her. I told her how upset I was with her for making me go to an orphanage, only to hear her apologize for that. But then she told me she was unhappy with my behavior. I'd not only disrespected her tonight, but I'd been disrespecting women for a long time. She said it was only a matter of time before my shit caught up with me.

I told her about all the good fortune I'd come into by doing things for myself. She said that she was proud, but acknowledged that I still had a long way to go before she could say I was the man she wanted me to be.

When I tried to touch her, I woke up, and found myself still lying in Mackenzie's bed. My mother had come to me in my dream and tried to get me to see how wrong I'd been.

I lay still for a while, thinking about Mackenzie. My heart ached for her. She was such a joy to be around, and I hoped like hell that I'd be able to see her soon.

It was damn near light outside, so I pulled off my shoes and lay across my own bed. I watched the early morning news and dozed off again.

When the phone rang, it woke me, and when I looked at the alarm clock, it showed that it was almost noon.

"Jaylin, Mr. Schmidt is really upset with you," Angela said on the phone. "Why haven't you come in or called or something?"

"Because I don't fucking feel like it. Tell your father-in-law to kiss my black ass." I hung up. She called back several more times and I finally picked up.

"Don't you hang up on me," she said. "Now, if you're having some problems, I'll cover for you, but don't go hanging up on me."

"Sorry, Angela. I'm not in the mood for doing any work right now. Tell Schmidt I'm taking a leave of absence. In the meantime, ask Roy to handle all my accounts for me. Don't call me unless it's a matter of life or death, please."

"Okay, but you know he's going to want to talk to you."

"Well, fine! Have the motherfucker call back soon so I can tell him to go fuck himself."

"Jaylin, calm down. I don't know what's bugging you, but nothing is worth losing your job over. So, when he calls, you'd better change your attitude."

"Yeah, yeah, yeah. Whatever. Just tell him to call me."

Ten minutes later, Schmidt called from his office on speakerphone. Roy was in the office with him, and in so many words, Schmidt said if I didn't come in to work, I didn't get paid. That was quite hilarious because I had plenty of money stashed aside from my inheritance. In addition to that, most of my salary came from Higgins and his buddies anyway. I told Schmidt I would be back in action within sixty days, and he agreed to let me come back then.

For the rest of the day, I chilled. Took my phones off the hook and watched Judge Joe Brown and Judge Judy as they showed no mercy for people in their courtrooms. My body was sore and bruised from fighting with Stephon; I relaxed in the tub about five times that day.

The only phone call I made was to my lawyer to find out if there was anything I could do about Mackenzie. When he said there wasn't, I told him to go fuck himself and then fired him. Then I called my other attorney. He said that if I could prove Scorpio was an unfit mother, I stood a chance of getting Mackenzie. I wasn't trying to hate on Scorpio, but I wanted Mackenzie back in my life. I told him I would think about it for a while and get back with him.

After lying across the bed most of the day, drinking shots of Martel, I found myself getting horny. Right about now, Scorpio was probably fucking a brotha and sucking him well. I kicked her out five days ago, and my dick was lonely. I looked in the pocket of the suit I'd had on at Aunt Betty's funeral and found that tender's phone number.

I called, and when she answered and realized it was me, she sounded excited. I talked to her for an hour and managed to coax her into coming over to see me. I jazzed myself up, knowing she'd give it up to me in a heartbeat.

When she rang the doorbell, I watched the last five minutes of a sitcom and made her stand outside and wait. Still, as she saw me approach the door, she was all smiles. I let her in and closely observed her to make sure I wanted to go there with her. She wasn't no dime, but her plump, juicy ass in a tight red mini-skirt kept her in the running. I noticed her fingernail polish was chipped, and I immediately knew she couldn't be in Jaylin's world if she tried.

She looked a bit young, too, but when I asked how old she was, she told me she was twenty-seven. As long as she was legal, that's all I cared about.

We chatted in the living room for a while, and then I escorted her upstairs to my room. I wasn't about to waste any time.

"Hey, uh, what did you say your name was?" I asked, already forgetting.

"Brashaney."

"Can I get you something to drink?" I poured myself another drink.

"No, no, thank you. But if you have some water, I'll take a glass of that."

"I don't serve water to my guests. If you'd like something else, let me know."

She cut her eyes and gave her lips a slight toot. "I'm good," she said. "Forget it."

I didn't care much for her attitude, but for the time being, it didn't even matter. I sat on the bed, put my drink on the nightstand, and asked her to stand in front of me.

"Why?" she asked. She was starting to become a pain.

"Because I want you to take off your clothes. You don't mind if I watch, do you?"

"No, but can we listen to some music or something before we get down like that? I mean, I haven't been here five minutes and you already talking about taking off my clothes."

"Brashaney, I don't have time for games, baby. I invited you here tonight because a brotha liked what he saw. If you want to sit here all night and listen to some music, fine, go right ahead. But I asked you to come here so I could make love to you. So, what's it gonna be?" I stood up and started to remove my pants. "Are you taking off your clothes too, or do I put on some music?"

Brashaney gazed at my dick and started to remove her clothes. I could tell she was young because she had young, fresh titties. Nipples didn't even look full-grown yet. But when she got on top of me and swallowed my dick like a shark, I wasn't really sure.

I slid on a condom and just lay there as she put an arch in her back and tried to put it on me. I rubbed her ass a few times, and that was it. She took my hand and tried to make me feel her insides, but I couldn't get with it.

The thought of Stephon fucking Nokea was fresh in my mind. I remembered how he told me he be airlifting women, sucking their pussies, and I wondered if he'd done that to Nokea. I also wondered if she liked the feel of him better than me. How or why would she bring that much hurt to me, I didn't know. And Stephon's backstabbing ass . . . I couldn't believe he had gone there when I simply told him not to.

I was unable to focus on my intimate time with Brashaney, so I stopped.

"What's wrong, Jaylin? Aren't you enjoying yourself?" She took my hand and rubbed it on her pu-tain.

"Yeah, I'm enjoying myself. Let's get this over with so I can go to sleep, though."

I flipped her over and took control. I banged her insides just so I could have the pleasure of hearing her say my name. When I heard that, I cut her short and came. She was all up on me, trying to get close, but I wasn't having it.

"Say, baby, I'm really tired. Why don't I call you tomorrow?" I said.

She yanked the covers back, put on her clothes, and left without saying good-bye. I wasn't sweating it. I got what I wanted, so what the hell?

I woke up in a sweat as I dreamed about my Aunt Betty. The thought of her beating me and throwing me into a dark closet wouldn't leave my memory for shit. I prayed many nights for Mama to come back and save me, but she never came. I hoped Daddy would come and take me from that orphanage, but he never showed.

I lay across my bed, and a few tears rolled from my eyes as I thought about my horrifying past. To me, women were only good for one thing.

Wasn't no telling when I would feel any differently, but I knew that treating them the way I'd been doing wasn't going to make my life any better. Changes had to be made, but I knew that would take some time and willingness on my part.

29

NOKEA

Stephon and I cleaned up the place as best as we could, and then I asked him to leave for a short while so I could be alone. He called J's Towing Service to come get his car, and they took it to Al's Body Shop.

I was sick about what happened. I didn't intend for things to go as far as they did, and I really didn't expect Jaylin to be that upset with me about Stephon. I mean, they'd shared plenty of women before. Wasn't no biggie for them, so why did he trip with me? And the baby? I wondered if he believed me when I told him the baby was his. Maybe that's why he calmed down a bit and left. Either way, I'd never seen him that angry with anyone. I didn't like the fact that it was me who had upset him the most. I was the one who was always there for him, and for him to call me those names was extremely hurtful.

Ever since I saw him, I couldn't get him off my mind. The look on his face was terrifying, and I could only imagine how hard he was still taking the news. I'd known Jaylin for a long enough time, and I knew he'd shut out everyone and try to face this all alone. And, even though I enjoyed being with Stephon, I hated to admit that Jaylin still had my heart. Just having him in my presence did something to me.

Maybe I was wrong for sleeping with Stephon, but I had needs. I'd already gone thirty whole years without sex, and nine years without having sex with the man I loved. Since Stephon was the one who was willing to take care of my needs, I had to go with the flow. And Stephon wasn't there only for my physical needs. He'd been there for me: cooking for me, massaging my body for me, listening to the baby's heartbeat, and feeling the baby move—everything Jaylin should have been doing. It wasn't my fault he decided to be with Scorpio.

I guess things didn't work out with them either. He sounded upset with everybody, and I guess in his mind, he had good reason. But when was Jaylin going to realize that life didn't revolve around him? He had his own little messed-up world, and if people didn't do things his way, then there was no other way.

I sat around the house for hours and thought about Jaylin as I took care of the damage he'd done to my place. I had insurance on my piano, so the company I bought it from would replace the smashed one. I told them someone had vandalized my place. This guy Pat knew came over to fix my drywall. Pat came with him because I didn't feel comfortable letting a strange man in my house.

When she walked in, she was stunned. Plaster was all over the floor, my table was without glass, and one of my expensive statues was without a head.

"Girl, what the hell went on up in here? I know you told me it was bad, but this is ridiculous. You need to send that motherfucker a bill. He can't be coming over here tearing up your stuff like this."

"Come on in the kitchen, girl. I know Jaylin was wrong for messing up my things, but if you think about it, he kind of had good reason to."

"So if he would've punched you in your stomach, he would've been right? Is that what you're saying, Nokea?"

"No. All I'm saying is maybe, just maybe, I was wrong for not telling him about the baby to begin with. And then, to turn around and sleep with his cousin? I . . . I'm not sure if that was the right thing to do."

"I can't believe that after all this fool done to you, you still sticking up for him. Now, you did what you had to do based on the situation he put you in. If Stephon—or anybody, for that matter—stepped in and picked up where Jaylin left off, then, hey, his loss. Although I'm not saying sleeping with Stephon was the best thing to do, because I think he's a bigger ho than Jaylin is. If not a bigger ho, a smoother ho. He knows how to charm a woman out of her panties just like Jaylin. The problem is both of them fine, and now you've got your work cut out for you."

"Pat, I don't care about who's the finest between them. All I care about is who's going to be there for Nokea and her baby. That's it. I want a man to love me and to be a good father to this baby. And if that's Stephon, then that's what I want. I can't see Jaylin loving me like I want him to. Right about now, he wants to kill me. I don't stand a chance with him anymore," I said, looking down at the floor.

"Nokea, forget about who loves you for a minute. Where's your heart? Who do you really want to be with? Now, I can tell you who I want you with, but you got your own mind. So, make your own decision."

"What's your recommendation, Pat? Who do you want me with?"

She looked at me and crossed her arms. "Neither one. Personally, I think you were wrong for sleeping with Stephon, but since I hate Jaylin so much, I went with the flow. You need to get rid of both of them and move on. You're a beautiful person, Nokea, and you won't have a hard time finding somebody who will truly love you."

"But it's not that easy. Being with Stephon is great, but my heart still belongs to Jaylin. I don't know what I'm going to do. I'm worse off than I was before." I put my head on the table and started to tear up again. Pat tried to console me, but even she said I messed up.

She stayed with me until Stephon came back over. When she opened the door, she stared him down like a bloodhound. I rushed her out the door because I didn't want her starting no mess with him over the baby.

"Look, Nokea, you don't have to push me out the door. I'm leaving. He already knows he ain't got no business over here."

"Pat, I got plenty of business over here. I got a baby on the way and a woman I love. If that ain't enough business, then I don't know what is."

"Negro, please. Spare me the lies. Save them for Jaylin so he won't whip your butt again. You know darn well that ain't your baby, because if it was, you wouldn't even be here. And love? Don't make me sick. I think it's pathetic and absurd you call yourself in love with Nokea. Too bad she don't know any better because—"

"Pat, stop! I've had enough drama already. Please don't do this, okay?" I asked politely.

She threw up her hands and walked out. "Call me later, girl. Love ya," she said.

I shut the door and looked at Stephon. "Tell me, am I wrong for wanting to be happy? Can't I be with who I want to and leave it at that?"

Stephon came over and held me in his arms. "Nokea, we both knew this wasn't going to be easy. Baby, this is just the beginning. Hell, you haven't even told your parents about us yet. I'm not sure how they're going to feel about me being Jaylin's cousin. But when all is said and done, we're going to be happy together. You, her, and me," he said, rubbing my belly.

"No, it's a boy. I want a boy so I can name him—" I was about to say Jaylin, but I stopped myself. "A girl will be just fine."

"No, no, now, if you want a boy then pray for one. And when he comes, we'll talk about what to name him then."

"I said a girl would be fine, Stephon. I don't ever want to think about having a boy again."

We sat on the outside deck and chilled. Stephon wanted to make love to me, but I couldn't get with it. After he jumped in the swimming pool naked, I joined him. I wasn't even ashamed of my fat stomach because he made me feel like I was the most beautiful woman that walked the earth. He held me in the swimming pool and rubbed every part of my body.

Mentally, I still wasn't there yet, so I hopped out of the pool and lay on a beach towel while looking up at the sky. Stephon played in the water and did all kinds of crazy flips, trying to make me laugh. Thoughts of Jaylin, however, continued to occupy my mind. When my laughter went away, Stephon came over and laid his wet body beside me.

"Shorty, I know you're still in love with Jaylin, but just think about what we could have for a minute. I'm not trying to make you love me, but it would be nice if you would be willing to give it a try."

I turned on my side and looked into Stephon's eyes. "I'm trying, Stephon, but it's so hard. Seeing Jaylin again kind of brought out some feelings I didn't know were still there. So, be patient with me. I want this to work between us too, but it's going to take some time."

He kissed me from the top of my forehead to the bottom of my feet. He made love to me like this was the beginning of something special.

I was there with him for a while, but when I closed my eyes, I saw Jaylin. I squeezed my eyes tighter until he went away. I didn't expect him to be away for long, and deep down, I knew the drama was just beginning.

30

JAYLIN

Since I'd last seen Mackenzie, it had been almost a month. I felt like I was in another world, and only left the house when it was necessary. I knew that leaving a lot would cause me to do something stupid, like go cuss out Nokea some more or provoke another fight with Stephon. I decided to let them have each other. As a matter of fact, they deserved each other. I was truly bothered by their relationship, but with Nokea being pregnant by him, what could I do? I knew her parents would want her to be with the father of her child, and Nokea would do anything to please her parents. It was over, and once again in my life, I had to accept the situation and move on.

Other than Nokea, Scorpio's pussy was heavy on my mind. Since I was occasionally sticking my dick in

Brashaney, the rest of my time was dedicated to finding Mackenzie. I'd been to one law firm after another trying to find out what I could do. After paying big dollars for consultation after consultation, I gave up. The only thing I heard was that if I wasn't her biological father, there was little I could do.

I knew so little about Scorpio that I didn't even know where to begin to look for her. I called her house, but the number had been disconnected. I even drove around Olivette to see if I saw her car parked somewhere, but no luck. I was starting to give up on them too, but something wouldn't let me.

Since Ray-Ray's party was that night at his cousin's house, I stopped searching and went home, showered, and changed into one of my favorite outfits. I put on my black Armani wide-legged linen pants and my black-and-off-white, thick-striped linen shirt with an oversized collar. Then I put on my off-white gangster hat and tilted it to the side. GQ magazine didn't know what they were missing. My Rolex was on one wrist, and my thick gold diamond bracelet was on the other. I was bling-blinging, ready to get my party on as soon as I sprayed on some Clive Christian. I put twenty hundred-dollar bills in my money clip and stuck it in my pocket.

I knew I'd see Stephon tonight, but I wasn't backing down on going. I had other friends who would surely be in attendance, so what the hell?

As I was preparing to leave, Brashaney rang my phone. I had to answer because I didn't want to mess up the only booty I had lined up, especially since I knew I'd probably want to get my sex on after the party tonight. The last time we hooked up, I went through her purse while she was asleep and peeked at her driver's license. She didn't lie about her age; she was actually twenty-seven, and I felt bad about not being able to trust her.

I ended the conversation with Brashaney and told her to meet me at my place around two in the morning. I told her to make sure she called before she came just in case I wasn't home; then I jetted. I was looking good, feeling good, and smelling good, so I decided to drive my Boxster for the night.

Ray-Ray's bachelor party was at his cousin's mansion in Ladue, and I was dying to see what it looked like. I heard it was banging, but I also heard it wasn't banging better than mine; however, when I pulled up, I knew I had some competition. I had to push an intercom button and announce who I was just for the gates to open up and let me in. Then, I drove up a long driveway lined with waterfalls and marble rock landscaping. And when I saw the house—damn! It was bad. Twice the size of mine, with big white columns in the front, like the White House.

When I pulled up in my car, one of the valets parked it alongside the other cars. Mercedes, Lexus, Lincoln, Cadillac, Jaguar, BMW; you name it, they were there. But I was glad to see that nobody had a Boxster like mine. The lot also had its share of fucked-up cars too, so I knew it wouldn't be all good inside.

There was already about a hundred brothas outside, so I imagined what the inside would look like. As I walked past them, they checked me out from head to toe.

As soon as I hit the door, I saw men all over the place. There was a double staircase cluttered with people. Mesmerized, I stood in the spacious foyer, which was covered with green, black and white marble. Shit was off the hook. Whoever told me I could compete lied like hell.

I walked through the place, and I tried to figure out how I could be down. I was more infatuated with the house than I was with all the ladies that flounced around butt naked.

I finally stepped into the room where the party was actually happening. It was packed with fellas and the music thumped loudly. I bounced my head to the rhythm and quickly scoped one of my boys from the barbershop. He came up with a couple other brothas and asked what took so long for me to get there.

"Man, I was trying to hurry, but perfection takes a little more time," I boasted. Yes, I was arrogant, but I didn't give a damn.

"I heard that," Ricky said, giving me five. "You know you clean, though, bro. And where in the hell did you get that shirt? It's off the chain."

I usually didn't tell anybody where I got my things from because I didn't want nobody trying to look like me. So, I pretended that I didn't hear him, and then stepped away from Ricky to get a drink at the bar.

My eyes searched the crowded room for Ray-Ray, but I didn't see him. I knew he was somewhere, but finding him would be like searching for a needle in a haystack.

As I tipped the bartender, the lights flashed on and off, which meant a stripper was on her way out to entertain. The floor cleared, and all the brothas, including me, watched as this dark chocolate sista came in and danced her way to the middle of the floor. She danced to a slow, funky song as we all watched and waited for her to take it off. She had on a red leather shorts outfit with her cheeks hanging out. Her tinted blonde hair was straight and hung down to her butt.

When she sat in a chair and stretched her legs straight out in the air, motherfuckers hollered like they ain't never seen no pussy before. She grabbed this one brotha on the floor and put her goods all in his face. She tied him to a chair and pulled out some whips. Now, this shit was too damn freaky for me. Not interested in that kind of action at all.

I went back over to the bar and got another drink. A stripper who had already performed came over and stood next to me. She wiped the sweat from her forehead and threw a towel over her shoulder.

"So, are you having a good time?" she yelled over the music.

"Yeah, it's cool. But have you seen the groom yet?" I asked, leaning down toward her.

"The last time I saw him, he was in one of the rooms upstairs." Made sense. That was probably where I should've looked for him.

I smiled. "So, what's your name?"

"My real name or my play name?"

"Whichever one you want to give me," I said, really not giving a fuck.

"It's Nicola. What's yours?"

"Jaylin," I said, looking to see what all the hype was about because the men were hollering again.

"So, Jaylin, did you see my performance tonight?"

"No, I just got here. I must have missed you."

"Well, if you'd like, I can give you a private show in one of the rooms upstairs."

I looked down at Nicola and set my drink on the bar. No doubt about it, she was fine. But since I thought about how Scorpio probably put herself out there like this, I had to turn her down. She didn't give up, though. She stood next to me while we both watched Dark and Lovely entertain.

When she finished, Dark and Lovely walked out of the room with all kinds of dollars stuck in her thong, down her top, and in her hand. I slid her a hundred-dollar bill because the sista worked hard for it. She smiled and whispered that she would be back.

Since Nicola hadn't budged, I had no choice but to pay her some attention. I wasn't sure what she wanted from me. Even though I knew I had it going on, I hadn't paid her nearly much attention as the other men.

I finally saw Ray-Ray. He came down the steps with two chicks on his side. Stephon was right behind him with another chick.

I excused myself from Nicola and met Ray-Ray at the bottom of the steps.

"What's up, playa?" I said and then gave him five. Stephon and me eyeballed each other as he walked away.

"Man, man, man! Where have your ass been? I've been bragging to all the ladies about my partna Jaylin and you just now getting here?"

"Fool, I've been here. Since I got here, I've been trying to find your ass. But you can't be found if you don't want to be," I said, looking at the two chicks.

He laughed and took his arms from around them. "Hey, let me holla at you for a sec," he said.

We walked into another room where just a few people were chilling on a circular leather sectional.

"Listen, I'm not going to hold you up, because I know you want to get your party on like I intend to do. But Stephon told me what happened. I'm hurt, man, because we've been boys for too damn long. And to let some bitch come between y'all, that ain't even cool. I'm not saying y'all need to squash things tonight; all I'm saying is y'all need to get together and talk this shit out."

"Ray, thanks for trying, but this shit goes beyond him fucking my woman. If you're worried about me tripping tonight, then don't. I don't have any intention of messing up your night."

I grabbed Ray-Ray on his shoulder and we walked back into the other room. Stephon stood next to Nicola, but when he saw her look my way, he moved away. She walked back to me. I looked at him and he looked at me.

"Say, did I interrupt something?" I asked.

"Naw, seemed like she was hollering at you, dog."

"Right, right. But you know if you want her, I ain't got no problem with that."

"Nope, ain't interested. Besides, I already got my shit off tonight with that sista right over there," he said, pointing to the chick he came down the steps with.

"Well, I got two questions for you: Was it good, and how much did it cost you?"

He laughed.

"Aw, it was good. And so was I. So good that once I was finished, she gave me fifty dollars back."

"That's how you do it," I said, giving him five.

Stephon reached over and gave me a hug. I hugged him back. It felt good that my cousin and I were on our way to settling our differences. He was way more important to me than Nokea, and we were both wrong for allowing a woman to come between us.

Nicola had worked her way over to some other

brothas, so Stephon and I talked about the fine-ass ladies in the house.

"Man, I can't believe you haven't found you no shit up in here yet," Stephon said, gawking at everything that walked by. "It's all kind of ass floating around here."

"Yeah, I've been checking it out. I'm just waiting for the right one. There ain't no question in my mind that I'll be fucking tonight. The only question is, who's going to be the lucky woman?"

After scoping the set with Stephon, I thought about my conversation with Brashaney tonight. I pulled out my cell phone and called to cancel our plans. I knew there was no way I would be home by two. She didn't answer, so I left a message and told her I'd call her tomorrow.

When Stephon walked to the bar and got another drink, the lights flashed again. The floor cleared and the fellas made room for the next stripper. Nelly's "It's Gettin Hot in Here" started playing.

I almost spilled the drink in my hand when I saw Scorpio enter the room. She had on some thigh-high white leather boots with spurs and a cowboy hat. As she walked in and swayed her hips from side to side, the white tassels that covered her ass also moved from side to side. Her top was a strapless corset with fringes dangling from it. She strutted around the floor in circles and the men went crazy. She tossed her hat into the crowd, and her long, wavy hair fell down over her face.

Stephon looked at me and pointed. "Is that—?"

"Yes, but don't say anything. I don't want her to see me." I ducked behind Stephon and this other dude who seemed to really be enjoying the show. I had to see for myself how nasty she could get at one of these parties that she claimed were so innocent.

When the music changed, she pulled not one brotha, but three on the floor. She sat one in a chair and the other two on the floor. As they eagerly waited for her to get busy, she removed her top, swung it in the air, and tossed it on the floor. Then the money came in. It covered the floor. When she slid out of her shorts and draped them across the face of the brotha in the chair, I cut my eyes and shook my head with disgust. She was left standing with a sheer white thong. Wasn't no need for her to have it on because everyone could see the smooth hairs on her pussy.

My heart raced as I continued to watch. Stephon and the rest of the motherfuckers stood with their mouths open, mesmerized by her trashy performance. And when she flipped her body over the chair and wrapped her legs around the brotha's face, with her face in his lap, I had seen enough. I put my drink down and headed for the door.

Stephon grabbed me. "Come on now, Jay. You better than that, ain't you? Don't let this woman ruin your night. Besides, it ain't nothing but a show anyway."

He nudged me back into the room, and I sat in the corner where I couldn't see what went on. The noise was distracting because brothas screamed and hollered more for her than they did anybody. It was a good thing I'd ended it with her before tonight, because if I hadn't, I'd be up there kicking her ass.

After Scorpio racked up the dollars, she blew kisses at the fellas and swished her ass out of the room. I at least wanted to find out where Mackenzie was, so I told one of Ray-Ray's partnas to stop her when she came out of the bathroom. I told him to tell her that if she was willing to come to the room at the end of the hall, it would be worth a thousand dollars.

After I paid him a hundred dollars for doing it, he stood by the bathroom door and waited for her to come out. I ran upstairs to the room at the end of the hall and took off my shirt so she would think I was naked. Then I turned off the light, hopped in bed, and waited for her to enter.

The room had an odor like somebody had already been there fucking, but it was still a bad-ass room. It was gold and white, had mirrors on the closet doors that went from one end of the room to the other. The wallpaper was gold-and-white-striped, and the rugs matched the white curtains that draped the windows. The bed had gold satin sheets, but I threw them off, not knowing who had laid their ass on there tonight.

I sat up with my arms folded, my legs crossed, and leaned back on the soft pillows as I waited for her to come. It was dark as hell, but I could see a sliver of light coming through the window.

When the door creaked, I took a deep breath. She walked in and shut the door behind her. I knew it was her because I could smell her perfume.

"Excuse me, are you going to turn the light on?" she asked, still standing by the door.

"Are you going to take off your clothes for me like the song said?" I asked in a deep voice.

"No, I'm not. I came in here to tell you I don't go out like that. If sex is what you want, I'll set you up with one of my girlfriends out there. As a matter of fact, she's waiting for you now."

"I'm not interested in her." I made my voice deeper.

"I got a thousand dollars right now that got your name all over it."

"Sorry, sir, that ain't my style," she said, turning to open the door.

"I said that I'm not interested in her. I want you."

I reached for the lamp and turned on the light.

Scorpio turned and looked at me. She cracked a tiny smile. "Jaylin, what are you doing here?"

"I came here to party, just like everybody else," I said, admiring her sexy body in her tight blue jeans and matching jean halter-top.

"Did you try to set me up or something?"

"No. I liked what I saw downstairs and wanted a sista to come shake a brotha down. But since you don't get down like that, I guess I'll have to settle for your girlfriend."

She smiled. "Well, I can always make an exception for a man like you."

"Naw, naw. Don't do me any favors. I'm sure your girlfriend wouldn't mind shaking a brotha down tonight. Go ahead; call her in here so I can get this party started."

Scorpio turned around and threw her backpack with her clothes in it over her shoulder. "Well, you have a good time. I came here to do what I had to do tonight; now I'm going home."

I gazed at the gap between her legs where I used to lay my dick, and I couldn't let her walk out. I jumped out of the bed and put my hand on the door. As I pressed my body up against hers, I moved her hair to the side and whispered in her ear.

"Don't go, baby. Let's make love. I'm sorry for not trusting you. I'm sorry for putting my hands on you.

Just . . . just make love to me right now. I miss you beside me. I miss having your body next to mine."

She turned around and put her finger on my lips. "Jaylin, that all sounds good, but all I want to know is do you love me? If you miss me so much, then tell me. Do you love me?"

"Baby, you know how I feel."

"No, I don't. Tell me how you feel. If you can't tell me you love me, then ain't no sense in me hanging around."

I took a few steps back and looked her pretty self in the face. I rubbed her cheek and then kissed it. "Where's Mackenzie?" I asked, as I wasn't about to tell her something I didn't mean.

"Jaylin, is loving me that hard for you? All I'm asking is for you to love me, that's all."

"Scorpio, I can't. I'm sorry, but I don't feel that right now. Why do you want me to lie to you? What's the big deal? My feelings don't get any stronger for you than they are right now. If this ain't enough, I'm sorry. So," I said with a sigh, "where's Mackenzie?"

"She's at my sister's house."

"Bring her over to my house tomorrow so I can see her."

"I'll try. I have a few things I have to do tomorrow, but I'll try."

"Try hard." I opened the door so I could let her out.

She walked down the steps in front of me, and brothas pulled on her like she was a piece of gold. She smiled as a couple of them stuffed money into her back pocket, then she walked out the door.

Stephon stood in the doorway to the party room. I walked up to him. "Come by my house tomorrow so we can talk," I said.

"Fo' sho'. Any particular time?"

"Anytime. I'll be there."

He nodded.

I saw Dark and Lovely, who performed earlier, grabbed her hand, and told her let's go. Maybe a little whip action was what I needed. She put her drink down and didn't waste any time getting in the ride with me.

As we were leaving the premises, I saw Stephon getting into his car with the sista he'd been with earlier. He pulled his car in front of mine and cut me off. He got out of his car and came over to my window.

"Hey, man, I got something for you," he said, giving me a piece of paper.

I looked at it. It was a bill for $11,386 for the repairs to his car.

"Pay the bill, motherfucker," he said, laughing. "Pay the bill."

I ripped it up and threw it at him. "Negro, please. If I pay your bill, you can pay mine. If you don't move your car so I can go home and make love to this woman, I'm gonna tear your shit up again."

Stephon hopped back into his car, still laughing, and sped off. I zoomed past him on Lindbergh Boulevard and blew the horn.

I wanted to ask him about Nokea, but I was staying my ass out of it. She was sadly mistaken if she thought Stephon was any better than I was. He was slick, and knew how to charm the hell out of women. I didn't have time for that shit; I was the kind of man who always told it like it was, no matter what. I guess that's why Scorpio couldn't stand me. But I wasn't changing my ways for nobody.

When I pulled into the driveway, Brashaney's car

was parked in front of my house, even though I'd told her not to come over. Either that or she didn't get my message. I stepped out of the car, and Sexy Chocolate got out on her side. Brashaney stormed up the driveway and positioned her hand on her hip.

"Say, baby, I called and left you a message. You didn't get it?" I asked.

"No, Jaylin, I didn't. So, what's up? Are you kicking it with her tonight or what?"

"Well, since both of y'all here, why don't we make the best of it?" I said, thinking that a threesome would be good right about now.

Brashaney smacked the shit out of me. I could only smile. I knew I'd played her, but, hey, at least I tried to call.

She walked back to her car, threw her purse inside, and sped off.

I worked Sexy Chocolate all night long. I put on a condom three different times, but my big dick and all the pressure I put into this woman caused the damn thing to slide off. I hated using condoms, but under these kinds of conditions, what else could I do? I wished they'd come out with something special for a brotha packing it like me, because this shit just wasn't working.

I pulled off the torn condom and we went at it again. After I made her come six times, I'd had enough. All the moaning and groaning, screaming and hollering, and "oh, baby, this is the best dick I've ever had" bullshit worked my nerves. Been there and heard that shit before. I rolled my happy ass over and went to sleep.

After Sexy Chocolate left, I laid my lazy butt around

the house all day doing nothing. I waited for Scorpio to call and tell me when she would bring Mackenzie over, but it was almost three o'clock in the afternoon and she hadn't called yet. Brashaney did, though. She called and cursed my ass out for playing her like I did last night. I didn't say a word. Wasn't no sense in me getting all hyped up about a woman I really didn't care about. She went on and on about what a dog I was, and I just agreed and then ended the conversation.

Women are just some crazy creatures, I thought. They like too much fucking drama if you ask me. If somebody played your ass like I did last night, why would you still want to be bothered? She was actually calling here to try to fix shit. If I was such a dog, then why waste the time? Never could figure some women out—and wasn't trying to either.

I lay back on my bed with my hands folded behind my head and laughed at the situation. The phone rang again. This time, it was Stephon. He asked if I wanted to catch dinner at Applebee's, or order some Chinese food from the hood. Since I figured our conversation would get pretty deep, I suggested he come to my place and I'd have some Chinese food delivered. He agreed and said he was on his way.

I went downstairs to the kitchen and tried to find something to snack on because I was starving. When I realized I hadn't been grocery shopping in a while, I picked up the phone and quickly ordered our Chinese food. I ordered Stephon some house special fried rice with extra shrimp and bean sprouts because it was his favorite. Then I ordered the same for myself, minus the shrimp and sprouts. I threw in an egg roll just to be greedy.

My place seemed to be a bit out of order, so I called

Nanny B and arranged to have her come over the next day. I'd been in a slump, and cleaning up my place was the last thing on my mind. I asked the nanny to do everything from washing the windows to scrubbing the floors. She gave me a price, and I was all for it. Told her I'd see her at nine in the morning.

Finally, Stephon showed up, as beat as I was from last night's festivities.

"Damn, I thought I looked bad. What happened to your ass? Did you at least brush your teeth this morning?" I asked, laughing as he strolled in.

"Fuck you, man. I got a damn hangover. Hangover from drinking and from that pussy. Couldn't get that woman out of my place until an hour or so ago." He plopped down on the couch.

"Yeah, I know what you mean. After fucking, you just want to say, 'Get your shit and go,' don't you? But you know how it is; gotta play the nice, 'Oh, I really want you to stay here with me,' role just in case you might need that ass another time."

"You got that right," Stephon said, slamming his fist against mine.

"So, uh, I hate to ask you this, but do you want something to drink?"

"No, thank you! A tall, cold glass of ice water will be just fine."

"Just in case you forgot, I don't serve water to my guests, so get your ass up and go in the kitchen and get it yourself."

Stephon went to the kitchen, and I went to the bathroom. "Say, man, I ordered us some Chinese food," I yelled.

"Oh yeah?" Stephon yelled from the kitchen. "What did you order?"

"I ordered some—ahhhhh, shit!"

"Some what?" he yelled back.

"Damn! What the fuck!" I yelled, leaning my hand up against the wall and holding my dick in my other hand.

Stephon rushed to the bathroom door. "Man, what the fuck you doing?"

"My dick is on fire! This motherfucker straight up feel like fire shooting out of it."

"Fool, quit playing."

"Playing? Do I look like I'm—ahhhhh, shit!"

The doorbell rang, but I was in no condition to get it. "That's probably our food. Get the money off the table and give it to them. And don't give him a tip because he late. They told me ten minutes and it's been thirty-five."

"Man, you knew they weren't going to get here in no ten minutes."

"Well, they shouldn't be lying to people by telling them that bullshit all the time then. So, whatever you do, don't give him a tip."

Stephon laughed and went to the door. I slid my thang back into my pants, flushed the toilet, and washed my hands. I took slow steps entering the kitchen. Stephon cracked up as he removed the boxes from the bag.

"What in the fuck is so funny? I need a doctor," I said, looking for my doctor's number on the refrigerator.

"Jay, it ain't like you gon' die or nothing. I know her office closed today, so you might as well wait until tomorrow. Sounds like you might have a STD."

"Damn, does it have to burn like that? I ain't never

got caught up with no shit like this, and if that's what it is, I'm gon' mess somebody up."

"Yes, it burns like hell. Happened to me a couple of times. That's why I be telling you to strap your shit up. Ain't no telling where some of these tricks been."

"I do strap myself up—most of the time. I've had two women within the last week, and the condom ripped with the one I had last night. I guess I know where to place the blame."

"You gotta take responsibility too if you want to continue getting down like you do. It's too much shit out there now. And since motherfuckers dropping off like flies, I make sure I strap my ass up no matter what."

"I know, but the condoms just don't fit right. They be sliding off and everything. I'm gon' have to strap up my handler with a trash bag or something."

We cracked up, and then started to chow down on the rice.

"So, what was up with Scorpio last night? Are y'all still kicking it or what?"

"Nope. Not right now anyway. I'm sure I'll be back in them panties soon."

"What happened? I mean, I didn't even know she could get down like that."

"Shit, I didn't either. That's why I kicked her ass out of here. She lied, man, flat-out lied about her occupation. I didn't want to tell you the details because I didn't want your ass to tell me how badly I fucked up."

"Well, you did. But we all fuck up sometimes. I hope you learned from your mistakes.

"I feel you, though, about a woman who lies. Ain't nothing worse than a lying-ass woman. It's all right for us to lie, but women are supposed to be better than that."

"I agree. And it sure in the hell hurts when they do," I said, thinking about Nokea. I wanted to ask Stephon many questions about their relationship, but I didn't want him to think that I was interested in getting Nokea back. I knew if I said the wrong thing, that would spark an argument between us. I definitely wasn't trying to go there.

"So, my brotha, how are things going with you and Nokea? She's just about ready to pop that baby out, ain't she?"

"Yeah, just about. She's getting pretty big, but she looks good, though. Cutest little pregnant woman I've ever seen," Stephon said, smiling.

I chewed my food and tapped my fork on the table. "Tell me something . . . when was the first time you had sex with her?"

Stephon was silent for a while and then cleared his throat. "About two weeks after you did. Remember when you went over to her house—had sex with her after you had been with Felicia in your office?"

"Yeah, I remember. Remember well."

"Well, she found out about that. She was upset and came to me crying. After that, one thing led to another, and then she told me she was pregnant."

I thought about our conversation at the funeral. "So, you had sex with her before Aunt Betty's funeral? Don't you remember our conversation at the funeral? Just to jog your memory, I told you then that Nokea was the exception. You knew I didn't want the two of you getting involved. I made that quite clear."

"Yes, I remember what you said, but she was already pregnant by then."

I thought about how thick Nokea's breasts looked at the funeral, and my comments about her thickening up a bit. Stephon wasn't lying. "Okay, but if we had sex with her only a few weeks apart, how do you know it's your baby and not mine? After all, she did tell me it was my baby too."

"First of all, she lied. She told you it was yours because she didn't want you to punch her stomach. How I know it's mine is she had her period after she had sex with you. And when she went to the doctor, he pinpointed her delivery date nine months after the last time we had sex."

Stephon was unable to look me in the eyes. I wasn't sure if it was because he felt guilty or if he was bullshitting me.

I took a bite of my egg roll, folded my arms, and looked at him. "So, do you love her?"

"Yeah, I can pretty much say I do. Nokea's been in my heart for a long time, Jay. I know that's not what you wanted to hear, but she has. Problem was she could never get over you. But when she did, I saw an opportunity and took it."

"But the door wasn't open for you yet. She was still with me and still in love with me when you took it upon yourself to fuck her."

"I know, but she was tired of all the bullshit. After that incident with you and Scorpio in the shower, that did it. She might have slept with you after that, but her feelings were fading."

"Man, you know better than I do a nine-year feeling don't go away just like that. If I had to place a bet on it, I'd say you were the one who approached her and went after her vulnerability."

Stephon wiped his mouth with a napkin and

rubbed his hands together. "Actually, it worked both ways. She came to me, and I was willing to comfort her."

"If you don't mind me asking, how often do you comfort her now? I mean, since you out kicking it with other women and everything. Is she shaking a brotha down like she supposed to?"

Now, Stephon looked me straight in the eyes. "Not a day goes by that I don't make love to her. Lately, it's been here and there, but that's because of the baby. But I do enjoy every moment I spend with her."

My throat ached, but Stephon would never know it. Hurt me like hell that I'd waited nine years to make love to this woman, and now she was giving it to my cuz every time he wanted it. I wanted to stop torturing myself with the questions, but I had a few more for Stephon.

"So, if you say you love her so much, why you cheating on her?"

"Because I'm a man and that's what I do. These other bitches out here don't mean nothing to me. They just something to play with."

"Then why can't you be honest with Nokea and tell her you got other people in your life? I think she's going to be more hurt by you lying to her than she was by me being honest."

"And I disagree. I think what she don't know won't hurt. Eventually, I'm going to cut the shit anyway; and when I do, I'm going to be with her and only her. So why mess that up by bringing all this drama to the re-lationship?"

"Okay, man, your call. I just hope you can live by your words."

"You know I'll do my best," Stephon said. He reached out and slammed his hand against mine.

Stephon chilled at my place for a while. We played several games of pool and hooked up my video game to my theater-sized TV in the bonus room. He left the room a couple times and called Nokea. I don't think he told her he was at my house, but when I heard him tell her he loved her, his words stung like hell. Maybe asking him all those questions wasn't the right thing to do. I'd never felt jealous before when it came to him, but realistically, he had my woman.

As I thought about Nokea, my heart went out to her. Stephon and I were damaging her. Now, it was all up to him to make things right. At the rate he was going, I didn't think he was capable of doing that. I'd definitely have to sit back and watch him try.

When Stephon left, I hurried to the bathroom and almost fell to the floor, trying to stop the pain as I urinated. I couldn't wait to go see my doctor the next day. I truly had nobody to blame but myself for settling for any kind of woman and not fully protecting myself.

I finished up in the bathroom and the doorbell rang again. Damn, can a brotha get some peace around here? I thought. But when I saw who it was through the window, I jetted downstairs to open the door. It was Mackenzie and Scorpio.

When I opened the door, Mackenzie jumped right into my arms.

"Daddy!" she yelled as I hugged her and swung her around in my arms. She gave me a big wet kiss on the cheek, and then wiped the spit off with her hand. "Why did you make me leave?" she pouted.

"Mackenzie, I didn't make you leave. I had no idea your aunt was coming to get you that night."

"Mommy said you made us leave, and when I asked why, she said because you were mad at us."

I put Mackenzie down and cut my eyes at Scorpio. She shut the door and tooted her lips.

"You were mad at us, weren't you?"

"I was mad at you, not her." I dropped to one knee and looked into Mackenzie's eyes. "I'll tell you the truth one day when you're old enough to understand. But for now, please know I would never do anything to hurt you."

She smiled. "Can I go up to my room? When I left, I forgot to get something."

"Sure, baby. It's just like you left it. And you don't have to ask if you can go up to your room because it's yours. Okay?"

She ran up the steps to her room. I looked at Scorpio, wanting to kill her for lying to Mackenzie.

"Why did you tell her I put her out?"

"Because you did. When you kicked me out, you kicked her out. You didn't think I would let her stay here, did you?"

"I don't know, but I'm glad you brought her over here to see me. Has she missed me?"

"Jaylin, I don't know what you've done to my baby. She's been crying almost every day, talking about how much she misses her daddy. Why did you tell her that you were her father? You're gonna confuse her, and she don't need that right now."

"Scorpio, look, I explained to Mackenzie what type of daddy I was. She knows I'm not her biological father."

"No, she doesn't. She really thinks you're her daddy."

"Well, I'll have a talk with her later so she'll understand."

Scorpio walked further into the house and her eyes scanned the rooms. "Have you had company lately?"

"Why?"

"Because I want to know who's been getting what I've been thinking about every day since I left. Excuse me; I mean since you kicked me out."

"Get over it, would you? You know damn well that I had a good reason. Now, I was wrong for putting my hands on you, but I apologized for that."

"Naw, baby, you apologized with your mouth. I was hoping you would apologize with something else." She stepped up to me and placed her hand on my goods. Burning and all, my thang had the nerve to get hard. Since she had come over looking very good, I had to make sure I kept my distance and focused my mind elsewhere.

She placed her arms on my shoulders. "You know what I want?" she said, kissing me on the lips.

I eased my arms around her waist. "What do you want, baby?"

"I—you don't mind if I'm blunt, do you?"

"Be as blunt as you'd like."

"I want to have sex—fuck you, right now. I'm in no mood for those five-minute, I'm-too-tired fucks, but one of those all-nighters you give me when you're at your best."

My dick throbbed. I wanted to strip her ass naked and wear her out right then and there. But how was I going to get myself out of this one tonight?

I kissed her forehead and massaged her butt. She

backed away and led me up the steps. When we got to my room, I was able to come up with an excuse to save me for a while.

"Scorpio, you know I can't make love to you with Mackenzie still wide awake."

"Oh, I know. But I kept her up all day so she'd go to bed early tonight."

"You just knew you could bring your little sexy self over here and use a brotha for his thang when you wanted to, huh?"

"Yeah, pretty much; especially when he be using a sista for hers when he wants to."

I hopped off the bed and got me a shot of Remy. I almost didn't care about my little problem. All I wanted was to feel her insides. I tried to talk about something else, but no matter how hard I tried, she kept bringing me back to having sex with her tonight.

I was glad to see Mackenzie come through the doors. But when she came in and rubbed her eyes, I knew what time it was. She yawned and walked over to the chaise to sit next to me.

"Daddy, I'm sleepy. Can I stay the night with you?"

"Of course, Mackenzie. You can spend as many nights as you want with me."

"So, what about Mommy? Can she stay the night too?"

Scorpio interrupted. "Just one night, Mackenzie. After that, we're going back home. I'll bring you over next weekend to see Jaylin."

Mackenzie leaned on me and started to cry. "Daddy, why can't I live with you? I don't want to go to Aunt Leslie's house anymore."

I held her in my arms and wiped her tears. I knew how she felt; it was the same way I felt when I had to go live with my Aunt Betty. I carried Mackenzie to her room and told Scorpio we'd have to talk when I came back.

I read a bedtime story, and Mackenzie was asleep in less than ten minutes. I was so glad she was back in my life. There was no way she was leaving me again.

Scorpio had already taken off her clothes and was under the covers. She had the blankets up to her neck and a wide smile across her face. I sat on the bed and pulled the covers off her, just to get a good look at what I couldn't have. I rubbed my hand on her breasts and wiggled my hand down to her belly button. I lowered my hand again and cupped her pussy. My fingers slowly entered her wetness, and she widened her legs so I could dig deeper. Touching her excited the hell out of me, but I could go no further.

"Baby, I can't . . . I can't make love to you tonight," I said, removing my fingers.

She squeezed her legs together. "And why not, Jaylin?"

I thought about telling her the truth, but I couldn't.

She picked up Sexy Chocolate's business card on my nightstand and tossed it to me. "Is this why you can't make love to me? Did you have sex with her last night?"

I crumbled the card in my hand and massaged my goatee. "Yeah, baby. I did. I, uh, brought her here last night and things happened."

"Do you want to tell me about it, or should I call her so she can tell me like she told me this morning after she left?"

Damn, I was busted. Didn't even dawn on me that she knew Scorpio. She used what Sexy Chocolate told her to attack me even more.

"Yeah, that's right, Jaylin. I knew she wasn't lying to me because she described your 'million-dollar mansion' to a T. She told me how you fucked her so good and made her come six times. Said you banged her on the floor and in the shower. And then told me how good your dick tasted. When she gave me the exact measurements, I had to hang up on her. But the killing part about it was she said another bitch was waiting for you when you got here. When she described her, it didn't sound like anybody I'd ever seen you with, so I guess she's a new bitch, huh?"

I couldn't do nothing but sit there and listen. Seemed like if I tried to open my mouth, she already had the answer.

"You really don't waste no time, do you?" she said. "Kicked my ass out just to let somebody else in. I offered my body to you tonight because I wanted to see if you'd allow me to go behind her. I guess I should pat you on your back for backing out of it, huh?"

"Hey, it was the least I could do," I said in a sarcastic tone.

"You're pathetic, Jaylin, and you seriously need to cut the bullshit before it's too late." Scorpio pulled the covers back over her and turned away from me.

That was fine with me. I went downstairs and lay on the couch so I didn't have to hear any more of the drama. I knew I wasn't right, but I was my own man. I didn't have any ties anywhere and was free to do whatever I wanted.

But when I thought about it more, I realized Scorpio was probably hurting the same way I was when I found out about Stephon sleeping with Nokea. I went to sleep thinking of a way to make it up to her.

* * *

I sat on the examination table and waited for my doctor to come back in with the results. She was a beautiful and classy forty-seven-year-old woman, and if she wasn't married, I would have tried years ago to knock her. When she walked in, she lowered her glasses and peered over them.

"Jaylin Rogers, you know better. How many times have I told you to use a condom? You're a grown man and you shouldn't be naive when it comes to sex."

"Doc, I'm still a young man," I joked. "I'm only thirty-one."

"Well, you'll be thirty-two soon. You need to start being responsible for your actions. So, how many girlfriends do you have now?" she asked.

"None."

She chuckled. "You wouldn't be all messed up down there if you didn't have any."

"Doc, I've been waiting on you. Waiting for you to tell your husband you're marrying a young man like me."

"Jaylin, please. I wouldn't trade my monogamous husband in for something like you to save my soul. You need to get your mind out of the gutter and start putting it to work like you did when you landed your job. Leave all these fast-tail women alone and start taking care of yourself. This time, your test shows positive for gonorrhea. Next time, it might be something else. And you know what I mean."

"I got you, Doc. But you act like a brotha forever coming up in here with these kinds of issues. This was my first time, and as many—" I paused.

"Go ahead, say it. As many women as you've slept with, you could have had something a long time ago, right? But having unprotected sex one time is too many, Jaylin. And who says you don't have anything else? Gonorrhea is just the first sign. If you've been messing around for a long time without using protection, I suggest you get an AIDS test done. Today."

"Are you serious? You know I ain't got no damn AIDS."

"I'm just trying to be on the safe side. Go get yourself tested, and when your results come back, we'll talk about it then. Before I forget, make sure you contact your recent sex partners and tell them to get checked out. I know that might be a difficult task, but they need to be tested too."

"Trust me; I am not out there as bad as I could be. I've made some mistakes, and I promise you I'm going to correct them."

My doctor tooted her lips then gave me a shot and a prescription. When she said I couldn't have sex for two weeks, I placed my hand on my chest.

"Damn, you're breaking my heart. Two whole weeks? Why two weeks?"

"Because I said so. And you should be ashamed of yourself. Get your butt out of here and go take that test like I told you to," she said, hitting me on the butt with a towel.

"Now, you know that turns me on," I said, holding my behind.

"Jaylin, go! Get out of here before I have you arrested for statutory rape."

I was laughing as I left. I went down the hall and took the AIDS test. They told me my doctor would be in touch.

By the time I got home, Nanny B was there cleaning up the place—with the help of Scorpio and Mackenzie.

Just to make sure the cleaning was to my liking, I helped too. I was delighted to have Mackenzie back in my life, and seeing Scorpio again was a good thing too. I just wasn't sure if we'd work on our relationship again, and I didn't know if I had what it took to give it another try.

31

FELICIA

Jaylin didn't have to worry about me calling him any-time soon. After he choked and damn near killed me, I'd had enough. I hadn't called him since then, nor did I intend to.

My ex-boyfriend, Damion, was back in my life. No matter how hard I tried to pretend Paul had set it out for me, I couldn't. Dick just couldn't do the job I wanted it to do. And after I made it perfectly clear to Damion that I wasn't going to put up with his baby mama's drama, he promised me there would be no more confrontations.

When Friday night came, I didn't feel like being bothered with Paul or Damion. I took a hot shower and decided to go to The Loft. The women at work bragged about how they be having so much fun there, and this one lady named Shirley invited me to her birthday party. It was time for me to get out.

I put my braids in a bun and let two single ones dangle on the sides of my face. Then I slid into my coal black satin mini-dress with a V-dip in the back. I added silver accessories and put on my high-heeled black satin sandals that tied up just below my knees. My eyebrows were a little bushy, so I quickly arched them, and then put on the new Oh Baby lip-gloss I picked up at the MAC counter. I looked so good I couldn't even stand myself.

The Loft was packed, and I had a difficult time finding a parking spot. When I did, I looked in the mirror and slid some more lip-gloss on so I wouldn't have to keep running to the ladies' room. I made my way to the door, and this attractive older man with gray hair stopped me.

"Say, beautiful, are you coming in?" he asked.

"Yes, I'm supposed to meet some of my girlfriends from work."

"Then you don't mind if I escort you in, do you? I have a VIP card, so you don't have to pay. And when we get inside, if you'd like a drink, let me know."

He walked to the door with me, and I was flattered to enter the club with this man. He had clout, and, I could tell, money. Not only that, he must have had every woman up in there before, because when we walked in the door, many eyes started to roll.

I went right over to the bar with him, told him my name, and got my drink. After that, I planned to stay away from him for the rest of the night. I had enough drama already and definitely wasn't looking for more.

I found my girlfriends in the party room, sitting at some tables that were reserved for Shirley's party. When they saw me, the fakeness began. I'm sure they really didn't expect me to come.

"Felicia!" Shirley said, running up and giving me a hug like we were the best of friends. "I'm so glad you came. And your dress, girl, where did you get it? It is nice."

I hugged her back and didn't tell her nothing. She introduced me to some of her other phony friends, and they all gawked at me with jealousy in their eyes. I pranced over to the buffet table to get some chicken wings.

After I sat at a table and talked to this brotha whose breath smelled like garbage, I searched the dance area for a dance partner. The DJ was on point with the latest hits, and I was ready to get my party on.

The floor was crowded, hot, and musty. Somebody had definitely forgotten their deodorant before they came, but the brotha I danced with had his act together. He broke it down to the floor, and I tried to keep up with him. I could tell he was a little younger than I was, but what the hell? If he was setting it out like that on the dance floor, no telling how good he was in the bedroom.

We laughed with each other as we left the dance floor. I danced so well that several men grabbed on me for another dance. I declined, and my dance partner went to the bar to get us some drinks. I checked out his front and backside and surely thought about taking him home with me; however, when I looked up and saw Stephon at a table with some of his boys from the shop, my plans changed. He looked downright workable. He had it going on way more than my dance partner.

I went to the ladies' room to make sure everything was still in place. Then, I freshened my makeup and hiked my dress just a tad bit higher. When I came out of the bathroom, I headed straight to the table where Stephon sat. I swayed my hips from side to side and worked my ass so they'd be sure to notice when I walked by. I saw them checking me out from a distance, but I continued to look forward as if I wasn't paying them any attention. As I neared the table, one of Stephon's friends grabbed my hand.

"Say, baby, why you moving so fast? Why don't you have a seat and holla at a brotha for a minute?"

I smiled; my plan was already working well. I eased myself between Stephon and his friend.

"What's up, Felicia?" Stephon said, smiling at me with his pearly whites.

"Hey, Stephon," I said then quickly turned to talk to his friend.

His friend gazed at my breasts and licked his lips. "So, uh, what did your mama name such a pretty woman like you?"

"My mama wasn't the one who named me; my father was. My name is Felicia. Felicia Davenport."

"Felicia, huh? I got one question for you Felicia. How does a woman get as fine as you? You are definitely a sight for sore eyes."

He put it on too damn thick for me, and I tuned him out. I tried to figure out how I could kick up a conversation with Stephon, who sat next to me smelling like Gucci Envy. When he got up to dance with this other chick, I gazed at his tall, nicely cut body that showed through his light blue silk shirt and jeans. The front of his shirt was unbuttoned just enough to reveal the thickness of his chest. The light-blue round glasses that covered his hazel eyes had me melting like butter. Brotha had it going on, and it was hard for me, along with many other females, to keep our eyes off him.

The brotha I danced with earlier came over to the table and handed me my drink.

"I was looking for you. Why did you leave?" he asked. As I put the drink on the table, I noticed the crookedness of his teeth.

"Sorry, my man came in," I said, rubbing my hand on Stephon's partner's back. "I'll pay you for the drink if you'd like."

Cheap motherfucker stood and waited for me to give him the money. I reached into my purse and gave him ten dollars. He snatched it and walked away.

After he left, Stephon came back to the table with this light-skinned chick who wasn't giving him any room to breathe. There wasn't any room for her to sit, so she stood behind him and talked. I could tell he wasn't interested because he conversed with other women as she stood talking to him. She finally got the picture and walked away.

And when she did, I picked up where she left off.

"So, how have things been going, Stephon?" I crossed my legs so he could get a glimpse of my oily thighs. His eyelids lowered and he looked at them like he wanted to see what was between them.

"It's going pretty good, Felicia. Can't complain."

The DJ was right on time with a slow song. I interrupted as Stephon talked to one of his boys.

"Say, Stephon, would you like to dance?" I already

pulled the chair back because I knew he wasn't going to turn me down.

He didn't say yes or no. He just got up and followed me as I strutted with sexiness to the dance floor. When he put his arms around me and leaned his head down close to my shoulder, I lowered his hands and slid them over my butt.

"Felicia, what you doing?" he said, grinning and keeping his hands where I'd put them.

"What do you mean, what am I doing? I like for a man to have his hands on my butt when we're slow dancing. Not on my back."

He gave my butt a squeeze and pulled me closer to him. I felt his goodness press against me. "Is that better?" he asked.

"Yes. A whole lot better."

"So, what are you doing here? I didn't even know you still hung out at places like this."

"I don't. I came tonight because it's my coworker's birthday. And now that I'm here, I'm so glad I came."

"And why is that? Would it have anything to do with me?" he whispered.

"As a matter of fact, it does. I got a feeling something good . . . very good is going to happen to me tonight."

"Oh yeah? Something good like what?"

"Something good like my place or yours."

"Felicia, are you trying to get me in bed with you?"

"Stephon, I don't try at anything I do. I'm known for being a success. So, like I said, my place or yours?"

Stephon quit dancing before the music stopped and looked at me.

"Meet me outside in five minutes. I have to take

care of something before I leave, so just give me five minutes."

I walked off the floor and went back to the table to get my purse. His friend asked where I was going, but I kept on walking.

I was outside talking to a police officer when Stephon came out. "Come on," he said, grabbing my hand and walking me to his car.

When we got in, he sat for a minute before he started the engine.

"Are you having second thoughts or something?" I asked, rubbing his bald head.

"Nope. Just thinking about something." He backed up.

"May I ask what?"

"No, you may not. And put your seatbelt on. I don't like people riding with me without a seatbelt on."

I buckled myself in, just so I didn't have to hear his mouth. "I guess since you're driving, we're going to your place, huh?"

"Yes. Do you have a problem with that?"

"No. I didn't know if you wanted to take me there since Nokea might decide to show up."

"Felicia, if you came with me tonight to talk about Nokea or Jaylin, I can drop you right back off at the club. I don't play games like that, baby, all right?" he said sternly.

"I didn't come with you to talk about them, but I thought it was a legitimate question. And since you don't want to answer it, then so be it. Ain't no trip."

We were quiet the rest of the way to his house. By the time we got there, I'd almost fallen asleep from boredom. I hoped his sex was good, since he seemed to have no conversation.

As soon as we got inside, Stephon picked up his phone and checked his messages. I stood by the door until he gave me the go-ahead to have a seat. When he came back, he'd already taken off his shirt.

"Come on, let's go downstairs," he ordered. I walked behind him as he led me to the basement. He had the damn hook-up downstairs. It looked like a nightclub, unlike his upper level that wasn't much to brag about. His basement had a beige leather sofa that circled the room, huge beveled mirrors that covered the walls, a diamond-shaped bar in the middle of the floor with wine glasses that hung down above it, and beige leather bar stools surrounding the bar. The floor had shiny beige, black and white tile that looked like it had never been stepped on before. His entertainment center covered one complete wall. Jaylin's name was written all over his basement. I knew he had to be responsible for financing this sucker because there was no way Stephon could afford to live like this.

Stephon went over to the bar, while I took a seat on his sofa, which was amazingly comfortable. When he walked over to me, he didn't waste any time. He set a bottle of Moët on the floor next to him and kneeled in front of me. He rubbed his hands on my hips and slid them up the sides of my dress, raising it up a bit.

"Felicia, I don't want you having no regrets after I fuck you," he said, already pulling my dress over my head.

"No regrets for me." I started to unbuckle his pants. "And definitely no complaints," I said as I got a glimpse of what was seconds away from going inside me.

After he took off the rest of my clothes, he picked up the bottle of Moët and poured it all over my body. I trembled as he had the pleasure of slurping it up. It burned a little between my legs, but he had no problem cooling it off with his tongue. Then he stood up and flipped me upside down so he could get a better taste of me. I wrapped my legs around his head so I wouldn't fall. He held me tight around my back, and I worked his goodness on the other end.

My hair fell down as I pulled it from the excitement of him working me so well. He backed up to the couch and lay back with my legs still straddled across his face. He fondled me with his fingers, and I slid my body down and sat on top of him. I put my pussy to major work, which caused him to tighten up and grip my butt.

"Felicia! Slow down, baby. I ain't ready to come yet."

I cut Stephon no slack. I was showing him something Nokea probably didn't know how to. He squeezed my hips tightly and eased himself in deeper so I could feel the full effect of him. I was in another world with him. We seemed to click so well together, almost better than Jaylin and me—but right now, Jaylin was the last thing on my mind.

Stephon and I finished up in the middle of the bar. He had emptied the second bottle of Moët on his body, and this time, the pleasure was all mine. I hopped down off the bar and plopped my naked body on the couch. He joined me, and I lay against his chest.

"I had no idea you had it in you, Stephon. I mean, I've always noticed how nice looking you was, but I guess I couldn't see past Jaylin."

"Well, I knew it was good because Jaylin told me it was. But he didn't tell me you could work it like that."

"I only work it like that when I have to. Besides, it was hard for me to keep up with you. You're packing a load down there, aren't you?"

He laughed. "Something like that. But I'm sure you'll be good company for it."

"So, does this mean this isn't a one-night stand? I know you told me not to ask about Nokea, but won't this interfere with your relationship with her?"

"Nope. Because what she don't know won't hurt her. Besides, I'd like to keep this on the down low as much as I can. Jaylin and me kind of just settled things after what happened with me and Nokea, and I don't want to do anything to mess that up."

"Stephon, please. If Jaylin finds out we're sleeping together, he'd laugh. He doesn't care about me. He could care less who I have sex with.

"I was really surprised to find out you and him went at it over Nokea. Her friend Pat told one of my girlfriends at work what happened and she told me. I knew he liked Nokea, but I thought I'd never see Jaylin fighting over a woman."

"Well, he's happy now, Felicia. He's doing what he wants to right now, and so am I." Stephon rose up and moved me away from his chest.

"Where are you going?" I asked, not wanting him to leave my side.

He turned on his CD player and sang a song by Luther Vandross. He moved from side to side with his hands on his chest like he was slow dancing with himself. His eyes were closed as he sang, "Let me hold you tight, if only for one night."

I laughed, as his voice didn't sound anything like Luther's.

Stephon grinned and held out his arms for me. There was no way I could resist holding a sexy naked body like his. This man had charmed the hell out of me, and had me leaving with a serious smile on my face.

32

NOKEA

I was so ready to have this baby. My doctor had me on maternity leave early because my blood pressure was high. I tried to remain calm and focus on the positive things, but I was worried about how all of this would work out. Stephon told me about his conversation with Jaylin, and I felt bad that he had to lie to Jaylin about us having sex right around the same time as him. But I guess everything was for the best.

Jaylin's birthday was just around the corner, and no matter what, I always spent it with him. Even when he was in the orphanage. My mother and I went to see him and took him some toys both years he was there. As we got older, we met each year at Café Lapadero, where we laughed and talked about the crazy college life. When I turned twenty-one, he was all mine. I always went out of my way for his birthday, just to let him know that somebody still cared about him.

This year would be different. I wanted to do something nice for him, but things were going so well with Stephon that I didn't want to go behind his back and cater to Jaylin. Stephon was a charm, just what I needed during my pregnancy. He satisfied all of my physical and mental needs. Problem was I still didn't love him. Couldn't kick these feelings I still had for Jaylin.

Every time Stephon left, I cried. He had no clue how I felt, and I hoped to overcome my misery soon.

I wondered if Jaylin would show up at Café Lapadero on his birthday. Then I thought about calling him to make sure he would be there. Finally, I decided to just show up without calling.

The morning of Jaylin's birthday, I searched my closet for the nicest maternity outfit I had. I decided on a pink blouse with big white flowers on it. My white linen shorts were loose enough for me to get my fat butt into, but my hair wouldn't cooperate. I put on a white straw hat with a pink ribbon around the rim, and dolled up my face to perfection.

We usually met up around 1:00 P.M., and I got there an hour early so I could make sure we had a seat, just in case he showed up. The sun shone brightly, and there was a nice comforting breeze, so I asked the waiter if he could seat me outside. I asked him for an ice-cold glass of water with a lemon, and some rolls. I hadn't eaten anything all day because I was so nervous.

I pulled out the Black Expressions card I had bought him and signed the inside. I still had thirty minutes, so when the waiter came over again, I went ahead and ordered a salad. I didn't want to sit there doing nothing and looking anxious.

By quarter after one, my stomach felt queasy. Another fifteen minutes zoomed by, and I felt like a complete fool. Why would I think Jaylin would put forth any effort to meet with me? Especially on his birthday. He probably had plans with his other woman. Plans to make love to her and make her smile like he'd done to me over the years. Stupid me, I thought. Always setting myself up for disappointment.

Almost in tears, I stood and dug in my purse to pay the waiter. I wiped my eyes and hurried to leave the restaurant. Then, a soothing wind picked up and blew my hat off my head. I quickly turned to pick it up, and that's when I saw Jaylin behind me with my hat in his hand.

"Were you getting ready to leave?" he asked, handing my hat back to me.

"Yes," I said, still wiping my eyes.

"Why are you crying?" He put his hand on my cheek.

I sat down in the chair and chuckled. I wanted him to know how happy I was to see him, but didn't want to go overboard with expressing my feelings. "Since I've been pregnant, I tend to get emotional for no reason."

"I wouldn't know nothing about that." He pulled back a chair and sat next to me.

He looked amazing, and those addictive grey eyes behind his tinted glasses pierced my heart. His glowing tan was in full effect, and his tailored suit had me screaming naughty things silently to myself. No doubt, life seemed to treat him well.

I wasn't sure how well I looked; I'd gained so much weight since the last time I'd seen him. I hoped my look was to his satisfaction.

"So, were you getting tired of waiting for me?" he asked.

"Yes. I was about to leave. I . . . I didn't think you were going to come."

"I always come, don't I? No matter what, I always come." His eyes dropped to my stomach, and I saw his Adam's apple move in and out.

"I thank you for coming. You don't know what it means to have you in my presence right now," I said.

"No, I don't know. I really didn't think you were going to show. That's why I took my time."

"Really? I've been here. Been here since noon." I wanted so badly to talk about the baby, but I was unable to look him in the eyes. He lifted my chin and made me look up at him. My eyes watered.

"What's wrong, Nokea? Why do you keep crying? I thought you'd be happy to see me."

"I am very happy to see you, Jaylin. I just didn't think it would be this hard for me seeing you again. The last time we spoke, you were pretty upset with me, and—"

"And I'm sorry. I should have never come to your place and disrespected you like that. I was wrong. That's what I wanted to come here and tell you today. I'm sorry for everything I've ever done to you. I've had a lot of time to think about my mistakes, and what can I say other than I fucked up? I have to move on and stop thinking about what could have been."

"Do you ever think there might be a chance for us down the road? I mean, right now, you're happy, I'm happy." I was trying to convince myself that I was, and I didn't want Jaylin to know that I wasn't. "But . . . but is there a chance we can be happy together?"

"Honestly, Nokea, I don't think so. You have my cousin's baby on the way, and you shared something with him I will never be able to forget. And as much as I thought that maybe someday we would be, that dream ended when you got pregnant. I'm not saying I don't still have feelings for you. All I'm saying is I got to take my feelings elsewhere."

My throat ached; that was definitely not what I wanted to hear. I dug in my purse and pushed his birthday card to him. "I wanted to do something else special for you, but I decided to keep it simple."

He picked it up, read it, and smiled as he closed it. Then I reached into my purse and pulled out the teddy bear he'd given me on my birthday. The front of the shirt still said: HAPPY BIRTHDAY, NOKEA YOURS FOR-EVER, JAYLIN. Printed on the back was MY HEART BE-LONGS TO YOU FOREVER, LOVE NOKEA. He smiled again and gazed at the teddy bear.

There was silence for a while, and then he said, "I guess sometimes people who love each other just can't be together."

I blinked several times and fought back my tears with everything I had. "Yes, they can. Love can con-quer anything. If you love me, we can make this work, Jaylin." I reached my hand out to touch his. He eased his hand away.

"Nokea, it'll never work out. There's too much damage that's been done. And I'm not talking about with just you and Stephon. I'm talking about all the damage I've done to you, too. You deserve better. Much better than I can offer you." He stood up and reached into his pocket. He tossed a fifty-dollar bill on the table and put his teddy bear under his arm.

"Listen, I'm not going to be able to stay for lunch. I have some business to take care of. This should take care of lunch and then some. Good luck, baby. And I wish you and Stephon all the best."

He leaned down and gave me a lengthy kiss. Touching his lips and tasting his tongue felt so good to me, but he backed up as he felt me getting deeper into it. I took off his glasses so I could look into his eyes.

"I love you so much," I said, unable to get any other words to come out of my mouth.

He stared deeply into my eyes. "I know. And you know how I feel." He took his glasses from my hand and walked away.

I was crushed. I felt my entire body shake. I hurried to my car and cried like a baby. I cried because reality had set in. What we had was over, and I'd have to let go and make the best of my life with Stephon. Maybe Jaylin was right. After so much hurt, how could a relationship between us ever make any sense?

When I got home, Stephon was there. He was in the baby's room, putting up some wallpaper I'd picked out at Lowes. I felt guilty as I stood in the doorway and watched him go out of his way for a baby that wasn't even his.

He climbed down the ladder and came over to give me a kiss. "Hey, baby, are you okay? You don't look so good. Why don't you go lay down for a minute? I'm just about finished and then I'll go whip up something in the kitchen."

"I'm fine. I went to the mall with Mama this afternoon and she had me doing a lot of walking."

"Aw, so, what do you think? It looks good, don't it?" he said, looking around at the wallpaper.

"Yeah," I said dryly. "It looks perfect. Just how I imagi—" I grabbed my stomach because I felt a sharp pain.

Stephon reached over and held me. "Baby, go lay down. You look tired."

"All right. I think I will go lay down for a minute."

I felt faint walking into my bedroom. I lay across my bed and let out some more tears as I thought about Jaylin. When another pain hit me, I yelled for Stephon to come help me. If the pains were contractions, the baby was early.

Stephon ran into the room. "Nokea, are you okay?" he said, bending down on the bed to hold me.

"Stephon, I think I'm in labor."

"Well, come on! I . . . I'll go get your things for the hospital and get the car." He rushed around the room in a panic. "I'll come back and get you in a minute."

"Hurry!" I yelled as I continued to hold my stomach.

Stephon zoomed around the house and gathered my things. Then he came back into the room, picked me up, and carried me to the car. I couldn't tell who was more nervous. He kept asking me the same questions over and over: How are you doing? Can you feel the baby yet? By the time we finally got to St. John's Mercy Medical Center on Ballas Road, I actually could.

The emergency room crew rushed me to the delivery room and called my doctor to come immediately. I asked Stephon to call my parents to let them know. But when I told him to call Pat, he cut his eyes at me.

I begged until he said he would. Once everyone was notified, he came back in the room with me and held my hand for support.

I lay there in so much pain. My doctor asked me to push and I gave it everything I had—but I couldn't force this baby out for anything in the world. Stephon bent down and tried to coax me, and after he squeezed my hand tighter and yelled at me, I pushed harder and the baby came out.

It was a boy. A six-pound, five-ounce baby boy. After the nurses cleaned him, they put him in my arms. He was handsome, with a head full of curly, coal-black hair. He was light-skinned, and when he forced his eyes open, I saw that they were a beautiful grey like Jaylin's. The baby looked just like him.

I was filled with joy as I rocked my baby in my arms. Stephon reached out, and I gave my baby to him. He smiled and rubbed the baby's tiny fingers.

"You did good, Shorty. I'm really proud of you," he said, bending down to give me a kiss.

"Thank you. Thanks for being there for me. I don't know what I would have done without you," I said, feeling exhausted.

Stephon gave the baby back to the nurses. They cleaned me up and took me to the private room I'd arranged for.

Mama and Daddy rushed in a few minutes later, anxious to see the baby.

"Where is he . . . she?" Mama said.

"It's a boy," I said softly. "Your grandbaby is a beautiful, handsome boy."

Mama started crying, and Daddy held her in his arms. She made me cry. I knew from the beginning that my parents would be elated about this moment.

While Mama and I talked, Stephon stared out of the window like he was in deep thought. Daddy walked over to him and shook his hand.

"Man, thanks. Thank you for being there for my baby. I was a little worried about her, but I'm glad she and the baby have you."

"You're welcome, Mr. Brooks, but you don't have to thank me. I wouldn't have had it any other way."

I was relieved it was all over. I couldn't wait for the nurses to bring my baby in to see me. When they did, we spent the next few hours showing nothing but love to our new arrival.

I became tired and asked everybody for some alone time with my son. Stephon walked Mama and Daddy to the car, but Pat hung around so we could talk.

"Girl, you got yourself a fine young man there. He's got to be the cutest little baby I've ever seen, and I'm not saying that because you're my best friend."

"Isn't he beautiful? I feel so blessed to have a healthy, beautiful baby, especially since he came early."

"You mean especially since he came on Jaylin's birthday. Ain't that something? When Stephon called me, I damn near died because I remember you said you were going to see him today. So, how did that go? Did he even show up?"

"It was okay. He came, but he was late. Bottom line, he never said he loved me, and he said we could never be together." I started getting choked up.

"Nokea, let it go. Stop trying to chase him. If he hasn't come back to you in all this time, forget it. I hate to see you keep torturing yourself like this."

"I know. The baby is born now, so it's time. I think I'll be much better anyway knowing I have him in my life."

"Good. I'm glad to hear that." Pat gave me a kiss on the cheek. "Get some rest and I'll call you later."

As she headed out, Stephon came in with a balloon and some flowers he'd picked up at the gift shop.

"Take care of her for me," Pat said, stopping him at the door. "She's a good woman and she deserves a good man in her life."

"I got her back," Stephon said.

Stephon put the flowers and balloon on my windowsill. I smiled and scooted over so he could sit next to me. He rubbed my hair back with his hand.

"I know. It looks a mess, doesn't it?"

"Naw, Shorty, you look beautiful. I can't believe you're a mother now. You're going to be a good mother. I know you will. Your mama and daddy did a good job raising you. I would've given anything to have parents like yours." Stephon looked a bit sad.

"And you're going to be a good father. My baby is going to have a daddy in his life that he can be proud of. Thing is, I don't know what to call him. Would you help me name him?"

Stephon walked over to the baby and picked him up again. He brought him over to the bed and sat next to me. He stared at the baby for a minute and then looked at me.

"He really doesn't look like you, you know?" he said.

"Yes, he does. He's got my nose."

"No, he doesn't. Actually, he has my aunt's nose. He looks like his daddy. Don't you, little man?" He rubbed his nose up against the baby's nose. "You look just like your daddy. I say we name him after his daddy ... Why don't you call him Jaylin?"

"Stephon, I don't think that's a good idea. I mean, if we're going to raise him together—"

"Yes, we're going to raise him together, but ain't nothing wrong with me naming my child after my favorite cousin, is it?"

I wasn't sure about that. Naming the baby Jaylin could be full of future consequences. Too many people would question it, and I wasn't prepared to explain myself each and every time. Jaylin might see this as an insult.

Then again, what if he found out the baby was his? He would want his child named after him, so maybe I should go with what Stephon suggested.

"No, I guess there's nothing wrong with you naming your child after your cousin. Jaylin it is."

We both held little Jaylin in our arms for a while, and after he went to sleep, Stephon called the nurse to come get him. He climbed sideways in bed with me and stared into my eyes.

"I love you, Shorty. And every chance I get, I'm going to make you the happiest woman in the world. Your worries are over. So, no more tears, no more arguing, and no more disappointments." He reached over to hold me, and I kissed his cheek.

"You are so wonderful. Why couldn't you have come into my life before Jaylin? This would be a lot easier for me if you had. I just don't know when or how I'm going to be able to move on."

Stephon climbed out of the bed and stood next to me. "You're going to move on right now. It's time. I have always been in your life since you and I were kids. Unfortunately, you just recently started to notice. Life is so unpredictable, and sometimes we have to go wherever it takes us. I never thought in a million years

I would be here with you, loving you like no other man in this world can love you. And asking you to . . . to be my wife." He reached into his pocket and pulled out a small black box.

I was too nervous to open it, so he opened it for me. I looked at it and blinked my eyes so I wouldn't cry.

"Shorty, will you marry me? I don't want to waste any more time being without your love." He took the ring out of the box and waited for an answer. Must have paid a fortune for it because the diamond was huge!

"When, Stephon? When do you want to do this?"

"Whenever you want to, baby. I want to give you time to get things situated with the baby, and also time to share the news with your family. After that, I want you to be my wife, and I'm not taking no for an answer."

I took the ring from Stephon and slid it on my finger. I looked at it and smiled, as it weighed down my finger. Again, I wasn't sure about this, and my feelings for Jaylin definitely hadn't all gone away. I had feelings for Stephon, too, and it was now time to choose one versus the other.

I couldn't deny how much Stephon had made me happy, and being happy with a man that I loved had always been my goal. How could I walk away from something that felt so right?

"If I never loved you before, I love you now. There is no way I'm going to let you walk out of my life when you've been so good to me. Six months, Stephon. In six months, you'll have your wife." I gave him a juicy wet kiss that showed him just how excited I was.

Stephon stayed in the room with me all night. He fell asleep in the bed next to me. Feeling slight pain, I eased out of bed and sat in a chair by the window. I raised my hand several times and looked at my ring. I was happy, and all of this started to make sense to me. I stared up at the sky and thanked God for my healthy baby boy and my new handsome fiancé.

33

JAYLIN

Meeting with Nokea was one of the toughest things I had to do. I'd been thinking about her a lot, but I figured it was because I had been missing her so much. During our brief lunch, something else hit me and surprised the fuck out of me. I realized how much I loved her. Chills ran through my body, and when I looked deeply into her watery eyes, my heart felt as if it jumped out of my chest and into hers.

I wanted to tell her how I felt, but seeing her pregnant brought so much hurt to me. Damn, why did she have to be pregnant with Stephon's baby? She looked so beautiful. A part of me really wished she were having my baby.

I couldn't stop thinking about what she'd said during my fight with Stephon. Maybe she was truthful about me being the father, but if I dug deeper to find out, Stephon would think I was desperate to get his woman back. I didn't want either of them to think I was desperate. If the baby was mine, I was sure the truth would come to the light.

For now, Mackenzie and Scorpio had moved back in, and their presence helped me cope with my thoughts of Nokea and Stephon. I'd forgiven Scorpio for lying to me about stripping, and since I saw for myself that she was all about the money, I felt at ease with her. She promised me that she wouldn't lie to me again and said that taking off her clothes for men was history. That was fine with me, as there was no way for me to accept a woman with that kind of profession. She had to do better.

When I got back from meeting with Nokea, Scorpio and Mackenzie were in the kitchen baking me a Black Forest cake for my birthday. My favorite. It was a bit lopsided, but I appreciated their efforts.

I sat on a stool in the kitchen, and my mind drifted back to my day with Nokea.

"Baby, you seem kind of preoccupied today," Scorpio said. "Does your birthday always get you down? You know, since you're getting older and everything." She laughed.

"Don't go calling me old until this motherfucker here can't rise anymore," I whispered, grabbing my thang.

"Jaylin, watch it. Mackenzie's in here."

"I'm sorry," I said. I walked over to Mackenzie, who was standing by the counter and putting more icing on the cake, as if it didn't already have enough.

"Daddy, do you like it?" she asked, licking the icing off the spatula.

"Like it? I love it! And since you made it, I really love it."

She scratched her head and whispered, "I really didn't make it. Mommy did. If it's not good, blame her, not me."

We all laughed. One thing I like is an honest woman. She wasn't taking credit for what she did or didn't do. I gave Mackenzie a wet kiss on her cheek and she wiped it off.

"Yucky." She smiled.

Mackenzie put the finishing touches on the cake. She and Scorpio stuck some candles in it and sang "Happy Birthday" to me. It was so sweet, I damn near wanted to cry. Later, they took me to Morton's Steak House in Clayton and we got down on the food.

On the drive home, my cell phone rang. It was Stephon. I hadn't heard from him all day, which was quite unusual, since it was my birthday.

"Say, man, happy birthday. Sorry I just got around to calling, but I had a busy day," he said, sounding like he was out of breath.

"Well, I'm glad you found time in your busy schedule to call a brotha."

"Listen, what's on your agenda next weekend? I have something really important I want to holla at you about, but it'll have to wait until then."

"Nothing much. I signed Mackenzie up for ballet classes, but that's at nine A.M. After that, I don't have any plans."

"Good. Then I'll see you Saturday afternoon. I'll call before I come to make sure you're there because it's important that we talk."

"If it's that important, why don't you meet me at my house tonight? I'm on my way there now."

"Naw, it's got to wait until the weekend. Besides, I want you to enjoy the rest of your birthday with your fine-ass woman tonight."

"All right, man, if you insist. I'll be by the shop this week anyway to get my hair cut, so I'll see you before then."

"Okay, that's cool."

He wished me happy birthday again then hung up.

By the time we got home, Mackenzie was sound asleep. I carried her to her room and kissed her for at least the hundredth time that day. I really appreciated her and Scorpio as they tried to make sure I had a good birthday.

I shut Mackenzie's bedroom door, and I saw Scorpio standing in the foyer, looking upstairs at me.

"Oh, Jaylin," she said, smiling and motioning with her finger for me to come to her.

I smiled because I knew what time it was. She was butt-ball naked and headed toward the Jacuzzi.

By the time we got outside, I had dropped my shit off in the living room. And since I'd kept sex between us on the down low because of my burning ordeal, I was ready to tear into her. And that I did. Loving was so good, made me wanna fuck all night long. She was definitely in it for the long haul, and I couldn't find it in my heart to kick her out again, no matter what.

I'd hired a private detective to keep an eye on her, though. She promised me that she would leave all the bullshit behind. According to her, her main focus was finishing school, but I had to know for myself if she was still lying to me about stripping. I knew that if she was, it was all about the money. Had nothing to do with trying to get her fuck on with somebody, because

I was doing that and doing it well. It was strictly about making sure she kept some money in her pocket, and the best way she knew how to do it was by showing men her sexy body.

I wanted to believe that she wasn't doing it anymore, but last week, she came in at two in the morning. She claimed she and her sister went out with some of their girlfriends, but I wasn't no fool. I'd been a playa for many years and definitely knew the game. Knew it well.

After I finished making love to her, we went to the kitchen and warmed up our leftovers from Morton's. I sat on a stool while Scorpio stood her naked body in front of mine and stuffed some potatoes in my mouth with a fork. I rolled the potatoes around in my mouth and opened it wider for some more.

"Wow, I didn't know your mouth could get that wide," she said, holding back on giving me some more.

I slid her closer between my legs and held her ass with both of my hands.

"Where do you see us in five years?" I asked.

She laid the fork on the plate and gave me her full attention. "I see us standing here in the kitchen doing the same thing we're doing right now. I see me loving you more and more each day. And maybe, just maybe, some more kids in our future."

I nodded and thought about what she'd implied. Since Nokea had moved on with Stephon, that didn't sound too bad. But we still had some major work to do with our relationship. My feelings weren't quite there yet, and if Scorpio wanted that kind of future with me, she'd have to do a better job of showing me. Lying to me wasn't helping at all, and my gut told me she might still be up to her old tricks. My loving Nokea wasn't helping the situation much either, but I hoped to overcome my feelings.

"Do you ever think you'll get tired of making love to me?" I asked. "Not only that, do you think you'll always be willing to put up with my sometimes fucked-up ways? I can be a motherfucker when I want to, you know."

"Oh, trust me, I know. Probably better than anybody does, because I'm the one who lives with you. But it's okay, baby. I know how to deal with your mood swings. I just get out of your way when you don't want to be bothered. Eventually you come around, don't you?

"Anyway, why are you asking me about the future? Are you planning on keeping me forever?" She rubbed her fingers through my hair like she always did.

I took her hand and kissed it. "Baby, I have a serious problem with loving women. I told you this before, and I don't want it to damage our relationship. It's not that I don't love being with you, but something inside just won't let me get too attached. And every time I feel as if I'm getting too close, I back off. I go find myself another lady to occupy my time . . . to take away some of these feelings I have for you."

"But in due time, Jaylin, that will change. Those days I was away from you were hell for me. But it took that for me to realize I couldn't lie to you if I wanted to be with you forever. I had to do whatever it took to get you to trust me again, so now I'm working on proving to you that you can trust me. And once you realize you can, all of this is going to change. You'll be able to love me like I want you to. You'll see."

I turned Scorpio around and pressed her butt

against my thighs. "I'm tired, baby. I'm truly burned out. It's time for me to get back to business. I've focused too much of my time elsewhere. After tonight, I don't know how much time I can offer you. I mean, I still want us to be together, but my work is going to take priority over everything. By not working, I'm losing money, and I don't like to lose out on money. I'm a brotha who likes the finer things in life, and I know what I have to do to make sure that continues."

"Jaylin, by all means, handle your business. You've been more than a blessing to Mackenzie and me, and there's no way I'm going to stop you from doing what you need to do. All I ask is that you don't forget about us.

"Maybe some day I'll be able to handle it if you do, but Mackenzie won't. She's so crazy about you it scares me. I really think she wants to be with you more than she does me. But you do have a way with the women, so I know exactly where she's coming from."

"I will never forget about you and Mackenzie, especially her. She's made me realize so many things over these past months, and most of all, she's shown me how to love somebody. In a different way, of course. And for me, that's a big step. I don't think I've ever loved anybody in my entire life except for her and—" I paused.

"Her and who, Jaylin?"

"My mother. Her and my mother," I said, backtracking my thoughts, as Nokea was on my mind.

"Well, I'm at least glad you love a part of me. That makes me feel special right there. And I know if you can love my daughter the way you do, then you will eventually find a way to love me too."

"Maybe so," I said, kissing her on the neck and getting ready to stick my thang where it belonged.

Scorpio and I got down in the kitchen, but I cut it short because I had to get up early and get back to work.

Shit was crazy back at work. The market had dropped to a five-year low and everybody lost money—including me. Roy tried to keep up with all my accounts, but the way things were, we were all lucky to still have jobs.

No sooner had I plopped down in my seat than Schmidt rang my phone.

"Jaylin, welcome back. Hope you're ready to do some work around here today because you have it cut out for you."

I laughed and hung up on his ass. That's what I always did when I felt pressured: just laughed and did what worked best for me.

My first call was to Higgins. Even though I spoke to him occasionally from home, I wasn't really able to handle my business with him like I was at the office. He was cool. He understood the ups and downs of the market, and didn't blame me at all for losing thousands of dollars in his investments. My suggestion to him was to buy while the market was low and wait to see what happened. Didn't expect much to happen anytime soon, but a year or two from now, we could all be back on the right track. He went with the flow, and so did his buddies.

Roy had seen how well my plan worked for me, and he got on the phone and started to do the same. We called people we hadn't talked to in years and tried to get them to invest. Drove around visiting companies that didn't have pensions or any type of retirement plans set up and talked to them about investing.

By day's end, we cooled out in my office and I kicked off my shoes. Schmidt came in and congratulated both of us. He told me he was glad to have me back, and frankly, I was glad to be back. Not once did I think about my crazy life outside of work, because that's truly what it was—crazy.

After my conversation with Scorpio, I couldn't deny or ignore my love for Nokea. I knew she was the one who'd kept me going all these years. And as much as Scorpio fulfilled all my other needs, my heart was empty because Nokea hadn't been there to fill it for a long time.

But wasn't a damn thing I could do about it. I was hurt too badly by Nokea being with Stephon. I just had to play it cool and focus on the good things I had in my life, like Mackenzie and my job, my obscenely wealthy status, and even my woman's pussy. Now, that was getting better and better each day. Too bad it wasn't filling that emptiness inside of me.

Things started to calm down by the end of the week. Everybody went with the flow of the market and waited for something positive to happen. Angela and Roy became my right-hand team. They had things under control and made sure I didn't feel much pressure. It didn't dawn on me that the reason they were so close was because Roy was banging her.

When I left the office on Friday, I had to go back and get my keys because I forgot them. I passed by Roy's office, and the lights were out. I knocked because I could hear a bunch of rumbling going on. The only thing I could do was smile. I had been there and definitely done that before. Roy didn't know what he was getting himself into. And to think Roy and her husband were supposed to be friends.

It made me think about who I truly considered to be a friend of mine. I really didn't have many to begin with, because it was too many brothas hating when I inherited my grandfather's estate and started making money years ago. I had to leave behind motherfuckers I'd known for years. Not because I was big-balling, but because they tried to take me for everything. When they came to my house, shit always came up missing. I learned my lesson early on and decided to limit myself to just a few friends. Actually, Stephon was the only person who was close to me. Our bond was pretty tight, and no matter what, I wanted to keep it that way.

When I pulled in the driveway, Scorpio's car was gone, as usual. Sometimes she was there when I got home, but usually she wasn't. When I met up with the private detective yesterday, he said she was definitely still in school. He also confirmed there was nothing going on with her and the guy who read her scripts. But he did tell me she still stripped at parties. Showed me pictures of her entering and leaving motherfuckers' cribs—and even tried to defend her by saying she always left alone. I paid him for his time and thanked him for what he called "easy work."

The thing about it was I didn't even trip. Pretended like I didn't know nothing about it. And I for damn sure didn't ask her any questions when she came in that night. She always gave me excuses anyway, like she went somewhere to study, or she went out with her sister. Explanations sounded pretty good. And if I was a brotha who didn't know any better, I'd probably have believed her. Sad thing about it was she thought I was stupid enough to believe it. Puzzled the hell out of me, but what the fuck? Arguing about it wasn't even worth my time—for now.

The only reason I kept my mouth shut was because of Mackenzie. I told Scorpio I wanted to adopt her, and without any hesitation, she agreed. Once everything was settled with the courts, I would do what I had to do to have shit Jaylin's way. I didn't hate Scorpio for what she did because after all, money was the name of her game. I knew people out there who would do anything for money, no matter who got hurt. But she was a sista who could have had it all. Didn't even have to go out like she did. Looks, charm, personality, and a little more willingness on her part could have gotten her anything she wanted. She wasn't even smart enough to realize that, and that's why I couldn't love her like she wanted me to. Besides, there was no way to forgive a woman who lied to me as much as she did.

Mackenzie always waited for me when I came through the door. I sat her on my lap in the living room as I rummaged through my mail. There was a letter from my doctor, and I figured it had to be the results from my AIDS test.

I tapped the envelope against my hand and let out a deep sigh. I unfolded the letter, and when I saw the word "negative" and read the comments from my doctor, I jumped for joy. She'd tried to reach me by phone, but the letter said her attempts were unsuccessful. Either way, I was ecstatic.

Mackenzie jumped for joy with me. She didn't even know what I was happy about; she was just happy because I was. Now, that was the kind of love I needed in my life, I thought as I swung her around the living room.

I paid Nanny B for watching Mackenzie and called Stephon to make sure he still planned to come over the next day. When a very familiar voice answered his phone, I thought I'd dialed the wrong number.

"Felicia?" I said.

"Hold on," she said.

He immediately answered. "What's up?"

"Man, was that Felicia who answered your phone?"

"Yeah." He was to the point, but didn't have much else to say.

"I just called to see if it was still on for tomorrow. But, uh . . . what is Felicia doing answering your phone?"

"I was asleep. She wasn't supposed to answer it. If Nokea had called, I would have been fucked."

"No doubt. When did this shit between you and Felicia—"

"Man, I don't know. It's been a while. I was going to tell you, but I didn't want you tripping like you did when you found out about Nokea."

"Man, whatever. That's your prerogative, so, hey, go for what you know. It ain't nothing but some pussy." I heard Felicia in the background moaning.

"Right, right. But, uh . . . I'll give you a holla tomorrow. This one here just can't seem to get enough."

"Tell me about it. I definitely know how that is."

I hung up and didn't even sweat it; wasn't even worth me stressing over. Stephon was just too much, and so was Felicia. She was nothing but a whore who I'd seen through from day one.

I put Mackenzie on my shoulders and carried her up to my room. When I put her down, she laughed so hard that spit dripped down her face.

"Mackenzie, that's nasty," I said, tickling her.

"Daddy, stop tickling me," she said, grabbing her stomach. "Stop before I tickle you back."

"Okay, give it your best shot." I flexed my muscles and tightened my six-pack. She ran her hands across my stomach. When I didn't laugh, she stopped. She sat on the edge of my bed and pouted.

"Mackenzie, what did I tell you about pouting? What's wrong with you now?" I asked.

"When I tickled you, you wouldn't laugh. Mommy said when you stopped laughing and smiling, it was time for us to leave. I don't want to leave, Daddy. I want to live with you forever and ever and ever." Her eyes were watery.

"Mackenzie, how many times have I told you you're here to stay? Wherever I go, you go. I don't care what your mommy says to you. You're my little girl, and I'm never going to let you leave this house again until you get married."

"Like you and Mommy. She said you were going to marry her."

"No, no, Mackenzie. I told you that you'd be the first to know when I decided to get married. And right now, I'm not ready for a wife."

"But I'm ready for a husband. Can my husband come live here with us too?"

I laughed. "Baby, you're much too little to be talking about having a husband. Daddy's going to have to screen these young men out here for his baby." I hugged her. "When the time comes, he'd better have it going on like your daddy does, and be able to buy a beautiful home for you so you can live there with him."

Scorpio walked in. "Or be able to afford one yourself," she said, taking her jacket off and laying it on the bed. That still bugged the hell out of me, and after my look cut her in half, she picked up her jacket and hung it up in the closet. Mackenzie ran up and gave her a hug. She walked out of the room with Mackenzie, and I went into the bathroom and shut the door.

I ran some steaming hot bath water and hung my clothes neatly in the small closet in the bathroom. I slid my body deep down in the tub and closed my eyes. Scorpio knocked on the door and cracked it open so she could see me.

"Hey, you got a minute?" she said, walking into the bathroom.

I kept my eyes shut. "Always. I always got a minute for you."

"Jaylin, why do you be telling Mackenzie all that crazy stuff? I want her to grow up being independent. I don't want her to think the only way she can make it is if a man takes care of her. You're giving her the wrong impression about life."

I opened my eyes and looked up at her. "Scorpio, almost everything you tell Mackenzie has been a lie. If anyone is giving her the wrong impression about life, it's you, not me. I make a decent living; you don't. Well, in your mind it's decent, but I tell you what: I'm going to make damn sure, whether you like it or not, she don't have to go through life shaking her ass just for a fucking dollar. Now, if you got a problem with that, that's too bad. I'm not going to sit here trying to convince somebody who thinks I'm too damn stupid to realize she's still taking her clothes off for money." I closed my eyes again.

She left the bathroom. I could see her in the mirror as she sat on the bed with a sad look on her face. I wasn't in no mood to comfort her. After a few minutes, she took off her clothes and re-entered the bathroom.

"Can I join you?" she asked.

"I'd rather you didn't. I'd like to enjoy my bath alone, if you don't mind."

"Sure." She covered herself with a towel and went back into the bedroom, where she lay across the bed and turned on the TV.

When I got out of the tub I flaunted my big dick in front of her. I walked into the closet and pretended to look for something. I stood right in the doorway and dried my body with a towel, then grabbed the *St. Louis American* newspaper off the nightstand and pimped out the door. I lit up a Black and Mild, and after I closed the door to the bonus room, I laid my naked body on the floor with my feet propped up on my leather sofa. I opened the newspaper and started to read.

Scorpio came in and locked the door behind her. She took the newspaper from my hand, straddled my chest, and then squatted down on it. I got a glimpse of her good stuff, as every bit of it stared me in the face.

She rubbed her clitoris and allowed her fingers to find a way inside of her. As her juices flowed, she placed her fingers on my lips because she knew that was definitely how to turn me on. I sucked her fingers into my mouth and continued to lay there and watch. Her attempt was to fuck with my mind, and just for the hell of it, I took my fingers and teased her walls. But when she turned around and put her ass in my face, how could I resist?

I took a few puffs from the Black and Mild, and then blew the smoke out of my mouth. I scooted her down on my face so I could lick her at the right angle. I tore her insides up, but when I thought about my weakness—and my love for Nokea—I stopped the action. I moved her over to the side and stood up.

"Jaylin, why . . . why did you stop? Don't you want to fuck me?" she said, kneeling down in front of me.

I rubbed my hand on her cheek. "Baby, sex doesn't solve everything. I'm not saying that I don't want to have sex with you. All I'm saying is it's not going to keep us together forever. I find it funny how easily you can make me love your pussy. Question is, can you make me love *you*? For that is the only thing that counts." I picked up my paper, grabbed my Black and Mild, and went to sleep in one of the guest rooms.

Scorpio didn't bother me for the rest of the night. I actually got up in the middle of the night to make sure she hadn't left with Mackenzie. When I saw both of them still there, I went back into the guest room and went to sleep.

Mackenzie and I rushed to get her to ballet class on time. I bought her pink shoes instead of purple, and she made a big deal about it. We had to stop by the store to exchange them before we went to ballet class. I made it perfectly clear to her there would be no more complaining from her. And even though she cried again, I felt good about standing my ground for the first time.

After seeing her twirl around the floor, I couldn't do nothing but smile. I stood and watched as she took charge and learned everything that the instructors taught her. I was proud of my baby girl, and I didn't care what anybody said.

Before going home, we stopped at McDonald's on Olive Street Road. Mackenzie took about ten minutes deciding what she wanted, only to play with the toy that was inside of the Happy Meal. She didn't eat a thing. When I tried to make her eat it, she did her normal routine and pouted. I reminded her about our conversation earlier, and when I told her I would give her hamburger to Barbie, she ate it.

I was surprised to see Scorpio's car still in the driveway when we got home. Usually, on Saturday mornings, she'd find somewhere to go, like to the gym or to Chesterfield Mall or Saks Fifth Avenue to spend my damn money. But I guess that since I'd gotten a little tight with the money, she had fewer options.

She must have heard us pull up because she came outside and asked how Mackenzie's ballet class went. I felt like if she really wanted to know, she would have gotten her ass out of bed and gone with us. But what the hell? I couldn't make Scorpio do what she was supposed to do, and I wasn't trying to kick up an argument with her today.

I went into my office to turn on my computer. My intentions were to catch up on some work before Stephon came over, so I closed the door so I wouldn't be interrupted. No sooner had I taken off my jacket than I saw him pull in the driveway. He was outside talking to Scorpio—probably trying to coax her ass in the bedroom too. I did promise him a while back that as soon as I was finished with her, he could have her. Again, it didn't matter to me either way.

They were outside for a while, so I pulled my curtain to the side to see what was taking them so long. Scorpio held a baby's car seat in her hand, and Mackenzie jumped up and tried to look at the baby.

I guess Nokea finally had it. I was nervous about seeing the baby. My palms had already started to sweat. Stephon, of course, couldn't wait to bring it over here and throw it in my face.

I heard them coming through the front door, so I rushed to my desk and pretended to be occupied. Scorpio peeked in and told me Stephon was here to see me. When I told her to let him in, he walked into my office with the baby in his arms. I lowered my head and rubbed my goatee. Couldn't even get up enough nerve to look at the baby. Frankly, I didn't know how to respond.

Stephon cleared his throat and got comfortable in a chair. "Say . . . say, man, I know I'm a little bit early, but I figured you and Mackenzie was probably back from her ballet class."

I looked up. "I see you got yourself a son there, huh? Can't help but notice all the blue and white he got on."

"Yeah, I got myself a son."

"How's Nokea? I didn't think she was due so soon. I just saw her on my birthday and she didn't mention anything about her delivery date being so soon." I knew Stephon had no idea we'd seen each other on my birthday.

"So, you saw her last week? What did she say?"

"Nothing much. She just wished me well and we talked, that's all."

"Aw, okay. But, uh, she's doing pretty good. She hasn't been getting much sleep because of the baby, but her parents and me been trying to help out."

"So, I take it you had a night off last night since you were at your place with Felicia."

"Aw . . . yeah, that. Well, ah, Felicia and me, we cool. But I didn't come here to talk to you about her. We've shared plenty of women in the past—eight of them, to be exact—and we never made it a big issue, so I don't want to do it now."

"I'm with that. And actually, it was nine, to be exact. I counted them myself just last night."

"Well, nine then. I'm just glad we don't have a problem keeping it in the family."

"Naw, no problem. But there was one exception. One I sure regret not keeping to myself, and one that I told you was off limits," I said bluntly.

"I know, Jay, but things happen. I couldn't control my feelings for her, and I really thought you had moved on with Scorpio. So, what I'm about to tell you, I want you to listen and listen good. If you get upset with me, just know I didn't come here to fight with you again. I think it's time this came out in the open so we both can get on with our lives."

My voice rose. "That's all I'm trying to do. But every day, it's something new. I don't know how much of this back-stabbing, playa-hating bullshit I can take."

"It ain't even like that, Jay. Just . . . just let me start from the beginning. As you know, when we were growing up, I always liked Nokea. You never paid her any attention, and I think that's why she liked you so much. When y'all started dating, I was really disappointed. And then when you kept fucking around on her, seeing all these other different women, Jay, it bothered me. All I wanted to do was see her happy. Year after year, same ole shit.

"And each time y'all got into it, she came to me for

comfort. This past year was the first time she ever said it was over between y'all, and I truly felt it was time for her to move on."

Stephon moved around in the chair and tried to get more comfortable. He could see the daggers in my eyes ready to do damage. "So, anyway, I stepped up," he said. "I tried to show Nokea what a good man could really be like. But she still wouldn't love me like she loved you. No matter how hard I tried. So, a part of me felt like if I saw other women, maybe her feelings would change. Maybe she did want a bad boy in her life.

"She doesn't really know about the other women in my life, but I think she suspects something. And just that small suspicion is bringing her closer to me. Making her want to be with me more and more. I know it sounds crazy, but it's the truth. She seems to like me more because she thinks I'm a challenge for her now."

Stephon took a deep breath and looked at the floor. "When I told you I made love to her a few weeks after you did, I lied. When I told you we'd had sex before my mother's funeral, I lied about that too. I lied to you because I wanted this baby to be mine. I wanted Nokea to be happy and I wanted him to be raised by me. I know it was wrong, my brotha, but I felt like at the time, it was the best thing to do."

He stood up and carried the baby over to me. He took him off his shoulder and laid him in my arms. "This is your baby, Jay. I took it upon myself to name him Jaylin because he looks just like you."

Since I'd obviously been lied to so much, I hesitated before accepting the baby in my arms. Stephon laid him in my arms and pulled the blankets back so I could get a good look at him. My eyes searched his head full of curly black hair and the shape of his eyebrows. Stephon rubbed his cheeks, and as he squirmed around a bit, his eyes started to open. When they did, I could see myself written all over this baby. Wasn't no denying him. My baby picture was almost identical.

I looked at Stephon as a tear rolled down my face. I hadn't cried in a while, but having my son in my arms just did something to me.

"Did anybody ever think about what I wanted? This could have changed things for me a long time ago. Why in the hell would you and Nokea lie to me about something like this? Maybe I didn't have my head on straight, but . . . but this is something we could have worked through together. You all I got, man, and I thought our bond was much stronger than that."

"It is. That's why I couldn't go another day without telling you. It's been killing me not being able to, but I did what I thought was best."

"So, when was he born? She had to have just had him."

"He's a week old today. She had him on your birthday. I guess after seeing you last week, it was too much for her."

I smiled and thought about my conversation with Nokea at the restaurant. God surely had a way of making a way out of no way. As stubborn as I'd been, I didn't know why he was looking out for me. "On my birthday, huh? She had him on my motherfucking birthday?"

Stephon nodded.

"So, now what, Stephon? You seem to be the man with all the answers. Where do we go from here?"

The grin on his face vanished and he walked back over to the chair to take a seat. "I've asked Nokea to marry me. She accepted, and in less than six months, we're going to be married. I don't want to keep you from seeing your son, but I want to be a part of his life too. He will definitely know who his father is, and I will never do anything to keep him from you."

I could have damn near died. I shook my head and even had to chuckle a bit from the bullshit I'd just heard. "So, now you're going to marry her? Just like that. Walk her down the aisle knowing damn well that she still loves me. Man, that's crazy. How can you be with a woman knowing how she truly feels? Don't make any sense, and neither does your plans for my baby."

"Well, it makes sense to Nokea and me. She's different, cuz. Ever since she's had the baby. She wants a family. She wants to set the same good example her parents set for her. And personally, I think she is starting to love me. I didn't expect you to be happy for us, but this is what we want, and nobody is going to stop us.

"I would like for you to get on board and be my best man. I really wouldn't have it no other way."

I stood up and put the baby on my shoulder. Did the best I could, anyway, because I really didn't know how to hold him. "You have got to be out of your fucking mind," I said, not knowing any other way to put it.

I looked out the window to see where Scorpio and Mackenzie were. They were outside washing the cars, so I turned my attention back to Stephon. "You expect me to stand there and watch you marry the woman I love? Man, please. I don't even want to talk about this shit anymore. My mind is going a mile a minute."

"Jay, man, don't be like that. We can put this behind us. You can accept this situation for what it is and move on with Scorpio and Mackenzie. Please, man; be there for me. You don't have to give me an answer today, but think about it."

He picked up the baby's car seat. "We gotta go. I told Nokea I was taking him to the barbershop to brag on him, and she's going to be looking for us, so I'd better go. I didn't tell her I was coming by here, so if you could keep this quiet until you decide what to do, I'd appreciate it."

I looked at the baby and kissed his forehead. My mind was so messed up, I didn't know which way to turn. I gave him to Stephon and watched him lay the baby in the seat. This motherfucker had my son and my woman, and it was actually the first time in my life I felt as if I'd lost control.

He slammed his hand against mine, and before heading to the door, he turned. "Just think about being my best man. Call me when you've made up your mind."

"I already gave you my answer, Stephon. Ain't much to think about. But you can do me a favor and tell Nokea to call me. I want to talk to her and make sure this is what she wants to do."

"No problem. Will do. I'll ask her to call you later." He shrugged and left.

I figured Stephon wasn't going to tell Nokea nothing. I looked out of the window and watched him put my baby in the car.

I couldn't blame anyone but myself for fucking Nokea over like I did. I never thought my mistakes would cost me a son and the only woman I'd ever loved. But no question about it, what goes around definitely comes around, and I was starting to feel the effects of every bit of it.

I sat in my office all day long with the door locked. Scorpio hollered in and told me she and Mackenzie would be at her sister's house. I guess she figured I needed time to myself since I'd refused to come out of my office when she asked me to.

It had gotten late, and I was still in my office. I didn't turn on any lights, just lay on the couch, sleepy as ever, and thought about my child. During my dreams, I called on Mama and she encouraged me to go get my son. She said she was finally proud of me for realizing the mistakes I'd made. Claimed it was not always about me, even though I wanted it to be, and yelled at me for not respecting her wishes by being with Nokea. When I tried to touch her again, I woke up. It was another dream, but it seemed so damn real.

I wiped my face with my hand and the tears just kept on coming. No one was there to stop my pain. No one was there to hug me, and right about now I needed that more than anything in the world.

As I soaked in misery, there was a light knock at the door. "Daddy, are you in there?" Mackenzie whispered. "Come out. You haven't played with me all day."

I chuckled and felt a sudden sense of relief. I went to the door, and then locked it after Mackenzie came in. We remained in the dark because I definitely didn't want Mackenzie to see I'd been crying.

"Daddy, why are you in the dark?" she whispered as she hopped up on the couch and turned on the lamp next to it.

"Because, Mackenzie, I'm thinking." I held my head down.

She got off the couch and stood in front of me. Then she lifted my head like I did hers when she felt down.

"Have you been crying?" she asked. "It looks like

you've been crying." She wiped her hands on my face. "Don't cry. I'll take care of you."

I held Mackenzie tight and we rocked back and forth together.

"I love you, Mackenzie. I really and truly do love you."

"I love you too, Daddy. But I'm hungry. Would you make me some of those pancakes you made me last week?"

"Pancakes at this time of the night?" She nodded. "Sure, baby. You can have anything you want."

I led her into the kitchen. Scorpio came in and sat on one of the stools. Mackenzie didn't waste no time telling her I'd been crying, and she looked at me with sympathy in her eyes.

"I know that was your baby Stephon had today. I want you to know that I will be here for you if you need me. Whatever you decide to do, I'll back you all the way. Even if that means you want me to move out. I talked to my sister about moving back in with her, and she said it would be okay—"

"Scorpio, I don't know what I'm going to do. I haven't asked you to move anywhere, so don't go making plans to move out just yet, all right? Besides, I need you right now. Need you more than I ever have before. You and Mackenzie both."

Scorpio gave me a hug and placed her lips on my ear. "I love you," she whispered. "More than you will ever know."

34

FELICIA

Stephon and I were having sex every chance we got. He crept into my place and I crept into his. But when he told me he had proposed to Nokea, I was devastated. It didn't stop him from putting it on me, so I tried hard to get him to change his mind. But no matter how hard I worked him, he stood his ground. I thought about calling Miss Homebody and telling her the news about her so-called fiancé, but knowing her, she'd probably try to kill her damn self. I didn't want to be responsible for nobody taking their life over a man who wasn't worth it.

Nokea was a fool, though, a prime example of every stupid woman who puts all her trust in one damn man. I knew better. And even though I had kicked Paul and Damion aside for Stephon, the door was always open so they could come back. I made sure of that, because I kept our conversations going, the dinners going—and even the money. Anything I needed, they gave, even though I'd cut off all the sex.

Wasn't no need for me to be screwing three men when Stephon tore it up like he did. He was definitely better than Jaylin. Besides, who better to replace him with than his own cousin?

When Stephon and I got together, we were like two dogs in heat. He couldn't stay away from me for two days, so I didn't know how he thought this marriage thing would work itself out. If it did, he knew damn well that as soon as the honeymoon was over, he'd climb right back into my bed. When I brought that to his attention, he laughed. Laughed because he knew I wasn't lying. He knew Nokea couldn't satisfy his physical needs like I could. So, I knew for sure, if I wasn't a pain in her side now, I'd sure as hell be one in the future. The Jaylin drama was over for us, but her husband drama had just begun.

I'd called ole Jaylin a few times to try to explain the Stephon situation, but he blew me off. I knew he wouldn't trip, but to hear him say, "A dog might get into a little trash sometimes," it kind of messed me up. I asked if the dog in him was up for dinner, but he hung up on me. I was a little hurt even though I wasn't expecting him to embrace me with kindness.

Stephon promised to be at my place no later than nine o'clock. When he didn't show until eleven-thirty that night, I was pissed. I had put the food back in the refrigerator, and I sat on the couch with my arms folded while he tried to explain why he was so late.

"Look, Felicia, I told you I had to work late tonight. A couple of fellas called the shop and told me they needed their hair cut before going to this concert tonight. Since I need the money, I stayed and cut it for them."

"Well, you could have called. I cooked all this food for you and you didn't even have the decency to call and tell me you were going to be late. And when I left you a message, you didn't even call back."

"I don't know what else to tell you. I'm answering to you like you're my woman or something. Let's get an understanding now, before this shit starts to get out of hand. I don't answer to no motherfucker! I've told you where I've been out of the kindness of my heart. But if you don't believe me, that's your problem, not mine." He stood up and got ready to leave.

I grabbed his hand. I didn't want him to leave without giving me something hot and heavy. "Stephon, I'm sorry. You're right. You don't owe me an explanation. I appreciate what you told me, but I get upset when I think about you being with someone else."

"Well, ain't no need for you to think about it. I am with someone else—and not just Nokea—so make up your mind about us, because I'm not going to be dealing with this bullshit every time I come over here. Either you're with it, or you're not. If you decide to deal with it, then I don't want to hear anything else about my delays, Nokea, or anything else, all right?"

"Ain't no trip. You do you and I'll do me. But don't get upset with me when I get back to business with some other men who I've put on the back burner for you."

"I never asked you to put anybody on the back burner for me. You did that yourself. So, don't be mad at me about your own decision."

"I'm not. I thought it would make things easier for us, that's all."

"Naw, baby," he said, standing and unbuttoning his pants. "Do what you want and with who you want. I ain't got no control over it."

Stephon took off the rest of his clothes and undressed me as well. He fucked me good. So good that I realized even though his mouth said he had no control, his dick showed me he did.

After he left at three in the morning, I put on some clothes and took a late night drive by the St. Louis Riverfront. I parked close to the river and laid a blanket on the ground so I could sit and think for a while.

For a sista to have it going on like I did, I was a bit disappointed in myself for settling for less when it came to men. I knew I could have always had one to call my own, so I couldn't figure out why I had to have somebody else's man. Paul was the only man I ever had that I could call my own, and I treated him like a pest. I knew if we got together, there would be nothing in the world I couldn't have—with the exception of a big package. I was smart enough to know a good man when I had one, and going forward, I intended to focus on trying to improve my relationship with Paul. Stephon's good loving might set me back a few times, but since he had plans to move on with Nokea, it left me with few options.

35

NOKEA

Nobody in the world was happier than I was. I still had my moments of Jaylin withdrawal, but slowly but surely those were fading away. My baby showed me how important it was for me to be there for him. I loved him more than life itself. I wouldn't let anyone or anything stand in the way of our happiness.

Not even Stephon. I could feel something wasn't right with him, and when Pat told me she saw him at the Old Spaghetti Factory with another chick, it confirmed my suspicions. He said she worked at the shop with him, but I knew better.

I started having trust issues with Stephon, but didn't really give our relationship much attention. The only person who needed that kind of attention from me was little Jaylin.

I still planned to marry Stephon because there was no solid evidence that he had lied to me about who the other woman was. And after being with Jaylin for so long, a part of me felt that cheating was in a man's nature. Sometimes, no matter how hard they tried not to, they just couldn't be right, even if their lives depended on it. As long as Stephon gave little Jaylin and me what we needed, and he hadn't brought any chaos to our relationship, I accepted his explanation. Yes, he made a mistake by not telling me about his lunch plans, but I'd made plenty of mistakes too. He'd forgiven me for mine, so how could I not forgive him for his?

When the phone rang, I was rocking LJ to sleep. Mama was supposed to come over and watch him while I went to the gym, but she was late. The call was from Jaylin. I was surprised to hear from him, but it was good to hear his voice.

"Can I come by to see you?" he asked.

"That might not be a good idea."

"Please. I really need to see you."

I knew Stephon had told him about the engagement, so I figured I needed to explain my reasoning. "I'll come to your place. Mama should be here soon, so give me a few hours, okay?"

"All right," he said and hung up.

I put on my lime green fitted dress that hugged my petite body, which was already back in shape. My hair had really grown since I had the baby, but it was styled with one side short, and the other side long and swooped above my eye. I sprayed my body with perfume. When Mama got there, she could tell I wasn't headed for the gym.

"Actually, Mama, I'm going to see Jaylin. He called and wanted to talk to me about something."

"Do you feel comfortable about going to see him?"

"Yes. It's been a while since we've seen each other and we really need to have this discussion."

Mama didn't say another word. She gave me a hug and told me she loved me.

I was nervous about going to see Jaylin. My stomach turned in knots and my sweaty palms kept the steering wheel wet. Every time we got together, something always seemed to go wrong. This time, I kept my head up and prayed for God to give me strength.

I rang his doorbell and took deep breaths as I waited for him to answer. He opened the door wearing only his blue jeans with the top button undone. His body was so perfect. I had a vision of rubbing my hands all over his bare chest. When I saw that he wasn't smiling, though, my heart beat faster.

"Have a seat, Nokea," he said, leading me into the living room. "Can I get you anything?"

"Yes. Some water. My throat is very dry." I was testing him because I knew Jaylin never served water. He went into the kitchen and came back with a glass of water in his hand.

"Here," he said, sitting on the table directly in front of me. "Take your time because you won't be getting any refills."

"And why not?" I said, laughing.

"Because I ran out. Besides, there's something in that water I don't want anyone to have but you."

"Oh yeah, and what's that?" I asked, holding the glass to my lips.

"It's called love. And since I never knew how to love anyone before, it was hard for me to do. Today, there are so many things I want to tell you, but first I want to tell you I love you. I don't know what that means to you, but it means a whole hell of a lot to me."

I uncrossed my legs and put the glass of water on the table next to him. He took my hands and held them together with his.

"I want my son, Nokea. I want him in my life twenty-four/seven. Not only that, but I want my woman back."

I couldn't believe the words that were coming from Jaylin's mouth. Just to hear him say that he loved me released a lot of pressure that was inside of me. I'd waited years to hear those words, and I tried hard to fight my emotions.

His timing couldn't have been more off.

"What about your other women? What about Stephon? I can't turn my back on him after all he's done for me. I'm finally happy, and he's part of the reason why. Things can't just happen when you want them to."

"I know, but there aren't any more women. The only person I've been chilling with for a while is Scorpio. I realized that I couldn't ignore my feelings for you, and using her to help me cope with my situation wasn't fair. A few weeks ago, we decided that she would move out. I bought her a condo, had my interior decorator hook it up for her, and we're trying to move on. I won't lie to you and say that we've stopped having sex, because sometimes we do. But she and Mackenzie are all that I have right now.

"I love Mackenzie with all my heart, and she visits me twice during the week and spends the night here on the weekends. Right now, I'm going through the channels to adopt her. But it's still not enough, especially since I know I have a son now.

"You can't lie to me anymore, Nokea. I know that the baby you had is mine. I saw so for myself." Jaylin was almost in tears, and so was I.

"He is yours, but I—"

"Don't say a word. Let me show you something."

He led me up the steps and took me to one of the guest rooms he'd converted into a baby's room. It was to die for—better than the room I had for the baby at home. It was blue, yellow and white. The walls were painted with white clouds and had yellow birds drawn on them like they flew around the room. The crib was round, with a sheer blue canopy above it that draped to the floor. It was white and matched the dresser and the changing table that had light blue handles on them. The closet was filled with baby clothes: T-shirts, pants, jogging suits, and tennis shoes. Like LJ was really going to be able to wear all these things.

I was taken aback by the room, and observed it in tears. When I looked in the baby's bed and saw the teddy bear Jaylin and I had exchanged on our birthdays, I lowered my head. Jaylin came over and held my waist from behind.

"I feel you, baby. I know exactly how you feel." He turned me to face him. "Tell me, do you still love me?"

I wanted to scream "yes" at the top of my lungs, but I knew that revealing my feelings would be a big mistake. Jaylin would take my words and run with them, and Stephon would be left brokenhearted. I was confused and needed to speak with Stephon before I told Jaylin how I really felt.

"Jaylin, I can't answer that. I'm confused right now. I . . . I don't know how I feel. I need to go—"

He pulled me closer to him. "Then don't answer me right now. Just let me make love to you today. I know it hasn't been that long since you've had the baby, but I promise you this will feel different. It will be everything you always wanted it to be, and I'll show you nothing but love. And if you don't feel my love, then you leave here and go marry Stephon. I won't interfere with your relationship with him anymore."

I shook my head. "I . . . I do still love you, but I don't know if making love to you is going to solve my problems. I don't know if it's going to answer all these questions I have. In fact, I think it's going to complicate things more if—"

He put his fingers over my lips, and then took my hand and escorted me to his bedroom. He pulled my dress over my head and smiled at my naked body. At that moment, I wanted Jaylin just as much as he wanted me. Maybe even more. I eased back on the bed and watched as Jaylin removed his jeans. He got on the bed and held himself up over me.

"Don't think about anything else right now but me. Clear your mind right now, and think about how good I'm going to feel inside of you."

I got a jump start and stroked his dick as he circled his tongue around my nipples. He pressed my breasts close together and massaged them. When he went down and licked my navel, I ran my fingers through his wet, curly hair.

I couldn't wait for his tongue to enter me, and when it did, it caused a high arch in my back. My legs trembled, and I felt my insides vibrate. A tear rolled from the corner of my eye. The love he had for me was truly being displayed.

"I'm so sorry for causing you any hurt," I said. "You have to know that I never, ever stopped loving you."

Jaylin kept working between my legs, and I put my hand down there to stop him.

"I don't want to come like this. Give me what I really want and I'll come as much as you want me to."

Jaylin licked my taste from his lips and placed my legs on his shoulders. He went inside and worked me in a smooth, circular motion. My eyes shut, and I sucked on my bottom lip from the feeling. It had never felt like this before, and I came quickly. He turned me on my stomach, and a few minutes later, I came again. His hands rubbed almost every part of my body, and as I gave him a ride, his thickness rubbed against my clit.

I was on the verge on coming again, but Jaylin hadn't come yet, and I wanted to satisfy him as well. I lowered myself, but he stopped me.

"What's wrong?" I asked.

"Nothing. I'm just loving the feel of you, and I didn't realize how good you look from the back."

I took that as a hint and a compliment, and turned my backside to Jaylin. He lay on me and nibbled on my ear.

"That tickles," I said, moving my head so he would stop.

"Oh, I plan to make it tickle. But you aren't going to want me to stop."

He held himself up with his strong arms and rolled his tongue down my back. After he kissed my butt cheeks and massaged them with his hands, he reached underneath me and fondled my clit while inserting himself back inside. The only thing I could do was lay my head on his pillow and somehow prevent myself from trying to pull out my hair.

Minutes later, we released our energy together, and he continued to kiss the back of my neck.

"I love you, I love you, I love you," he repeated, and then turned me to face him.

I held his face with my hands. "You have no idea how good I'm feeling right now, but where do we go from here? Do we now just say to hell with everybody who's been there for us? Do I go home and tell my parents I've changed my mind, I'm not marrying Stephon?"

Jaylin rubbed his nose against mine and kissed my forehead.

"You don't tell anybody anything," he whispered. "You go home and hold our son in your arms and think about what's best for him. You think about who you want in your future. Think about where your heart truly is and where it's always been. And when you get your answer, you come over here again so I can make love to you like I just did and we can talk about putting our relationship back together.

"If there is any doubt in your mind, I want you to be honest with yourself. I don't want you to have any regrets. I would rather you stay with Stephon if you think I can't be everything you want me to be.

"The last time I told you to go home and think about it, I didn't give you time to think. I decided for you, and I quickly moved on. This time, I'm not. If you want to be with me, I'm here. I'm not going anywhere until I hear from you."

I nodded. "Just give me some time, okay?"

We got out of bed and took a shower. As he washed me, the thought of him being with Scorpio was in my mind. I wouldn't doubt if the thought of me being with Stephon had crossed his as well. But in an effort to clear my thoughts, I asked Jaylin to make love to me again. He honored my request.

Jaylin sat in bed and watched reruns of *Good Times* as I put on my clothes and got ready to go. He said Scorpio was on her way to bring Mackenzie over for the night, so I hurried as fast as I could. She was the last person I wanted to see. She had a way of making me feel insecure. I really felt none of this would have happened if she had never met Jaylin. But the more I thought about it, I realized someone else probably would have interfered.

Jaylin walked me to the door and kissed me good-bye. No sooner had I got in my car than Scorpio pulled up. Surprisingly, she spoke to me. I responded politely. She really was an attractive woman, and I knew Jaylin had to have a difficult time keeping his hands off her. As much as he claimed to love me now, a part of me knew she would remain in the picture no matter what.

I watched Mackenzie run up to Jaylin. He picked her up and was all smiles. As he waved good-bye to me, she waved with him. I waved back and swallowed the lump in my throat. I really knew that no matter how hard I tried, this just wasn't going to work out.

As I drove away, Luther sang on the radio "I'd Rather." I listened as he said he'd rather have bad times with the one he loved than good times with someone else. I didn't quite know why, but I definitely understood his message.

36

JAYLIN

I could see the hurt on Scorpio's face when Nokea pulled off. And when she went up to my room and saw the bed all messed up, she sat on the chaise and held her temples with the tips of her fingers.

"So, is it over between us, Jaylin? Is this as far as we go?" she asked.

I went over to the bed and sat in front of her. "For the most part, it is. But I don't want you to ever think you didn't mean anything to me. These past months with you have been the most exciting time in my life. You've taught me a lot—more than I thought you would when I met you. I was in it for one thing and you knew it. Never in my wildest dreams did I think I would come out of this with a daughter who I love with all of my heart—and with so much respect for you. So, don't walk away from this feeling empty-handed. If there's anything, and I mean anything, I can do for you, I will."

Scorpio sighed and moved her head from side to side. "I never wanted anything from you but for you to love me. I was so sure we were moving in the right direction, and I'm not giving up on you that easily. I was always taught to fight for what I wanted, and I intend to do just that."

"Baby, I'm in love with someone else. I don't know what good fighting is going to do you when my heart is with her. I'm just telling you this because I don't want to see you continuing to hurt yourself over something that will never be. Besides, I want my little girl growing up with a mother who is sure of herself. One who knows she shouldn't fight a battle that can't be won. Don't make her suffer through watching us tear each other apart because we can't get along, okay?" I handed Scorpio the Kleenex box on my nightstand and she wiped her watery eyes.

"So, what about just last week? You made love to me like you wanted to be with me forever. You can't tell me you were thinking about her when that was happening. You seemed to be right there with me. Am I wrong?"

"No, I was there. I'm always there, but . . . but there comes a time when sex just ain't everything. I know coming from me you might not believe that, but I'm at a point in my life when I want more than just sex. And I can't get that from you. Not to say you're not good enough, but there are some things that, over time, I know only Nokea can give me."

"See, Jaylin, this is all messed up, because I know

you. You're saying these things today, but tomorrow you'll have a new attitude. You'll come over to my place and strip me naked, convince me to make love to you, and be right back there again two days later.

"In the meantime, what do you want me to do? I love you and I don't want to deny you. I can tell you no all I want, but we both know I'm going to give in to you. How can I not set myself up for disappointment when I know this is going to happen?"

"I'm not going to deny what you're saying, because you're right. But I'm going to need you to stand your ground. Depending on what Nokea decides, the future might be different for me. I probably will be knocking at your door some lonely nights, but be woman enough to stop me, especially since I've told you where things stand.

"The harder you make things for me, the easier you're going to make things for yourself. I have a sexual passion for you that I've never had for any other woman. Including Nokea. But the love I have for her goes deeper than that. I admit that the passion we have for each other may send me your way again, but you know where my heart is, because I've told you."

Scorpio went into the bathroom and splashed water on her face. She looked at the shower that still dripped from my recent encounter with Nokea. I didn't like myself right about now because I'd for damn sure hurt too many women. I always thought it was about me, and felt bad for not recognizing their needs and feelings too.

I tried to make Scorpio laugh before she left by playing Twister with her and Mackenzie. Scorpio seemed out of it, and I saw her biting her nails. I knew she was hurt, and I was glad she agreed to remain friends. When she got ready to leave, I gave her a hug and patted her ass.

"If you decide to give yourself to someone else, don't let him get it from behind, because that's my place," I said, laughing.

She found no humor in my words. She gave me a blank stare. "See, Jaylin. That's what I'm talking about. Don't be saying things like that when you know the only person I want behind me is you."

"I'm sorry. I just wanted to see a big, bright smile before you left."

She gave me a fake grin and showed her pearly whites. We both laughed and rocked back and forth with a tight hug.

"Thank you for being you," I whispered in her ear.

"Anytime," she said and walked out the door.

The next several weeks were hell. I was sure I'd hear from Nokea, but she never called. Didn't come by either. I promised myself I would give her time to think about what she wanted to do; however, when Stephon called and asked again about being his best man, I knew the wedding was still on. I made it clear that it wasn't in my best interests to be there for him like that. When he mentioned they'd made plans for their honeymoon, my heart ached. I wasn't about to tell him how hurt I really was, and continued our conversation like I wasn't even tripping.

After our conversation was finished, I got back on the phone and called Scorpio. She'd made herself available to me whenever I called. And when she did try to tell me no, I went to her place anyway. I had a key to let myself in, and fucked her like fucking was going out of style. Couldn't help myself.

Nokea didn't even have the decency to call and tell me she'd made her decision. That really fucked me up. Tore me apart. I couldn't concentrate on anything but the last time we were together. She had to know how deep my love for her was. If not, she had to feel it.

I went into my son's room and looked around. He was growing fast, but he didn't even know I existed. He'd never spent one fucking night with me, and if it were left up to Stephon and Nokea, he never would. Stephon promised he wouldn't keep him from knowing who his real father was, but every time I talked about bringing LJ over to see me, Stephon always made excuses. If he did bring him by, they only stayed for a few minutes then he said they had to go.

The only reason I hadn't interfered was because I gave Nokea my word that I would allow her time to make her decision. It looked as if she'd already made it. I just truly wished it had been different.

37

JAYLIN

A week before the wedding, I had mentally prepared myself for it. Everything else in my life was going smoothly. The market was on its way back up and my adoption of Mackenzie had gone every bit of my way. I spent every moment I had with her, and tried to ease some of the pain I felt. Without a doubt, she helped me cope very well. I'd even changed the baby's room back into a guest room. Gave all his clothes to charity and sold his furniture to my neighbor who was pregnant.

Nokea never did call, and I made no attempts to call her. It was obvious who she wanted to be with, so I left things as they were. Scorpio didn't lie, though; she wasn't giving up on me. She did everything in her power to win me over. Even had Mackenzie begging me to get back together with her. But even though Nokea had moved on with her life, I still wasn't ready for the type of relationship Scorpio wanted.

One thing that fucked me up was that, according to Stephon, he still saw Felicia and screwed around with his ex-girlfriend. And if that wasn't bad enough, he planned to get up on this new chick that started working with him several weeks ago. So, Nokea really had her work cut out for her if she planned on marrying him.

Me, I pretty much chilled. I was down to fucking Scorpio about twice a week and that was it. That was good for us, considering the fact it used to be twice a day. I'd met a few other ladies from time to time, but wasn't nothing but phone conversations going on. I hadn't invited anyone over yet, and I wouldn't until I knew for myself this marriage was actually going to happen.

The night of Stephon's bachelor party was like any other night to me. He'd invited me and so did our other boys, but again, I told everyone that I didn't want to have anything to do with his marriage to Nokea. Stephon called during his party and mentioned all the fly women and fun they were having. I wasn't in the mood, so I told him to knock one out for me and hung up.

I lay my head back on the pillow, tired of flipping from channel to channel, when the phone rang. When I heard Nokea's voice, I quickly sat up on my bed.

"Jaylin, I'm sorry I haven't called you until now, but I felt this was the only way for me to figure out what I really wanted to do. I had to give Stephon at least six months to show me how much he wants this, and he's worked hard at proving himself. And even though I know there's a possibility he could be with someone else, I truly don't believe it's going to be any different with you. You told me if I had any doubts, to marry him, so . . . so that's what I'm going to do."

I had finally heard it straight from her mouth. I cleared my achy throat.

"So, have you prepared yourself for the storm? It's headed your way. And it's nothing like what I did to you either. It's worse." She didn't respond. "My storm is over. I'm not saying I've been celibate, but it's been different since I've been in love with you." I could hear her sniffles on the other end of the phone.

"Stop torturing yourself and let this happen with us, Nokea. I won't get another chance to ask you before the wedding tomorrow, but think about it. Close your pretty eyes tonight and think about us—"

She hung up the damn phone on me. As I sat for a moment and thought about why, she called back.

"Hey," she said, sniffling. "I love you."

"Right back at you," was all I could say before she hung up again.

I turned on the radio to listen to the Quiet Storm. The lights were off, and I lay in bed with my eyes closed and one hand resting on my chest. I hoped that Nokea realized how much I loved her. As Gerald Levert sang "Made to Love Ya," the words to the song took effect and tears welled in my eyes. My heart felt like somebody was squeezing it in their hands and wouldn't let go. To help ease my pain, I reached into my drawer, pulled out a picture of Nokea, and laid it in bed next to me. I couldn't really see her picture in the dark, but I rolled my fingertips around it thinking that I could feel her and wishing I could have her in my bed forever. I fell asleep hoping she'd do the right thing the next day.

While I was in a deep sleep, Mama stared down at me and smiled. She told me to go get my woman and my son, and told me to never lose them again. When I reached out to touch her, Aunt Betty touched my hand and told me it was too late for me. Said that Stephon was the one for Nokea and I didn't deserve her. As I started to dispute that with her, they both faded and I woke up.

The sunlight beamed through my room, and I sat up on the edge of the bed with my face resting in my hands. I wiped my hand down my face and got out of bed to prepare myself for a long day. I went back and forth about attending this wedding, and came to the conclusion that we all needed closure. I had to go, just to see if Nokea would truly go through with it. If I were there, it might be the perfect time for her to realize what a big mistake this was and come to her senses.

I was running late for the wedding messing around with Mackenzie. She acted like she was the one getting married. We dressed alike in our cream-colored outfits; mine a suit and hers a dress. We accented the cream with royal blue because that's what she wanted. But when I tied the bow around her waistline, she insisted it didn't look right.

She pouted all the way there because I wouldn't stop by the store and get her a new color. Once again, I reminded her of our previous conversation and she perked up.

Mackenzie and I sat in the last pew of the church because I really didn't want to be seen; however, Stephon spotted me and came over to take a seat.

"Thanks, man. Thanks for coming. I know how hard this is for you, but you always had my back when I needed you to—even when you refused to be my best man. It was foolish of me to ask, knowing how you felt. I just hope after today, this will all be over and we can go on being like brothers again. I miss kicking it with you, dog, talking about the ladies. . . . You missed a live party last night. I'll tell you all about it, but now ain't the time. You can't tell me you don't miss us kicking it, Jay, and—"

"Of course I do," I said, interrupting his bullshit talk. "After today, I'll be fine. I came here not only for you but for me as well. I need closure. I need to put this shit behind me and get on with my life. Today, after I see how happy you and Nokea are going to be, then maybe I'll be able to do that. So, go do your thang, man. Don't let me stop you."

Stephon smiled. "You don't know what it means to have your support. And Nokea and me are going to be very happy. She's made her choice, Jay, and all we can do is accept it."

It was a good thing that Mackenzie was by my side because this would have been the perfect opportunity for me to knock the shit out of Stephon. Instead, I reached for his hand and shook it. "Good luck. I hope everything works out for you."

"Likewise," he said and then made his way to the altar.

I looked at the front pew and saw Nokea's mother holding my baby. I wanted to at least hold him, but I was sure the last person her mother wanted to see right about now was me. So, Mackenzie held her arm around mine and held it tight. It was as if she could feel the pressure I was under and was acting as my support system.

When the music started, I took a deep breath and watched as the bridesmaids started making their way down the aisle. The maid of honor came in, and I knew it was just about Nokea's time. The slower the maid of honor walked, the better off I was, but it seemed like the music was on fast forward, and then the pianist broke out with "Here Comes the Motherfucking Bride."

I dropped my head. I couldn't even stand up to watch Nokea as she walked down the aisle. Mackenzie stood up because everybody else had, and she claimed the bride was the most beautiful person she'd ever seen. I finally stood up and straightened my jacket.

When Nokea turned the corner right by my pew, she looked at me with her eyes filled with water. Mine were filled too, but I nodded and gave her the go-ahead. She cracked a tiny smile and slightly nodded back.

As she got closer and closer to the altar, I looked at Stephon. He was all smiles. Definitely knew he was getting a jewel. He'd promised me he would make her happy. Did I believe him? Hell no, and I had to do something about it.

When Nokea made it to the altar, everybody took a seat. The minister prepared for the exchanging of vows. But before he did, he asked if there was anyone who knew any reason why these two should not be joined together in holy matrimony. "Speak now, or forever hold your peace," he said.

I dropped my head again, covered my face with my hands, and then watched my legs as they trembled. I knew damn well that this was something I couldn't let happen. I knew that Nokea loved me. If I put her on the spot, she'd have to come to her senses and do the right thing.

I cleared my clogged throat and stood up. I stepped into the aisle and watched as many heads turned to me. Mackenzie grabbed the back of my jacket and asked where I was going. I touched her soft cheek and whispered, "I always told you you'd be the first to know, baby. Daddy's going to get married."

As I proceeded down the aisle, I could feel Mackenzie close behind me. Everybody watched and whispered. I stood at the altar with my hands behind my back. The entire church was in disbelief. I looked directly at the minister.

"I, uh—" I cleared my throat again. "I don't mean you any disrespect, sir, but I have love for this woman who stands before you today."

I held out my hand for Nokea to take it and looked into her big brown eyes. "If love for me is not a good reason to stop this wedding, then I don't know what is." I continued to hold out my hand for Nokea.

Stephon's mouth was wide open, and Nokea's father looked like he wanted to tear me apart. Nokea, though, forced out a tiny smile as she dropped her bouquet on the floor. . . .

Later that night, I stood looking over my balcony, drinking a glass of wine. My eyes searched the stars as a south wind blew, and I lifted my glass to Mama. Nokea had made her choice, and now I'd made one. I'd come to realize how important it is to treasure the one you love, and I now knew how easily a good thing can slip away. I tilted my wine glass upside down, poured some of the wine over the balcony for the people who weren't with me, and thought about where I'd go from here.

A tear rolled down my face while I stared at the curvaceous silhouette lying sideways in my bed. She was my rock. From the day I met her, she had always been in my corner, and she was my shoulder to cry on. She put up with my attitude, and had loved me during the times I felt as if I had no direction and didn't love myself. I looked at her in bed, swallowed the huge lump in my throat and thought . . . *is this really what I want?*

Coming Soon

Naughty No More

By Brenda Hampton

Nokea

Who would've thought that my husband, Jaylin, and I would be celebrating our three year anniversary? After many years of trials and tribulations, a failed marriage to my now ex, Collins, and Jaylin being able to put Scorpio, a woman he was engaged to behind him, here we were. We had our children, Jaylene who was three years old, and Jaylin Jr., a.k.a. LJ, who was now five. Life couldn't be any better for us, and Jaylin's inheritance at the age of 16, from his grandfather's estate, and his numerous years as an Investment Broker, brought him to millionaire status. Several years ago he retired, and once we got married, he required that I retire too.

Our mansion in Florida was by the beach, offering us nothing but peace and quietness. I in no way missed

living in our hometown, St. Louis, but I often missed
being with my parents, and my best friend, Pat, who
still lives there. Our fifty . . . sixty something year old
nanny, Nanny B, takes good care of all of us and she
wouldn't have it any other way. It took a lot of work
for us to get to this point, and if anyone thought that
relationships were easy, they were fooling themselves.
I had known Jaylin since elementary school, and we
didn't become an item until I was in my twenties.
Still, his womanizing ways caused us many setbacks,
and when he finally asked me to marry him, it wasn't
until I was almost thirty-three. He was thirty-four and
time was starting to run out on us. I hadn't planned on
waiting this long for him to get his act together, but
through dealing with him for so many years, I knew
there was a step-by-step process.

Many people had to be cut from our lives, but the
one person that I hated to see distance himself was
Jaylin's best friend, Shane. Reason being, Shane and
Scorpio had started dating, causing Jaylin much
headache and jealousy. He hadn't admitted his an-
guish to me, but I knew my husband and knew him
well. When Shane declined Jaylin's offer to leave St.
Louis and start his architectural business here, their
friendship went down hill. I figured Jaylin had made
the offer to get Shane away from Scorpio, but as far as
I know, that hasn't happened, especially since Scorpio
was *supposed* to be pregnant. She and Jaylin had only
dated for a few years, but the way she had a hold on
him, you would have thought they'd known each
other for a lifetime. She was the kind of woman who
often toyed with a man's feelings, and her jaw dropping body and movie star attributes, made many men
eat out of the palms of her hands. More than anyone,

I'm so glad that Jaylin got away from her tight grip. To this day, it still bothers me that she has a beauty shop named Jay's, and that Jaylin still considers her seven year old daughter, Mackenzie, his child. He adopted her several years back, and I guess there's not much I can do about that. What I can do is make sure my man is the happiest man in the world. He's had his trust issues with me too. Once upon a time, I had been engaged to his cousin, Stephon, but he was set up and killed by someone. Stephon was Jaylin's right hand man, until Stephon got on drugs and started pursuing me and Scorpio. Needless to say, things got ugly, and even in Stephon's death, Jaylin has never forgiven him. It was only a few years later that I took it upon myself to kiss Shane. I was upset that Jaylin had gone to St. Louis for a visit, and he wound up having an encounter with Scorpio. I consulted with Shane about my concerns, and even though he convinced me that nothing had happened between the two, I still *offered* Shane my thanks.

At the time, I wanted to offer him more than that, and I'll never forget the way he looked that day with his shirt off. Jaylin's body was cut, but Shane's was cut to perfection. His carmel skin always had a smooth look to it, but what sent me into a trance that day were his light brown eyes. Scorpio was lucky to have him, but not as lucky as I was. Jaylin had swagga, and even a man like Shemar Moore didn't have nothing on him. His bedroom gray eyes had broken many hearts and his nine plus inches of goodness had injured many backs. He was great . . . excellent in the bedroom, and even though I hated it, there were plenty of women who could testify.

A while back, I told Jaylin about the kiss between

me and Shane, and for whatever reason, he wasn't that upset. He knew how much I loved him and him only. There was no other man in the world that I wanted, needed, or desired. Jaylin was my destiny, and no matter what, we will be together for the rest of our lives.

Today, on our anniversary, he surprised me. Instead of using our neighbor's yacht to sail the ocean, Jaylin had purchased a yacht for our family. Of course, he named it after our son, LJ, and once I saw how beautiful the huge 40 foot vessel was, I almost cried.

The majority of it was white with navy blue trimmings around the outside and LJ's initials were scripted on the front in big bold letters. Navy blue and white adorned the interior too and the floors were covered with soft white carpet. The walls were draped with silky navy blue fabric and cherry oak panels covered portions of the walls as well. In the saloon there were two navy swivel leather chairs and a leather sectional in front of them. A cherry oak bar was there for our drinking pleasures, and once Jaylin took me up the spiral staircase to the master stateroom, I couldn't believe my eyes. Multi-colored fiber-optic lighting lit up the queen-sized bed and made the room feel and look like the most exquisite place I'd ever seen. As Jaylin wrapped his arms around me from behind, all I could do was think about how lucky I was.

"Do you like it?" he asked while nibbling my earlobe.

"I . . . I don't know what to say, other than, yes, I love it! How did you have time to plan all of this and . . . and I know this must have cost you a fortune?"

"It took almost eight months to build and if I tell you how much I paid for this thing, you'd kill me. The company I purchased it from promised they'd have it ready for today and it's been docked for almost a week. A few days ago, I saw it for the first time, and baby, I almost cried my damn self. We are so fortunate and I am enthused about spending our anniversary weekend here."

I turned to face Jaylin, placing my arms on his broad shoulders. "Me too. I love you so much and my little present for you doesn't even compare to this. I'm almost ashamed to give it to you."

"You know, I'm a simple man, Nokea. All you gotta do for me is get naked and show off your sexy caramel skin. That's how I get excited. Whatever you have for me can't possibly compare to you walking around the entire weekend without any clothes on. No negligee, no thongs, nightgowns, swimming suit . . . nothing."

"Nothing?" I asked, rubbing my nose against his. "What about when I go outside and stand on the deck? You don't want me naked out there, do you?"

"Trust me. For the next few days, you won't even see the outside. I'm going to satisfy your every need in here, and whatever is on the outside won't even matter."

From experience, I knew Jaylin's statement to be all so true. I gave him a long, wet kiss and before things got heated, he continued to show me around.

After the rest of the yacht was shown to me, I decided not to show Jaylin the silver electronic picture book I'd gotten him for our anniversary. The inside screen displayed some of my favorite pictures of us and I'd only paid two-hundred bucks for it. I knew Jaylin wouldn't trip, but compared to the yacht, I felt as if my gift was rather cheap.

As our hype started to settle, Captain Jack, whom Jaylin had hired to navigate our vessel, took off. Being on the ocean always provided a sense of peace, and while dressed in our silk blue matching robes, we held up our glasses to give a toast.

"Here's to fifty more years," I said, while sitting at the bar. Jaylin stood tall behind it, clinking his glass with mine.

"Fifty? Fifty years is a long time to be with somebody, Nokea, and I . . . I can't drink to that kind of commitment. Besides, you looking like Nia Long today, but ain't no telling what you'll look like in fifty years," he joked.

With my glass in my hand, I walked away from the bar and took a seat in one of the swivel chairs. "Oh, well," I shrugged. "Then, I'll just have to find someone else to spend my life with."

Jaylin came from behind the bar and kneeled down in front of me. He removed my glass from my hand and set both of our glasses on the circular table between the chairs. As his hands roamed up and down my legs, he looked deeply into my eyes. "Fifty, sixty, seventy . . . whatever. We'll be together forever, Mrs. Rogers. You can count on that and when we toast again, make sure it's for a lifetime, alright?"

"Got it," I smiled. "Now, Mr. Lifetime, other than sex, what are we going to do to keep ourselves busy for the next few hours?"

Jaylin showed his straight pearly whites and snapped his finger. "Somehow, I knew you'd ask that question. And, trust me, your husband already got this shit planned out."

He went over to the bar, picked up two large bottles of Remy, and reached for two shot glasses. He placed them on the table, and invited me to sit on the floor with him. While looking at each other from across the table, Jaylin poured Remy in both glasses.

"What are you doing?" I laughed.

"We're about to play strip poker. If not that, truth or dare. Loser or liar must drink up!"

"Are you serious?" I laughed.

"Very. It'll be fun. Besides, you are always at your best when you're sloppy drunk."

"I've never been sloppy drunk. You must be thinking of someone else."

"Nokea, on your birthday last year, you were sloppy drunk. Remember? You passed out on the beach and I had to carry you . . .

"Yeah, yeah, I remember. I also remember what you did to me that night and if you think you're going to do *that* to me again, you might as well forget it."

"Please," he begged. "I . . . we had so much fun that night. Besides, I like taking advantage of you. Sometimes you be holding back, but when you get that oil in your system, baby, you get loose than a mutha."

My mouth hung open. "Loose? That's terrible to say about your wife. You should be ashamed of yourself, but for the record, the liar will lose and that will be you. I'm feeling truth or dare and I get to ask the first question, right?"

"Shoot," he said, and then shot the full glass of Remy to the back of his throat. He coughed, then smiled. "Just trying to prepare myself."

I chuckled and reached across the table to rake my fingers through Jaylin's thick, naturally curly hair. "Do you miss Shane? It's been a long time since you last spoke to him, and you seem kind of, at times, in another world."

"I talk to him every now and then, Nokea. And, I would be lying if I said I didn't miss our friendship, but for the time being, things must stay as they are."

"You mentioned that things were just okay with him and Scorpio. By now, I thought you and Shane would eventually put the past behind and work on your friendship?"

Jaylin gulped down another glass of Remy. Obviously, because of Shane and Scorpio's relationship, Jaylin had put his friendship with Shane on the back burner. I could tell it bothered him, but for months, he hadn't said much about it.

"Why are we talking about this right now?" Jaylin asked, rubbing his goatee that suited his chin well. "I thought we're supposed to be playing truth or dare?"

I traced my finger along the side of smooth, light skinned faced. "I know, baby, but I worry about you. I want you to be happy and you seemed bothered by Shane declining your offer to come here and live."

"I was bothered, but I'm over it. The only thing I'm bothered by is you leaving next weekend to go to St. Louis. And even though it's only for the weekend, you know I'm going to miss you, right?"

"Then go with me. You don't have to go to Pat's baby shower with me, but you can still go to St. Louis. Maybe even stop by Shane's place and kick it with him? I know he'd be happy to see you."

"Naw, I'm cool. You need this time to yourself and I want you to have a good time."

"Okay," I said, taking a sip from the glass. Jaylin took another one as well. "Your eyes are getting really, really glassy looking. You might want to slow it down a bit, Jaylin."

"You know I'll have to consume a lot of alcohol before I get drunk. A few shots of Remy ain't gon' do nothing to me."

"If you say so. But, uh, let's continue our game. Since you already know everything about me, I'll take a dare."

"Yes, I do know everything about you. So, I dare you to drink this whole bottle of Remy without stopping."

"If I drink that whole bottle of Remy, I will be sick as a junk-yard dog. You don't want me to be that sick, do you?"

"Then drink half."

"Some," I smiled. "And whatever I drink, you must do the same."

"Bet," he said.

I opened the other bottle of Remy, taking a few sips from it. Jaylin carefully watched me, but when he walked over to the bar to get some ice, I filled the wineglasses with Remy and placed them behind the chair so he wouldn't see them.

"Damn," he said, looking at the half empty bottle upon his return. "You doin' it like that, huh?"

I smiled and nodded. "I sure am. Now, it's your turn."

No questions asked, he turned up the bottle, guzzling down the alcohol. "Okay, stop," I laughed, trying to pull the bottle away from his mouth. I showed him the wineglasses I'd filled. "I'm sorry, baby, I cheated."

Jaylin slammed the almost empty bottle on the table and looked a bit dazed. "Why you cheat like that?" he softly said. His gray eyes were watered down and there was no smile on his face.

"Are you okay?" I asked. "You look . . ."

"Look like I'm fucked up, don't I?"

"Very."

Jaylin sucked in a heap of air, wobbling a bit as he stood up. With a universal remote, he dimmed the lights and turned on some music. Seeming awfully woozy, he fell back in the chair, staring at me as if he wanted to eat me alive. I'd seen that look many times before and I definitely knew what it meant. It was show time and I had no problem standing up to show my handsome husband everything I had.

I kept my eyes connected with his and removed my robe. Naked, I pulled him up from the chair and held his muscular frame close to my petite body. He worked his way out of his robe too and as the tunes of Prince kicked in with "Purple Rain" we embraced each other.

"Your body . . . damn your body feels good," he slurred. "I . . . I remember that time I rented a boat for your birthday and you . . . you couldn't hang with my ass . . .

I moved my head away from Jaylin's buffed chest and looked up at him. "Are you talking about when I was thirty and still a virgin?" I asked.

With his eyes closed, he smiled and nodded.

"What in the world would make you think of that?" I asked. "That night, I was so hurt and not because your dic . . . penis was too big for me either. If your memory serves you correctly, you left that night to be with Scorpio."

"I ain't do no shit like that, quit lying!" he challenged, opening his eyes.

"Yes you did! And, I drove to your house and found the two of you in the shower, remember?"

Jaylin smirked, knowing darn well he remembered. "Damn, you did, didn't you? I was tripping, wasn't I?

You should have kicked my butt and left my ass alone. Instead, you gave me another chance, didn't you? A few weeks later, I tapped that ass and we made our son, didn't we?"

"Yes, we did."

We stood in silence and I could feel Jaylin's body getting heavy.

"Baby, I think you might want to sit down for a minute."

"I'm guuud," he said, and then backed up to the leather sectional. He dropped his head back and tightened his eyes. I could tell he was doing his best to shake off the liquor. He opened his eyes, and licked his lips as his eyes dropped between my legs. "Come here girl, wit yo sexy lil self. I got myself a bad-ass wife and I got to be the luckiest man in the world."

"Wit yo what?" I said, appreciating his compliments. I headed his way and straddled his lap. Jaylin touched my inner thighs, and when he slid his finger deep in my tunnel, he snickered.

"Wit yo slippery and wet, sexy, horny-ass self."

"Very horny self. Now, sober up and take care of this slippery wet mess, would you?"

Jaylin sucked his finger, tasting my sweetness on his finger. I slid off his lap and laid back on the couch to get into position. He sat up, rubbing his temples with his eyes closed.

"Baaaby," he slurred. "I'm fucked up. Would you, uh, go get me some ice cold water, please?"

I hurried off the couch to get his water. Within a few minutes, I came back and handed the water to him. He was laid back with his eyes closed.

"Jaylin, wake up," I said. "Here, drink some water."

He threw his limp hand back at me while massaging his goods. "I'm tired, baby. Real tired. Come on, though, so I can make use of my nine, plus some."

"We have all night for you to make use of that. Just drink some water and let's go lay down, okay?"

"Nooo," he ordered. "Lay back down on the couch. I . . . I wanna taste that pussy.".

Those words always excited me, but it was obvious that Jaylin had too much to drink. I tried to get him to drink some water, but he pulled me down on the couch. He parted my legs, and with his curled tongue, he took slow licks up and down my already juiced slit. I pressed his head in closer, inviting him to sink his tongue further inside of me. As he did, he widened my coochie lips, exposing my clit that he knew how to stimulate so well. He licked along the furrows of my walls, causing my breathing to become intense. I sucked in a heap of fresh air, releasing it to calm the electrifying feeling that was stirring between my legs. I was so ready to start raining cum, but the excitement Jaylin was bringing came to a screeching halt. I sat up on my elbows, looking down at him.

"Jaylin?" I questioned. There was no answer, so I called his name again.

"Hmmm," he moaned.

He slowly sat up and leaned to the other side of the couch. Wanting so badly to continue, I laid on top of him. I kissed his lips, sucking them with mine. I licked down his chest, giving special attention to his sexy abs by pecking them. His stomach heaved in and out, and when I picked up his hardness to put it into my mouth, it had no reaction.

I sighed, making my way back up to him. "Honey, are you okay?" I whispered. He slightly cracked his eyes and nodded. His eyes faded again and true disappointment was written on my face. "You know I'm going to make you pay for bailing out on me like this, don't you?" He slowly nodded again and I continued. "I am horny as hell and how dare you do me like this on our anniversary. You do know it's your anniversary, don't you?"

At that point, he was out. Didn't provide any gestures and started to snore. What else could I do but go down low again. That didn't do me much good at all because *it* was as limp as he was. "Darn-it!" I yelled and tried to wake him again. I lightly shook his shoulder. "I got your present, baby, look." He didn't budge. I huffed, realizing that this night was over. I cut my eyes at the almost empty bottle of Remy and frowned.